QUALITY OF SERVICE CONTROL IN HIGH-SPEED NETWORKS

T0324821

QUALITY OF SERVICE CONTROL IN HIGH-SPEED NETWORKS

H. JONATHAN CHAO

XIAOLEI GUO

A Wiley-Interscience Publication
JOHN WILEY & SONS, INC.

Copyright ©2002 John Wiley & Sons, Inc., New York. All rights reserved.

Published simultaneously in Canada.

For ordering and customer service, call 1-800-CALL-WILEY

Library of Congress Cataloging in Publication Data:

Chao, H. Jonathan, 1955-
  Quality of service control in high-speed networks/H. Jonathan Chao and Xiaolei Guo.
  p. cm.
  Includes bibliographical references.
  ISBN 0-471-00397-2 (cloth : alk. paper)
  1. Computer networks--Management. 2. Quality control. I. Guo, Xiaolei. II Title.

TK5105.5.C459 2001
004.6--dc21

2001026943

10 9 8 7 6 5 4 3 2 1

CONTENTS

PREFACE

This book addresses the basics, theory, architectures, and technologies to implement quality-of-service (QoS) control in high-speed networks, including asynchronous transfer mode (ATM), Internet Protocol (IP), and multiprotocol label switching (MPLS) networks. For the last few years, we have witnessed an explosion of Internet demands, yielding rapid development of the corresponding technologies in the backbone network, such as terabit routers, dense wavelength division multiplexing (DWDM) transmission equipment, and optical cross-connect switches (OXCs). The success of telecommunications in supporting a wide variety of Internet services, such as multimedia conferencing and video-on-demand, depends greatly on, (in addition to high-speed transmission and switching) the reliable control in the underlying high-speed networks to provide guaranteed QoS.

QoS provision in a network basically concerns the establishment of a network resource sharing policy (including link bandwidth and buffer space) and then the enforcement of that policy. As the line speed increases to 10 or 40 Gbit/s and the number of connections in each line increases to several hundreds of thousands, implementing QoS control, under the constraints of the timing and memory requirement, becomes very challenging. Unlike other books in the area, this book not only gives a comprehensive overview of the existing QoS control techniques, but also describes practical approaches to implement the techniques in the high-speed network.

Most of the book is based on the material that Jonathan has been teaching to the industry and universities for the past decade. He taught a graduate course "Broadband Network" at Polytechnic University, NY, and used the draft of the book as the text. The book has incorporated feedback from both industry people and college students. We believe this book is timely to meet the demand of industry people who are looking for the solutions for meeting various QoS requirements in the high-speed network.

AUDIENCE

This book surveys the latest technical papers that readers can refer to for the most up-to-date development of control strategies in the high-speed network. The readers are assumed to have some knowledge of fundamental networking and telecommunications. Some of this book may require readers to have some knowledge of probability models and college-level mathematics. Since each chapter is self-contained, readers can easily choose the topic of interest for both theoretical and practical aspects. A comprehensive list of references follows each chapter. This book should be useful to software, hardware, and system engineers in networking equipment and network operation. It should be useful as a textbook for students and lecturers in electrical engineering and computer science departments who are interested in high-speed networking.

ORGANIZATION OF THE BOOK

Throughout, IP and ATM networks are used as examples. The book is organized as follows:

- Chapter 1 presents a systematic overview of QoS control methods, including admission control, traffic access control, packet scheduling, buffer management, flow and congestion control, and QoS routing. It also introduces ATM, Internet integrated services (Intserv), Internet differentiated services (Diffserv), and MPLS. It also describes the traffic characterization and QoS parameters.
- Chapter 2 explores admission control and its process of deciding whether a new connection between a source–destination pair can be accepted across a network. The decision is made subject to this new connection's QoS requirement (e.g., loss and delay) without affecting the committed resource provision for the existing connections at each node on its path.
- Chapter 3 focuses on traffic access control and its mechanism of enforcing certain criteria regarding an established connection's traffic load and burstiness, for instance, at the network ingress/entry point of this connection. If the connection crosses two subnetworks under different administration, similar control (also called traffic policing) may also be needed at the egress point of the first subnetwork.

- Chapter 4 presents a historical overview of different packet scheduling algorithms. It also describes under each scheme how to regulate packet transmission orders among a set of connections multiplexed at a network node so that their QoS requirements can be satisfied.
- Chapter 5 is dedicated to discussing the implementation of packet fair queuing, an intensively studied packet scheduling algorithm due to its desirable capability of providing QoS guarantees (e.g., on delay) for connections with diversified QoS requirements. Practical examples covered in this chapter include designs based on sequencers, priority content-
  addressable memory (PCAM), and random access memory (RAM).
- Chapter 6 presents buffer management, which controls the access of incoming packets to the buffer space and decides which packet should be discarded when, for instance, the buffer is full or a threshold-crossing event happens.
- Chapter 7 explains flow control and congestion control. Flow control addresses the needs of speed matching between a source–destination pair or any two nodes on the path of a connection. Congestion control addresses the regulation of traffic loading across a network for congestion avoidance and recovery.
- Chapter 8 covers QoS routing and describes its process of deciding which path (in a connection-oriented network like ATM) or which next-hop node (in a connectionless network like IP) should be chosen for a new connection among multiple physical paths across a network. The decision may depend on the resource availability along the path and whether the QoS requirement of the new connection can be met.
- Chapter 9 describes basic architecture and conceptual model of Diffserv in details. It describes the network boundary traffic conditioning and per hop behaviors functions used to support Diffserv.
- Chapter 10 covers MPLS technology. It includes the basic concepts, such as the label stack, route selection, penultimate hop popping, and label-switched-path (LSP) tunnel, Label Distribution. It also describes the MPLS mechanism to support Diffserv and two applications of MPLS: traffic engineering and virtual private networks (VPNs).
- The Appendix briefly describes Synchronous Optical Networks (SONET) and ATM for readers who need to attain a basic understanding of the physical layer and link layer protocols of the high-speed network.

ACKNOWLEDGMENTS

We are indebted to all who contributed to this book. Especially, we thank Professor Sugih Jamin (University of Michigan), Dapeng Wu (Carnegie Mellon University), Dr. Jun-Shik Hong, Dr. Li-Sheng Chen (Alcatel),

Dr. Necdet Uzun (Cisco), Dr. Yau-Ren Jenq (Fujitsu), Dr. Soung-Yue Liew, Dr. Zhigang Jing (Polytechnic University), Dr. Roberto Rojas-Cessa, and Dr. Taweesak Kijkanjanarat. We thank Dr. Jing and Dr. Liew for contributing Chapters 9 and 10, and Mike Barrett and Dr. Mukesh Taneja for reviewing the chapters.

Jonathan wants to thank his wife, Ammie, and his children, Jessica, Roger, and Joshua, for their love, support, encouragement, patience and perseverance. He also thanks his parents for their encouragement. Xiaolei would also like to thank his wife, Mei Feng, for her love and support.

We have done our best to accurately describe the QoS control methods, technologies, and implementation architectures. If any errors are found, please send email to chao@poly.edu. We will correct them in future editions.

H. JONATHAN CHAO
XIAOLEI GUO

September 2001

CHAPTER 1

INTRODUCTION

The broadband integrated services digital network (B-ISDN) with standardized asynchronous transfer mode (ATM) is envisaged to support not only current services, but also new services with varying traffic characteristics and quality-of-service (QoS) requirements [1-7, 11]. In simple terms, ATM is a connection-oriented packet switching and multiplexing technique that uses short fixed-size cells to transfer information over a B-ISDN network. The short cell size of ATM at high transmission rates is expected to offer full bandwidth flexibility and provide the basic framework for guaranteeing QoS requirements of applications with a wide range of performance metrics, such as delay and loss.

Meanwhile, the advent of broadband networking technology has dramatically increased the capacity of packet-switched networks from a few megabits per second to hundreds or even thousands of megabits per second. This increased data communication capacity allows new applications such as video conferencing and Internet telephony. These applications have diverse QoS requirements. Some require stringent end-to-end delay bounds; some require a minimal transmission rate; others may simply require high throughput. As use of the Internet diversifies and expands at an exceptional rate, the issue of how to provide necessary QoS for a wide variety of different user applications is also gaining increasing importance [8, 11, 15-20]. This book attempts to clarify the QoS issue and examines the effectiveness of some proposed network solutions.

In short, QoS depends on the statistical nature of traffic. An appropriate service model should be defined and some network QoS control methods should be engineered to meet a range of QoS performance requirements

1

(e.g., throughput, delay, and loss), which are usually represented as a set of QoS parameters associated with the service model. Section 1.1 describes the nature of traffic. Network technologies are presented in Section 1.2. Section 1.3 describes QoS parameters. QoS control methods for traffic management are discussed in Section 1.4. A summary is given in Section 1.5.

1.1 NATURE OF TRAFFIC

There are two main traffic types: delay-sensitive traffic and loss-sensitive traffic. Delay-sensitive traffic is characterized by rate and duration and may need real-time transmission. Examples include video conferencing, telephone, and audio/video on demand, which usually have stringent delay requirements but can accept a certain loss. Loss-sensitive traffic is characterized by the amount of information transmitted. Examples are Web pages, files, and mail. It usually has stringent data loss requirements but no deadline for completing a transmission.

There are other traffic types, such as playback traffic, multicast traffic [e.g., conferences, distributed interactive simulation (DIS), and games], and traffic aggregation [e.g., from local area network (LAN) interconnection]. Observations of LAN traffic [12] reveal its *self-similar*, or *long-range dependent*, behavior: The rate is variable at all time scales; it is not possible to define a duration over which the traffic intensity is approximately constant. These observations have been confirmed repeatedly. A plausible explanation for self-similarity is that LAN traffic results from a superposition of bursts whose duration has a heavy-tailed distribution [7, 12].

1.2 NETWORK TECHNOLOGIES

The network has been evolving to provide QoS guarantees to the users. For instance, ATM, widely adopted in the backbone network, can reserve the bandwidth and buffer for each virtual connection. Similarly, the Internet integrated service (Intserv) can also provide QoS for each flow (see definition in Section 1.2.2) in the Internet Protocol (IP) network. Internet differentiated service (Diffserv) provides different treatment for packets of different classes, instead of on a flow basis, so that it has better scalability than Intserv. Multiprotocol label switching (MPLS) allows the network providers to have better control and provision of QoS through traffic engineering policies.

1.2.1 ATM

The ATM Forum and the International Telecommunication Union (ITU) have different names for some service classes (known to the ITU as "ATM transfer capabilities"), while some classes are standardized by only one body. Hereinafter, unless otherwise stated, we use ATM Forum terminology for illustrations [5, 11, 13, 14]. Interested readers are referred to the Appendix for further details about Synchronous Optical Network (SONET) ATM protocols. The following summarizes the traffic parameters used in the ATM networks. Detailed definition can be found in Chapter 3.

The *constant bit rate* (CBR) service category applies to connections that require cell loss and delay guarantees. The bandwidth resource provided to the connection is always available during the connection lifetime, and the source can send at or below the *peak cell rate* (PCR) or not at all. A CBR connection must specify the parameters, including PCR or peak emission interval ($T = 1/\text{PCR}$), cell delay variation tolerance (CDVT), maximum cell transfer delay, and cell loss ratio. The standard defines the rate in terms of an algorithm making necessary allowances for jitter (cell delay variation) introduced between a terminal and the network interface. The chosen algorithm, called the virtual scheduling algorithm or the continuous state leaky bucket, is now standardized as the generic cell rate algorithm (GCRA).

The *variable bit rate* (VBR) service category is intended for a wide range of connections; it includes real-time constrained connections (rt-VBR) as well as connections that do not need timing constraints (nrt-VBR). (Note that CBR is normally for a real-time service.) The VBR is basically defined by its PCR, sustainable cell rate (SCR), and maximum burst size (MBS). The SCR indicates the upper bound for the mean data rate, and the MBS indicates the number of consecutive cells sent at peak rate.

The *available bit rate* (ABR) standard specifies how users should behave in sending data and resource management (RM) cells in response to network feedback in the form of explicit rate and/or congestion indications. An application using ABR specifies a PCR that it will use and a minimum cell rate (MCR) that it requires. The network allocates resources so that all ABR applications receive at least their MCR capacity. Any unused capacity is then shared in a fair and controlled fashion among all ABR sources. The ABR mechanism uses explicit feedback to sources to ensure that capacity is fairly allocated.

At any given time, a certain amount of the capacity of an ATM network is consumed in carrying CBR and the two types of VBR traffic. Additional capacity may be available for one or both of the following reasons: (1) not all of the total resources have been committed to CBR and VBR traffic, and (2) the bursty nature of VBR traffic means that sometimes less than the committed capacity is used. Any capacity not used by ABR sources remains available for *unspecified bit rate* (UBR) traffic as explained below.

The UBR service is suitable for applications that can tolerate variable delays and some cell losses, which is typically true of *transport control protocol* (TCP) traffic. With UBR, cells are forwarded on a first-in, first-out (FIFO) basis, using the capacity not consumed by other services; delays and variable losses are possible. No initial commitment is made to a UBR source, and no feedback concerning congestion is provided; this is referred to as a *best-effort* service.

The *guaranteed frame rate* (GFR) is intended to refine UBR by adding some form of QoS guarantee. The GFR user must specify a maximum packet size that he/she will submit to the ATM network and a minimum throughput that he/she would like to have guaranteed, i.e., a MCR. The user may send packets in excess of the MCR, but they will be delivered on a best-effort basis. If the user remains within the throughput and packet size limitations, he/she can expect that the packet loss rate will be very low. If the user sends in excess of the MCR, he/she can expect that if resources are available, they will be shared equally among all competing users.

1.2.2   Internet Integrated Services (Intserv)

The Internet, as originally conceived, offers only point-to-point best-effort data delivery. Routers use a simple FIFO service policy and rely on buffer management and packet discarding as a means to control network congestion. Typically, an application has no knowledge of when, if at all, its data will be delivered to the other end unless explicitly notified by the network. A new service architecture is needed to support real-time applications, such as remote video, multimedia conferencing, with various QoS requirements. It is currently referred to as the *integrated services internet* (ISI) [15–18].

The concept of flow is introduced as a simplex, distinguishable stream of related datagrams that result from a single user activity and require the same QoS. The support of different service classes requires the network, and routers specifically, to explicitly manage their bandwidth and buffer resources to provide QoS for specific flows. This implies that *resource reservation, admission control, packet scheduling*, and *buffer management* are also key building blocks of the ISI.

Furthermore, this requires a flow-specific state in the routers, which represents an important and fundamental change to the Internet model. And since the Internet is connectionless, a *soft state* approach is adopted to refresh flow states periodically using a signaling system, such as resource reservation protocol (RSVP). Because ATM is connection-oriented, it can simply use a *hard state* mechanism, in that each connection state established during call setup remains active until the connection is torn down. Since it implies that some users are getting privileged service, resource reservation will also need enforcement of policy and administrative controls.

There are two service classes currently defined within the ISI: *guaranteed service* (GS) [18] and *controlled-load service* (CLS) [17]. GS is a service characterized by a perfectly reliable upper bound on end-to-end packet delay. The GS traffic is characterized by peak rate, token bucket parameters (token rate and bucket size), and maximum packet size. GS needs traffic access control (using the token bucket) at the user side and packet fair queuing (PFQ) at routers to provide a minimum bandwidth. (The QoS control mechanisms, such as token bucket and PFQ, will be further explained later.) Since this upper bound is based on worst case assumptions on the behavior of other flows, proper buffering can be provided at each router to guarantee no packet loss.

The CLS provides the client data flow with a level of QoS closely approximating what the same flow would receive from a router that was not heavily loaded or congested. In other words, it is designed for applications that can tolerate variance in packet delays as well as a minimal loss rate that must closely approximate the basic packet error rate of the transmission medium. CLS traffic is also characterized by (optional) peak rate, token bucket parameters, and maximum packet size. The CLS does not accept or make use of specific target values for control parameters such as delay or loss. It uses loose admission control and simple queue mechanisms, and is essentially for adaptive real-time communications. Thus, it doesn't provide a worst-case delay bound like the GS.

Intserv requires packet scheduling and buffer management on a per-flow basis. As the number of flows and line rate increase, it becomes very difficult and costly for the routers to provide Intserv. A solution called differentiated services (Diffserv) can provide QoS control on a service class basis. It is more feasible and cost-effective than the Intserv.

1.2.3 Internet Differentiated Services (Diffserv)

Service differentiation is desired to accommodate heterogeneous application requirements and user expectations, and to permit differentiated pricing of Internet services. Differentiated services (DS or Diffserv) [10, 19, 20] are intended to provide scalable service discrimination in the Internet without the need for per-flow state and signaling at every hop, as with Intserv. The DS approach to providing QoS in networks employs a small, well-defined set of building blocks from which a variety of services may be built. The services may be either end-to-end or intradomain. A wide range of services can be provided by a combination of:

- setting bits in the type-of-service (TOS) octet at network edges and administrative boundaries,
- using those bits to determine how packets are treated by the routers inside the network, and
- conditioning the marked packets at network boundaries in accordance with the requirements of each service.

According to this model, network traffic is classified and conditioned at the entry to a network and assigned to different behavior aggregates. Each such aggregate is assigned a single DS codepoint (i.e., one of the markups possible with the DS bits). Different DS codepoints signify that the packet, should be handled differently by the interior routers. Each different type of processing that can be provided to the packets is called a different *per-hop behavior* (PHB). In the core of the network, packets are forwarded according to the PHBs associated with the codepoints. The PHB to be applied is indicated by a Diffserv codepoint (DSCP) in the IP header of each packet. The DSCP markings are applied either by a trusted customer or by the boundary routers on entry to the Diffserv network.

The advantage of such a scheme is that many traffic streams can be aggregated to one of a small number of behavior aggregates (BAs), which are each forwarded using the same PHB at the routers, thereby simplifying the processing and associated storage. Since QoS is invoked on a packet-by-packet basis, there is no signaling, other than what is carried in the DSCP of each packet, and no other related processing is required in the core of the Diffserv network. Details about Diffserv are described in Chapter 8.

1.2.4 Multiprotocol Label Switching (MPLS)

MPLS has emerged as an important new technology for the Internet. It represents the convergence of two fundamentally different approaches in data networking: datagram and virtual circuit. Traditionally, each IP packet is forwarded independently by each router hop by hop, based on its destination address, and each router updates its routing table by exchanging routing information with the others. On the other hand, ATM and *frame relay* (FR) are connection-oriented technologies: a virtual circuit must be set up explicitly by a signaling protocol before packets can be sent into the network.

MPLS uses a short, fixed-length label inserted into the packet header to forward packets. An MPLS-capable router, termed the label-switching router (LSR), uses the label in the packet header as an index to find the next hop and the corresponding new label. The LSR forwards the packet to its next hop after it replaces the existing label with a new one assigned for the next hop. The path that the packet traverses through a MPLS domain is called a *label-switched path* (LSP). Since the mapping between labels is fixed at each LSR, an LSP is determined by the initial label value at the first LSR of the LSP.

The key idea behind MPLS is the use of a forwarding paradigm based on label swapping that can be combined with a range of different control modules. Each control module is responsible for assigning and distributing a set of labels, as well as for maintaining other relevant control information. Because MPLS allows different modules to assign labels to packets using a variety of criteria, it decouples the forwarding of a packet from the contents of the packet's IP header. This property is essential for such features as traffic engineering and virtual private network (VPN) support. Details about MPLS are described in Chapter 10.

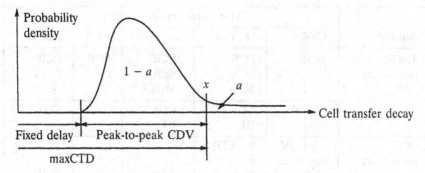

Fig. 1.1 CTD probability density model (real-time service model).

1.3 QoS PARAMETERS

In general, QoS is a networking term that specifies a guaranteed throughput level, which allows network providers to guarantee their customers that end-to-end delay will not exceed a specified level. Consider ATM as an example [13, 14]. The QoS parameters include peak-to-peak cell delay variation (CDV), maximum cell transfer delay (maxCTD), and cell loss ratio (CLR). As Figure 1.1 shows, the maxCTD for a connection is the $1 - a$ quantile of CTD.

The peak-to-peak CDV is the $1 - a$ quantile of CTD minus the fixed CTD that could be experienced by any delivered cell on a connection during the entire connection holding time.

The CLR parameter is the value of the CLR that the network agrees to offer as an objective over the lifetime of the connection. The CLR objective applies to either CLP = 0 cell flow or to the aggregate (CLP = 0) + (CLP = 1) cell flow, where CLP stands for cell loss priority and is indicated by a specific CLP bit in the cell header. The CLR is defined for a connection as

$$CLR = \frac{\text{lost cells}}{\text{total transmitted cells}}.$$

Figure 1.2 gives the ATM service category attributes, where DGCRA, short for dynamic GCRA, is an extension of the GCRA. Detailed definition of CDVT and BT (burst tolerance) can be found in Chapter 3. The DGCRA may be used to specify conformance to *explicit-rate* (ER) feedback at the public or private user-network interface (UNI), the private network-to-network interface (NNI), or the broadband intercarrier interface (BICI) for the cell flow of an ABR connection. It is dynamic because of the closed-loop flow control of ABR so that sources can be informed dynamically about the congestion state of the network and are asked to increase or decrease their input rate or sending window.

Attribute	ATM Layer Service Category				
	CBR	rt-VBR	nrt-VBR	UBR	ABR
Traffic parameters	PCR	PCR SCR MBS	PCR SCR MBS	PCR	PCR
	CDVT	CDVT BT	CDVT BT		CDVT
QoS parameters	p-p CDV max CTD CLR	p-p CDV max CDT CLR	CLR		CLR
Conformance definitions	GCRA	GCRA	GCRA		DGCRA
Feedback	Unspecified				Specified

Fig. 1.2   ATM service category attributes.

The goal of traffic management is to maximize network resource utilization while satisfying each individual user's QoS. For example, the offered loading to a network should be kept below a certain level in order to avoid congestion, which, in turn, causes throughput decrease and delay increase, as illustrated in Figure 1.3. Next we highlight a set of QoS control methods for traffic management.

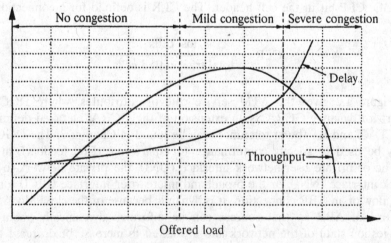

Fig. 1.3   Effects of congestion.

1.4   QoS CONTROL METHODS

Consider the data communication between two users in a network who are separated by a network of routers or packet switches, called nodes for brevity.[1] If the source has a message that is longer than the maximum packet size, it breaks the message up into packets and sends these packets, one at a time, to the network. Each packet contains a portion of the message plus some control information in the packet header. The control information, at a minimum, includes the routing information (IP destination address for the Internet, or virtual channel identifier for FR and ATM networks) that the network requires to be able to deliver the packet to the intended destination.

The packets are initially sent to the first-hop node to which the source end system attaches. As each packet arrives at this node, it stores the packet briefly in the input buffer of the corresponding port, determines the next hop of the route by searching the routing table (created through a routing protocol) with the routing control information in the packet header, and then moves the packet to the appropriate output buffer associated with that outgoing link. When the link is available, each packet is transmitted to the next node *en route* as rapidly as possible; this is, in effect, statistical time-division multiplexing. All of the packets eventually work their way through the network and are delivered to the intended destination.

Routing is essential to the operations of a packet switching network. Some sort of adaptive routing technique is usually used. The routing decisions that are made change as conditions on the network change. For example, when a node or trunk fails, it can no longer be used as part of a route; when a portion of the network is heavily congested, it is desirable to route packets around rather than through the area of congestion.

To maximize network resource (e.g., bandwidth and buffer) utilization while satisfying the individual user's QoS requirements, special QoS control mechanisms should be provided to prioritize access to resources at network nodes. For example, real-time queuing systems are the core of any implementation of QoS-controlled network services. The provision of a single class of QoS-controlled service requires the coordinated use of *admission control*, *traffic access control*, *packet scheduling*, and *buffer management*. Other techniques include *flow and congestion control* and *QoS routing*, as briefly explained below. Each of them will be further explained with detailed references in the rest of the book.

1.4.1   Admission Control

Admission control limits the load on the queuing system by determining if an incoming request for new service can be met without disrupting the service guarantees to established data flows.

---

[1] In this section we use the term *packet* in a broad sense, to include packets in a packet switching network, frames in a FR network, cells in an ATM network, or IP datagrams in the Internet.

Basically, when a new connection request is received, the *call admission control* (CAC) is executed to decide whether to accept or reject the call. The user provides a source traffic descriptor—the set of traffic parameters of the ATM source (e.g., PCR, SCR, MBS, and MCR), QoS requirements (such as delay, delay variation, and cell loss rate), and conformance definition (e.g., GCRA or DGCRA for ABR). The network then tries to see whether there are sufficient network resources (buffer and bandwidth) to meet the QoS requirement.

Given that most real-time queuing systems cannot provide QoS-controlled services at arbitrarily high loads, admission control determines when to generate a "busy signal". Proper resources may be reserved for an accepted request based on its QoS specification, such as a minimum bandwidth and buffer space. Chapter 2 introduces some examples of CAC for ATM and packet networks.

1.4.2 Traffic Access Control

Traffic access control (e.g., GCRA) shapes the behavior of data flows at the entry and at specific points within the network. Once the connection is accepted to the network, its emitting traffic to the network should comply with the traffic descriptor. If not, the excess traffic can be either dropped, tagged to a lower priority, or delayed (i.e., *shaped*).

Different scheduling and admission control schemes have different limitations on the characteristics (e.g., rate, burstiness) of traffic that may enter the network. Traffic access control algorithms filter data flows to make them conform to the expectations of the scheduling algorithms. Chapter 3 details GCRA for ATM networks and an integrated traffic shaper for packet networks.

1.4.3 Packet Scheduling

Packet scheduling specifies the queue service discipline at a node—that is, the order in which queued packets are actually transmitted. Since packets of many users may depart from the same outgoing interface, packet scheduling also enforces a set of rules in sharing the link bandwidth. For example, if a user is given the highest priority to access the link, his/her packets can always go first while packets from others will be delayed; and this privileged user can have his/her packets marked through some traffic access control algorithm when they enter the network.

In other words, packet scheduling prioritizes a user's traffic in two categories: delay priority for real-time traffic and loss priority for data-type traffic. One major concern is how to ensure that the link bandwidth is *fairly* shared between connections and to protect the individual user's share from being corrupted by malicious users (i.e., put a firewall between connections). In this respect, PFQ is very promising. Chapter 4 introduces various kinds of scheduling algorithms, including the PFQ family, targeted at different goals.

1.4.4   Packet Fair Queuing Implementation

There is a challenging design issue in that PFQ's packet reordering and queue management impose increased computational overhead and forwarding burden on networks with large volumes of data and very high-speed links. Chapter 5 presents the implementation of a PFQ scheduler and demonstrates how to move this kind of queuing completely into silicon instead of software so as to greatly reduce its effect on forwarding performance.

1.4.5   Buffer Management

The problem of buffer sharing arises naturally in the design of high-speed communication devices such as packet switches, routers, and multiplexers, where several flows of packets may share a common pool of buffers. Buffer management sets the buffer sharing policy and decides which packet should be discarded when the buffer overflows. Thus, the design of buffer sharing strategies is also very critical to the performance of the networks. Since variable-length packets are commonly segmented into small and fixed-length units for internal processing in routers and switches, the per-time-slot processing imposes difficulty in handling large volumes of data at high-speed links for both buffer management and PFQ, as mentioned above. Chapter 7 describes buffer management and its implementation issues in more depth.

1.4.6   Flow and Congestion Control

In most networks, there are circumstances in which the externally offered load is larger than that can be handled. If no measures are taken to restrict the entrance of traffic into the network, queue sizes at bottleneck links will grow and packet delays will increase. Eventually, the buffer space may be exhausted, and then some of the incoming packets are discarded, possibly violating maximum-delay–loss specifications. *Flow control* (some prefer the term "congestion control") is necessary to regulate the packet population within the network. Flow control is also sometimes necessary between two users for speed matching, that is, for ensuring that a fast transmitter does not overwhelm a slow receiver with more packets than the latter can handle. Chapter 8 describes several flow control mechanisms for ATM and packet networks.

1.4.7   QoS Routing

The current routing protocols used in IP networks are typically transparent to any QoS requirements that different packets or flows may have. As a result, routing decisions are made without any awareness of resource availability and requirements. This means that flows are often routed over paths that are unable to support their requirements while alternate paths with sufficient resources are available. This may result in significant deterioration in performance, such as high call blocking probability.

To meet the QoS requirements of the applications and improve the network performance, strict resource constraints may have to be imposed on the paths being used. *QoS routing* refers to a set of routing algorithms that can identify a path that has sufficient residual (unused) resources to satisfy the QoS constraints of a given connection (flow). Such a path is called a *feasible path*. In addition, most QoS routing algorithms also consider the optimization of resource utilization measured by metrics, such as delay, hop count, reliability, and bandwidth. Further details are provided in Chapter 9.

1.5 SUMMARY

Realizing QoS relies both on the implemented service models and on the way the network is engineered to meet the demand [4–10]. The central point is multilevel congestion control and bandwidth allocation [21, 22, 23]. The top level (on the time scale of a day, a week, or longer) involves capacity planning and network design (which we haven't addressed in this book). Since network congestion can be caused by lack of network resources or uneven distribution of traffic, the best method for networks that are almost always congested (or persistently overloaded) is to install higher-speed links and redesign the topology to match the demand pattern. Congestion pricing such as *peak-load pricing* is an additional method to decrease the peak load in the network. Interested readers are referred to [24] for further details on this subject.

Given a network topology, *traffic engineering* is the process of arranging how traffic flows through the network so that congestion caused by uneven network utilization can be avoided [25]. QoS routing is an important tool for making the traffic engineering process automatic and usually works on the time scale of seconds. The private network–network interface (PNNI) and QoS-based open shortest path first (QOSPF) are two examples for ATM and Internet routing, respectively. The goal of QoS routing is to select a path that can meet certain QoS requirements while increasing the network utilization. A QoS routing table is dynamically constructed at each network node (router or switch) by using the link state advertisement (LSA) messages generated from neighboring nodes periodically (e.g., every 30 s).

On the time scale of the duration of a session, CAC routes a flow according to load level of links (e.g., the QoS routing information) and rejects new sessions if all paths are highly loaded. The "busy" tone in telephone networks is an example of CAC. CAC is effective only for medium-duration congestion, since once the session is admitted, the congestion may persist for the duration of the session. The QoS provision for an accepted session is ensured by proper bandwidth and buffer allocation at each node along the path. The bandwidth allocation is ensured, in turn, by proper packet scheduling algorithms.

For congestions lasting less than the duration of a session, an end-to-end control scheme can be used. For example, during the session setup, the minimum and peak rates may be negotiated. Later, a leaky bucket algorithm (traffic access control) may be used by the source or the network to ensure that the session traffic meets the negotiated parameters when entering the network. Such traffic flow control and congestion control algorithms are open loop in the sense that the parameters cannot be changed dynamically if congestion is detected after negotiation. In a closed-loop scheme, on the other hand, sources are informed dynamically about the congestion state of the network and are asked to increase or decrease their input rate or sending window. The feedback may be used hop by hop (at the data link or network layer) or end to end (at the transport layer, as in TCP). Hop by hop feedback is more effective for shorter-term overloads than end-to-end feedback.

QoS and some sort of traffic engineering have long been provided by ATM networks [25, 4–7]. In an ATM network, QoS can be provided by allocating a certain amount of bandwidth for a specific virtual connection (VC). Traffic engineering is usually done by computing the routes offline and then downloading the configuration statically into the ATM switches on an hourly or daily basis.

Implementing differentiated service in the MPLS network allows Internet service providers to offer an end-to-end connection with guaranteed bandwidth. Users' traffic will be policed at the network edge based on a service level agreement. A label-switched path is established through traffic engineering in the MPLS network, which is described in detail in Chapter 10.

Looking forward, multicast communications pose a bigger challenge for QoS provisioning in wide area networks. Because multicast is the act of sending a message to multiple receivers with a single transmit operation, the solutions adopted will depend on application-level requirements, the size of the multicast group, and network-level support [26]. The key issue is how to handle heterogeneity of QoS requirements as well as network paths and end-system capabilities among participants in a scalable and efficient manner. As V. Cerf, one of the founders of the Internet, said [26],

Together Internet broadcasting and multicasting are the next chapters in the evolution of the Internet as a revolutionary catalyst for the Information Age.

Intensive research and study on this subject are still underway, and interested readers are referred to [26] for a complete list of references.

REFERENCES

1. J. P. Coudreuse, "Les réseaux temporels asynchrones: du transfert de données à l'image animée," *Echo Rech.*, no. 112, 1983.

2. J. S. Turner, "New directions in communications (or which way to the information age?)," *IEEE Commun. Mag.*, vol. 24, no. 10, pp. 8–15, Oct. 1986.

3. J. Kurose, "Open issues and challenges in providing quality of service guarantees in high-speed networks," *ACM Comput. Commun. Rev.*, pp. 6–15, Jan. 1993.

4. D. Bertsekas and R. Gallager, *Data Networks*, Prentice-Hall, Englewood Cliffs, NJ, 1992.

5. A. S. Acampora, *An Introduction to Broadband Networks*: *LANs, MANs, ATM, B-ISDN, and Optical Networks for Integrated Multimedia Telecommunications*, Plenum Press, New York, 1994.

6. R. O. Onvural, *Asynchronous Transfer Mode Networks: Performance Issues*, Artech House, 2nd Edition, 1995.

7. W. Stallings, *High-Speed Networks*: *TCP/IP and ATM Design Principles*, Prentice-Hall, Englewood Cliffs, NJ, 1998.

8. P. Ferguson and G. Huston, *Quality of Service*: *Delivering QoS on the Internet and in Corporate Networks*, Wiley, 1998.

9. R. Braden, D. Clark, and S. Shenker, "Integrated services in the Internet architecture: an overview," RFC 1633, Internet Engineering Task Force (IETF), Jun. 1994.

10. S. Blake, D. Black, M. Carlson, E. Davies, Z. Wang, and W. Weiss, "An architecture for differentiated services," RFC 2475, Internet Engineering Task Force (IETF), Dec. 1998.

11. J. Roberts, "Quality of service within the network: service differentiation and other issues," presented at IEEE INFOCOM'98, Mar. 1998.

12. H. J. Fowler and W. E. Leland, "Local area network traffic characteristics, with implications for broadband network congestion management," *IEEE J. Selected Areas Commun.*, vol. 9, no. 7, pp. 1139–1149, Sept. 1991.

13. ATM Forum, *ATM User-Network Interface Specification*, Version 3.0, Prentice-Hall, Englewood Cliffs, NJ, 1993.

14. ATM Forum, *Traffic Management Specification*, Version 4.1, AF-TM-0121.000, Mar. 1999.

15. D. D. Clark, S. Shenker, and L. Zhang, "Supporting real-time applications in an integrated services packet networks: architecture and mechanism," *Proc. ACM SIGCOMM*, pp. 14–26, Aug. 1992.

16. R. Braden, L. Zhang, S. Berson, S. Herzog, and S. Jamin, "Resource ReSerVation Protocol (RSVP)—Version 1 Functional Specification," RFC 2205, Internet Engineering Task Force (IETF), Sept. 1997.

17. J. Wroclawski, "Specification of the controlled-load network element service," RFC 2211, Internet Engineering Task Force (IETF), Sept. 1997.

18. S. Shenker, C. Partridge, and R. Guerin, "Specification of guaranteed quality of service," RFC 2212, Internet Engineering Task Force (IETF), Sept. 1997.

19. D. Clark and J. Wroclawski, "An approach to service allocation in the internet," IETF Draft, *draft-diff-svc-alloc-00.txt*, Jul. 1997.

20. K. Nichols, V. Jacobson, and L. Zhang, "A two-bit differentiated services architecture for the Internet," IETF Draft, *draft-nichols-diff-svc-arch-00.txt*, Nov. 1997.

21. J. Y. Hui, "Resource allocation for broadband networks," *IEEE J. Selected Areas Commun.*, vol. 6, no. 9, pp. 1598–1608, Dec. 1988.

22. R. Jain, "Congestion control and traffic management in ATM networks: recent advances and a survey," *Comput. Netw. ISDN Syst.*, vol. 28, no. 13, pp. 1723–1738, Oct. 1996.

23. J. Roberts, "Quality of service within the network: service differentiation and other issues," IEEE INFOCOM'98 Tutorial,1998.

24. S. Keshav, *An Engineering Approach to Computer Networking: ATM Networks, the Internet, and the Telephone Network*, Addison-Wesley, 1997.

25. X. Xiao and L. M. Ni, "Internet QoS: a big picture," *IEEE Netw. Maga.*, vol. 13, no. 2, pp. 8–18, Mar./Apr. 1999.

26. J. Kurose, "Multicast communication in wide area networks," IEEE INFOCOM'98 Tutorial, 1998.

CHAPTER 2

ADMISSION CONTROL

The statistical multiplexing nature of packet-switching networks refers to the fact that several variable-bit-rate connections (or flows) can share a link with a capacity less than the sum of their peak (maximum) bit rate requirements. Typical examples are provided by video, where the rate may depend on the scene being viewed, and traffic interconnecting LANs, which are bursty in nature. Since network resources (such as buffer and link capacity) are finite and are commonly shared by connections, the role of an admission control algorithm is to determine proper resource allocation for a new flow (if it is admitted) such that service commitments made by the network to existing flows won't be violated and the QoS requirement of the new flow will be satisfied. Here, the resource allocation can be considered [3] at a call level (i.e., the lifetime of a connection), a burst level (i.e., a short time interval when information is sent in bursts), or a packet level (i.e., packets are transmitted bit by bit at line rate). The service commitments can be quantitative (e.g., a guaranteed rate or bounded delay), or qualitative (e.g., a "low average delay").

According to [48], there are two basic approaches to admission control: parameter-based admission control (PBAC) and measurement-based admission control (MBAC). The PBAC computes the amount of network resources required to support a set of flows given *a priori* flow traffic characteristics. It is better suited for service models that offer a bounded-delay packet delivery service to support real-time applications. When a flow requests real-time service, it must characterize its traffic so that the network can make its admission control decision. Typically sources are described by either peak and average rates or a filter like a token bucket; these descriptions provide upper bounds on the traffic that can be generated by the source.

The MBAC relies on the measurement of actual traffic loads in making admission decisions. Given the reliance of MBAC algorithms on source behavior that is not static and is largely unknown in general, service commitments made by such algorithms can never be absolute. Measurement-based approaches to admission control can only be used in the context of service models that do not make guaranteed commitments. Thus, higher network utilization can be achieved at the expense of weakened service commitments. In this chapter, we investigate a number of PBAC and MBAC algorithms. Sections 2.1, 2,2, and 2.4.2 are based on [48] (interested readers are referred to [48] for further details). Section 2.1 explains the concept of deterministic bound. Probabilistic bounds are described in Section 2.2. Section 2.3 presents CAC algorithms that can be adopted for ATM VBR services. Section 2.4 discusses the CAC algorithms for Integrated Services Internet.

2.1 DETERMINISTIC BOUND

Traditional real-time service provides a *hard*, or absolute, bound on the delay of every packet. A deterministic guaranteed service provides for the worst-case requirements of flows. These requirements are usually computed from parameterized models of traffic sources. The source models used for this computation may be very complex, but the underlying admission control principle is conceptually simple: does granting a new request for service cause the worst-case behavior of the network to violate any delay bound?

The admission control algorithms proposed in [2, 5, 26] require sources to provide peak-rate characterizations of their traffic. The algorithms then check that the sum of all peak rates is less than the link capacity. If sources are willing to tolerate queuing delay, they can use a token bucket filter, instead of the peak rate, to describe their traffic. The network ensures that the sum of all admitted flows' token rate is less than link bandwidth, and the sum of all token bucket depths is less than available buffer space. This approach is proposed in [29]. In [37], the authors present an admission control algorithm for deterministic service based on the calculation of the maximum number of bits, $b(\tau)$, that can arrive from a source during any interval τ:

$$b(\tau) = \min\{\lceil p(\tau \bmod T) \rceil, \lceil \rho T \rceil\} + \left\lceil \frac{\tau}{T} \right\rceil \lceil \rho T \rceil, \qquad (2.1)$$

where T is the averaging interval for ρ, the source's average rate, and p is the source's peak rate. Queuing delay per switch is then calculated as

$$D = \frac{\max_{\tau \geq 0}\{\sum_{i=1}^{n} b_i(\tau) - rT\}}{r}, \qquad (2.2)$$

where n is the total number of flows and r is the link bandwidth. The admission control checks that D does not violate any delay bounds. This algorithm performs better than those requiring peak-rate characterizations and can achieve acceptable ($> 50\%$) link utilization when sources are not very bursty (peak-to-average rate ratio < 4) and the delay bound is not too tight (> 60 ms per switch). When flows are bursty, however, deterministic service ultimately results in low utilization. In [18, 19], the authors propose reshaping users' traffic according to network resources available at call setup time. While reshaping users' traffic according to available resources may increase network utilization, the reshaped traffic may not meet users' end-to-end quality requirements. Instead of imposing a traffic shaper at call setup time, [40, 45] propose characterizing different segments of a real-time stream and renegotiating the flow's resource reservation prior to the transmission of each segment. Renegotiation failure causes traffic from the next segment to be reshaped according to reservations already in place for the flow. This scheme may be applicable to video-on-demand applications where the entire data stream is available for *a priori* characterization prior to transmission.

2.2 PROBABILISTIC BOUND: EQUIVALENT BANDWIDTH

Statistical multiplexing is the interleaving of packets from different sources where the instantaneous degree of multiplexing is determined by the statistical characteristics of the sources. Contrast this with slotted time division multiplexing (TDM), for example, where packets from a source are served for a certain duration at fixed intervals and the degree of multiplexing is fixed by the number of sources that can be fitted into an interval. The probability density function (pdf) of statistically multiplexed independent sources is the convolution of the individual pdf's, and the probability that the aggregate traffic will reach the sum of the peak rates is infinitesimally small (10^{-48}), much smaller than the loss characteristics of physical links. The ATM network, for example, has a loss probability of 10^{-8}, in which case guaranteeing a 10^{-9} loss rate at the upper layer is sufficient [6]. Hence, networks that support statistical multiplexing can achieve a higher level of utilization without sacrificing much in QoS.

Probabilistic guaranteed service, such as the controlled load service (CLS), exploits this statistical observation and does not provide for the worst-case sum-of-peak-rates scenario. Instead, using the statistical characterizations of current and incoming traffic, it guarantees a bound on the probability of lost packets:

$$\Pr\{(\text{aggregate traffic} - \text{available bandwidth})\tau > \text{buffer}\} \le \varepsilon, \quad (2.3)$$

where τ is a time interval and ε is the desired loss rate. The aggregate traffic of the statistically multiplexed sources is called the *equivalent bandwidth* (or *effective bandwidth* or *equivalent capacity*) of the sources [10, 30]. The admission control using equivalent bandwidth is similar to that in a conventional multirate circuit-switching network. If there is enough bandwidth to accept the new flow, the flow is accepted; otherwise, it is rejected. The advantage of the equivalent bandwidth is its simplicity.

We now look at different approaches used to compute equivalent bandwidth. Let $X_i(\tau)$ be the instantaneous arrival rate of flow i during time period τ. Assume that $X_i(\tau)$'s are independent and identically distributed (i.i.d.). Let $S(\tau) = \sum_{i=1}^{n} X_i(\tau)$ be the instantaneous arrival rate of n flows. We want $S(\tau)$ such that

$$\Pr\{[S(\tau) - r]\tau > B\} \leq \varepsilon, \tag{2.4}$$

where B is the buffer size.

2.2.1   Bernoulli Trials and Binomial Distribution

In [11, 17], $X_i(\tau)$ is modeled as Bernoulli random variables. The aggregate arrival rate is then the convolution of the Bernoulli variables. The number of arrivals in a sequence of Bernoulli trials has a binomial distribution. Assuming sources are homogeneous two-state Markov processes, the previous convolution reduces to a binomial distribution [7, 14, 21]. This computation results in overestimation of actual bandwidth for sources with short burst periods, because the buffer allows short bursts to be smoothed out and the approximation does not take this smoothing effect into account [16].

In [35], instead of a single binomial random variable, the authors use a family of time-interval-dependent binomial random variables; that is, associated with each time interval is a binomial random variable that is stochastically larger than the actual bit rate generated. This method of modeling bit rates was first proposed in [23]. It allows a tighter bound on $S(\tau)$. The main drawback of modeling $S(\tau)$ with a binomial distribution is the cost of convolving the arrival probabilities of heterogeneous sources. In [35], for example, the authors suggest using the fast Fourier transform (FFT) to calculate the convolution. FFT has a complexity of $\Theta(n\hat{B} \log \hat{B})$, where $\hat{B}$ is the size of the largest burst from any source. Furthermore, when the number of sources multiplexed is small, this approximation of equivalent bandwidth underestimates the actual requirement [14].

2.2.2   Fluid-Flow Approximation

A fluid-flow model characterizes traffic as a Markov-modulated continuous stream of bits with peak and mean rates. Let c be the equivalent bandwidth of a source, as seen by a switch, computed using the fluid-flow approximation. In [16], c is computed by

$$c = \frac{\alpha b(1 - \rho)p - B + \sqrt{[\alpha b(1 - \rho)p - B]^2 + 4\alpha bB\rho(1 - \rho)p}}{2\alpha b(1 - \rho)}, \quad (2.5)$$

where b is the source's mean burst length, ρ is the source's utilization (average/peak), p is the source's peak rate, $\alpha = \ln(1/\varepsilon)$, and B is the switch's buffer size. The simple way to estimate the equivalent bandwidth of aggregate sources is to use the sum of the equivalent bandwidths of all sources. Although this scheme is very simple, it is effective only when the sources are not very bursty and have short average burst periods. When flows do not conform to this assumption, the bandwidth requirement is overestimated because the statistical multiplexing is not taken into account [16]. Equivalent bandwidths for more general source models have also been computed [1, 4, 13, 22, 27, 28]. Computing the equivalent bandwidth of a source using such a method depends only on the flow's fluid-flow characteristics and not on the number nor characteristics of other existing flows. The computation of equivalent bandwidth for general sources is, however, computationally expensive, with time complexity $O(n^3)$ [4].

2.2.3 Gaussian Distribution

In [6, 12, 16, 24, 33, 34], $S(\tau)$ is approximated with a Gaussian distribution. Given a desired loss rate, the equivalent capacity of the aggregate traffic is computed as $m + \alpha'\sigma$ according to [16], where $m = \sum_{i=1}^{n} \rho_i$, the mean aggregate bit rate; $\sigma^2 = \sum_{i=1}^{n} \sigma_i^2$, the variance of the aggregate bit rate (ρ_i and σ_i^2 are for source i); and $\alpha' = \sqrt{-2\ln \varepsilon - \ln 2\pi}$. This approximation tracks the actual bandwidth requirement well when there are many sources (e.g., more than 10 homogeneous sources) with long burst periods. When only a few sources are multiplexed, this approximation overestimates the required bandwidth. It also does so when sources have short bursts, because short bursts are smoothed out by the switch buffer. The approximation does not take this into account. [31] uses the minimum of the fluid-flow and Gaussian approximations, $C = \min\{m + \alpha'\sigma, \sum_{i=1}^{n} c_i\}$, in making admission control decisions, where c_i is the equivalent bandwidth of source i. When the sources have short bursts, $\sum_{i=1}^{n} c_i$ is adopted. Otherwise, $m + \alpha'\sigma$ is adopted.

2.2.4 Large-Deviation Approximation

Originally proposed in [3], an approximation based on the theory of large deviations was later generalized in [22] to handle resources with buffers. The theory of large deviations bounds the probability of rare events occurring. In this case, the rare event is $S(\tau) > r$ from (2.4). The approximations in [3, 22] are based on the Chernoff bound, while the one in [46] is based on the Hoeffding bound. The method is based on the asymptotic behavior of the tail of the queue length distribution. Further approaches to admission control based on the theory of large deviations are presented in [41, 42, 43].

2.2.5 Poisson Distribution

The above approximations of equivalent bandwidth all assume a high degree of statistical multiplexing. When the degree of statistical multiplexing is low, or when the buffer space is small, approximations based on Gaussian distribution and the theory of large deviations overestimate required bandwidth [16, 34, 46], while approximations using both fluid-flow characterization and binomial distribution underestimate it [8, 14, 15]. In such cases, [8, 14] suggest calculating equivalent bandwidth by solving for an $M/D/1/B$ queue, assuming Poisson arrivals.

2.2.6 Measurement-Based Methods

Each approach to computing equivalent bandwidth can be approximated by using measurements to determine the values of some of the parameters used [17, 34, 39, 41, 46, 47]. Two methods to estimate equivalent bandwidth are specifically suited to the measurement-based approach. The first is based on the Bayesian estimation method. From a given initial load and a set of recursive equations, one can estimate future loads from successive measurements. This approach is presented in [36, 44]. [36] further describes a hardware implementation of the measurement mechanism.

The second method is a table-driven method. An *admissible* region is a region of space within which service commitments are satisfied. The space is defined by the number of admitted flows from a finite set of flow types. The first approach to compute an admissible region uses simulation [9, 25, 32]. For a given number of flows from each flow type, simulate how many more flows of each type can be. admitted without violating service commitments. Running such simulations repeatedly with a different set of initial flow mixes, one eventually maps out the admissible region for the given flow types. The admissible region is encoded as a table and downloaded to the switches. When a prospective flow makes a reservation, the admission control algorithm looks up the table to determine whether admittance of this flow will cause the network to operate outside the admissible region; if not, the flow is admitted. The major drawbacks to this method for doing admission control are: (1) it supports only a finite number of flow types, and (2) the simulation process can be computationally intensive.

In [44], a Bayesian method is used to precompute an admissible region for a set of flow types. The admissible threshold is chosen to maximize the reward of increased utilization against the penalty of lost packets. The computation assumes knowledge of link bandwidth, size of switch buffer space, flows' token bucket filter parameters, flows' burstiness, and the desired loss rate. It also assumes Poisson call arrival process and independent, exponentially distributed call holding times. However, it is claimed in [44] that this algorithm is robust against fluctuations in the value of the assumed parameters. The measurement-based version of this algorithm ensures that the measured instantaneous load plus the peak rate of a new flow is below the admissible region. [20, 38] use a neural network to learn the admissible region for a given set of flow types.

2.3 CAC FOR ATM VBR SERVICES

In an ATM network, users must declare their source traffic characteristics to facilitate call admission control (CAC) and policing [49]. The source traffic can be characterized by the *usage parameter control* (UPC) descriptor, which includes the peak cell rate (PCR) λ_p, the sustainable cell rate (SCR) λ_s, and the maximum burst size (MBS) B_s. The CAC then tries to see whether there are sufficient resources to meet the individual user's QoS requirement. There has been extensive research on this subject [3, 16, 50–57]. Below are some examples for variable bit rate (VBR) service, where the QoS metric considered is the cell loss ratio.

2.3.1 Worst-Case Traffic Model and CAC

Deterministic rule-based traffic descriptors offer many advantages over statistical descriptors for specifying and monitoring traffic in ATM/B-ISDN [58]. The UPC descriptor, also called the dual-leaky-bucket-based descriptor, is a popular rule-based traffic descriptor. The CAC for rule-based sources raises the following issues: (1) characterization of the worst-case traffic compliant with the traffic descriptor (and hence allowed by the UPC to enter the network); (2) characterization of the combinations of sources that can be supported without violating the QoS requirements, assuming that each source submits its worst-case traffic to the network (compliant with the traffic descriptor).

In an integrated network, it would be nice to characterize the uniformly worst-case behavior in the general heterogeneous environment. Consider the departure process from the dual leaky buckets with the UPC parameters (λ_p, λ_s, B_s). This process can be modeled as an extremal, periodic on–off process with on and off periods given by

$$T_{\text{on}} = \frac{B_s}{\lambda_p}, \qquad T_{\text{off}} = \frac{B_s}{\lambda_p} \cdot \frac{\lambda_p - \lambda_s}{\lambda_s}; \qquad (2.6)$$

the probability of such a source being on is given by

$$P_{\text{on}} = \frac{T_{\text{on}}}{T_{\text{on}} + T_{\text{off}}} = \frac{\lambda_s}{\lambda_p}. \tag{2.7}$$

In [58], this process has been proven to be the worst-case source model for a bufferless multiplexer, but not always in the buffered case. However, in many situations this model can be considered as the worst-case source model in the sense of maximizing steady-state cell loss probability in an ATM node [50, 58–61]. In [58], it is shown that the CAC for such sources can be simplified by applying the equivalent bandwidth concept [refer to (2.5)].

2.3.2 Effective Bandwidth

The concept of effective bandwidth was first introduced in [3], where it reflects the source characteristics and the service requirements. In [50], Elwalid and Mitra show that, for general Markovian traffic sources, it is possible to assign a notional effective bandwidth to each source that is an explicitly identified, simply computed quantity with provably correct properties in the natural asymptotic regime of small loss probabilities.

According to [50], the source is assumed to be either *on* for an exponentially distributed length of time with a mean length $1/\beta$, emitting data at the peak rate λ_p, or *off* for another exponentially distributed length of time with a mean length $1/\alpha$, emitting no data, as shown in Figure 2.1.

Given the UPC parameters $(\lambda_p, \lambda_s, B_s)$, we can obtain α and β as follows, using the worst-case traffic model described in Section 2.3.1:

$$\alpha = \frac{1}{T_{\text{off}}} = \frac{\lambda_p \lambda_s}{B_s(\lambda_p - \lambda_s)}, \qquad \beta = \frac{1}{T_{\text{on}}} = \frac{\lambda_p}{B_s}. \tag{2.8}$$

Suppose a number of such sources share a buffer of size B, which is served by a channel of variable capacity c. The acceptable cell loss ratio is p. Define $\eta = (\log p)/B$. Then the effective bandwidth of all the sources (as if they

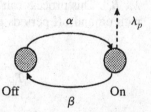

Figure 2.1 On–off source model.

constituted a single source in the system), e, is given by [50]

$$e = \text{MRE}\left(\Lambda - \frac{1}{\eta}M\right),\tag{2.9}$$

where MRE( ) represents the maximum real eigenvalue of a matrix, and

$$\Lambda = \begin{bmatrix} 0 & 0 \\ 0 & \lambda_p \end{bmatrix}, \qquad M = \begin{bmatrix} -\alpha & \beta \\ \alpha & -\beta \end{bmatrix}.\tag{2.10}$$

Based on the effective bandwidth, the admission criterion is satisfied if $e < c$, and violated if $e > c$. In [50], it is shown that the effective bandwidth is bounded by the peak and mean source rates, and is monotonic and concave with respect to a parameter of the admission criterion (such as CLR).

2.3.3 Lucent's CAC

To take advantage of the statistical multiplexing effect of a large buffer, another CAC was proposed in [53] for allocating buffers and bandwidth to heterogeneous, leaky-bucket-regulated traffic in an ATM node. This scheme has been implemented on Lucent's GlobeView- 2000 ATM switches [54], and thus we refer it to as Lucent's CAC.

The system model used in the algorithm is an ATM multiplexer with buffer size B, link capacity C, and dual-leaky-bucket-regulated input traffic sources, each specified by parameters $(\lambda_p, \lambda_s, B_T)$, where λ_s is also the token rate and B_T is the token buffer size of the regulator. The relation between B_T and the MBS B_s is given by

$$B_T = B_s\left(1 - \frac{\lambda_s}{\lambda_p}\right).\tag{2.11}$$

In addition, this model assumes that the outputs of the traffic regulators are extremal, periodic on-off processes with independent, uniformly distributed phases (i.e., the worst-case traffic model explained above).

The key idea behind the algorithm is a two-phase approach for solving the two-resource (i.e., bandwidth and buffer) allocation problem. In the first phase, it considers lossless multiplexing only, and a simple but efficient buffer-bandwidth allocation policy is used to make the two resources exchangeable. This step reduces the original two-resource problem to a single-resource problem and leads to the concept of lossless effective bandwidth.

The algorithm first considers the base case where there is zero cell loss. The lossless effective bandwidth e_0 for each homogeneous source is:

$$e_0 = \begin{cases} \dfrac{\lambda_p}{1 + \dfrac{B/C}{B_T}(\lambda_p - \lambda_s)} & \text{if } \lambda_s \leq \dfrac{B_T}{B/C}, \\[4ex] \lambda_s & \text{if } \dfrac{B_T}{B/C} \leq \lambda_s < \lambda_p. \end{cases} \quad (2.12)$$

In the second phase, the algorithm analyzes the effect of statistical multiplexing for the single-resource system by using established techniques, such as the Chernoff large-deviation approximation. In particular, a very small loss probability is allowed to extract statistical multiplexing gain from the time-varying unused portions of the resources allocated to the connections. The constraints are

$$\Pr\left\{\sum_i r_i(t) \geq C\right\} \leq \text{CLR}, \qquad \Pr\left\{\sum_i b_i(t) \geq B\right\} \leq \text{CLR}, \quad (2.13)$$

where $r_i(t)$ and $b_i(t)$ are the instantaneous rate and buffer occupancy of connection i, respectively. Using the Chernoff bound, one can find that

$$\Pr\left\{\sum_i r_i(t) \geq C\right\} \leq \exp[-F_K(s^*)], \quad (2.14)$$

where

$$F_K(s^*) = \sup_{s \geq 0} F_K(s), \quad (2.15)$$

$$F_K(s) = sC - \sum_{j=1}^{J} K_j \log\left(1 - \frac{\lambda_s^j}{e_{0,j}} + \frac{\lambda_s^j}{e_{0,j}}\exp(se_{0,j})\right), \quad (2.16)$$

J is the number of different traffic types, K_j is the number of connections that belong to traffic type j, $e_{0,j}$ is the lossless effective bandwidth for traffic type j, and λ_s^j is the sustainable cell rate for traffic type j.

According to [53], the lossy effective bandwidth e_j of traffic type j is given by

$$e_j = \frac{\log\left(1 - \dfrac{\lambda_s^j}{e_{0,j}} + \dfrac{\lambda_s^j}{e_{0,j}}\exp(se_{0,j})\right)}{s^* + \dfrac{\log L}{C}} \quad (j = 1, 2, \dots, J) \quad (2.17)$$

subject to the boundary condition $\sum_{j=1}^{J} K_j e_j = C$, where s^* is where $F_K(s)$ is maximized, and L is the CLR.

The statistical multiplexing gain g compares lossy effective bandwidth with lossless effective bandwidth:

$$g_j = e_{0,j}/e_j \quad (j = 1,2,\ldots,J).\qquad(2.18)$$

Intuitively, if C is large enough, there is a statistical multiplexing gain (i.e., $g > 1$); if C is small enough, there is none. Thus, there must exist a critical bandwidth C_c that is the smallest quantity such that the statistical multiplexing gain $g > 1$ for all $C > C_c$.

In other words, C_c determines whether statistical multiplexing gain can be obtained when connections are multiplexed. In a homogeneous system, if C is larger than the critical bandwidth of that traffic type, then a statistical multiplexing gain greater than unity can be achieved; otherwise, no more connections can be admitted than those determined by lossless allocation. Therefore, by comparing the critical bandwidth of a traffic type with the system bandwidth, one can classify the traffic type as either statistically multiplexible (S-VBR) or non-statistically-multiplexible (NS-VBR).

In general, the critical bandwidth of traffic type j is given by [53]

$$C_{c,j} = \frac{e_{0,j}\log(1/L)}{\log\left(e_{0,j}/\lambda_s^j\right)} \quad (j = 1,2,\ldots,J).\qquad(2.19)$$

Through numerical investigation, it has been shown in [53] that if connections of the same property (i.e., either all S-VBR or all NS-VBR) are multiplexed, the admission region boundary is almost linear. On the other hand, when S-VBR and NS-VBR connections are multiplexed together, the admission region boundary is nonlinear, but composed of piecewise linear segments.

2.3.4 NEC's CAC

To support multiple classes of services in an ATM switch, Ramamurthy and Ren proposed a multi-class CAC scheme [57], which we refer to as NEC's CAC, since the authors are with NEC. Since the CAC is executed for each class at each port of the switch, the multiclass connection admission control can be formulated as the following problem:

Suppose there are n existing class k (say, VBR) connections feeding to a single-server queue with an allocated bandwidth or service rate of C_k^{old} Mbit/s and a buffer space of B_k cells. The specified QoS guarantees (e.g., CLR) of these n connections are currently met by the system. When a new class k connection request arrives, the CAC has to determine the amount of new bandwidth C_k^{new} that is needed to ensure that the same QoS guarantees for all the $n + 1$ connections (including the new arrival) are met. The amount ΔC_k ($= C_k^{\text{new}} - C_k^{\text{old}}$) is the additional bandwidth required for class k if this new connection is to be admitted.

For VBR service, the NEC's CAC uses two multiplexer models based on the UPC descriptor. One is a lossless multiplexing model that performs best when the number of multiplexed sources, n, is small; the second is a statistical multiplexing model that performs best when n is large. Fluid-flow processes are used in both models. These two models are combined when the CAC has to make decisions on VBR sources.

2.3.4.1 Lossless Multiplexing

Assume that at an ATM node, the overall buffer allocated to VBR traffic is B_{vbr}, the maximum bandwidth allocated to VBR traffic is $C_{\text{vbr}}^{\text{max}}$, and input traffic sources are dual-leaky-bucket regulated. The ith connection has UPC parameters (λ_p^i, λ_s^i, B_s^i), $i = 1, 2, \ldots, n$. Let $C_{\text{vbr}}^{\text{old}}$ be the bandwidth currently allocated to the VBR class, supporting the n VBR connections.

The NEC's CAC first maps the departure process of each dual leaky bucket to the worst-case traffic model described in Section 2.3.1, with on and off periods given by

$$T_{\text{on}}^* = \frac{B_s^*}{\lambda_p^*}, \qquad T_{\text{off}}^* = \frac{B_s^*}{\lambda_p^*} \cdot \frac{\lambda_p^* - \lambda_s^*}{\lambda_s^*}. \tag{2.20}$$

The probability such a source is in the on state is given by

$$P_{\text{on}}^* = \frac{\lambda_s^*}{\lambda_p^*}. \tag{2.21}$$

When a new connection request is received with its UPC parameters (λ_p^*, λ_s^*, B_s^*), the CAC computes the new bandwidth $C_{\text{vbr}}^{\text{new}}$ for all $n + 1$ connections subject to zero loss probability as follows:

$$C_{\text{vbr}}^{\text{new}} = \max\left\{ \left(\lambda_p^* + \sum_{i=1}^{n} \lambda_p^i \right)\left(1 - \frac{B_{\text{vbr}}}{B_s^* + \sum_{i=1}^{n} B_s^i} \right)^+ , \lambda_s^* + \sum_{i=1}^{n} \lambda_s^i \right\}, \tag{2.22}$$

where $(x)^+ = \max(x, 0)$. Therefore, the additional bandwidth required to support the new connection is given by

$$\delta_1 = C_{\text{vbr}}^{\text{new}} - C_{\text{vbr}}^{\text{old}}. \tag{2.23}$$

2.3.4.2 Statistical Multiplexing

The above model is overly conservative when n is large. To take into account the statistical multiplexing gain, the NEC's CAC uses an alternative approach to achieve better statistical multiplexing gain.

It is well known that the cell loss rate of a multiplexer depends on the maximum burst length generated by the source with respect to the amount of buffering. To consider this, the algorithm modifies the above on–off model as follows. It defines a basic time interval

$$T_N = \frac{B_{vbr}}{2C_{vbr}^{max}}, \quad (2.24)$$

which can be viewed as the time to drain buffers that are half full. Then the CAC constructs a new two-state model for the source. The two states are a *high*-rate state with rate λ_H^*, and a *low*-rate state with rate λ_L^*. These states occur with probability P_{on}^* and $1 - P_{on}^*$, respectively. The high and the low rates are given by

$$\lambda_H^* = \min\left(1, \frac{T_{on}^*}{T_N}\right)\lambda_p^* + \max\left(0, 1 - \frac{T_{on}^*}{T_N}\right)\lambda_s^*$$

$$\lambda_L^* = \max\left(0, 1 - \frac{T_{on}^*}{T_N}\right)\lambda_s^*. \quad (2.25)$$

Note that the modified source has the same average rate λ_s^* as the original one. Thus, sources with burst sizes that are small compared to the buffer size are modeled by their average characteristics, while sources for which T_{on} is greater than T_N are modeled closer to the worst-case model and not significantly altered by this model.

Based on the modified source model, the algorithm approximates the rate of the aggregate traffic stream $R(t)$ of all VBR connections by a Gaussian process. In [57] it has been shown that, if $R(t)$ is approximated by a Gaussian process with mean μ_{new} and variance σ_{new}^2 given by

$$\mu_{new} = \mu_{old} + \lambda_s^*, \quad (2.26)$$

$$\sigma_{new}^2 = \sigma_{old}^2 + (\lambda_H^* - \lambda_L^*)^2 P_{on}^*(1 - P_{on}^*), \quad (2.27)$$

then by solving the inequality

$$CLR \stackrel{def}{=} \frac{E[R(t) - C]^+}{E[R(t)]} < \varepsilon$$

for the bandwidth allocated to VBR connections, C, we can get the new overall bandwidth, C_{vbr}^{new}, required to support all VBR connections (including the new request) with QoS guarantee of CLR $< \varepsilon$ (i.e., a required limit on cell loss rate), as

$$C_{vbr}^{new} = \mu_{new} + a\sigma_{new}, \quad (2.28)$$

where

$$\eta = \frac{\mu_{\text{new}}\sqrt{2\pi}}{\sigma_{\text{new}}} \cdot \varepsilon, \tag{2.29}$$

$$a \approx 1.8 - 0.46 \log_{10}\eta. \tag{2.30}$$

Therefore, the required additional bandwidth based on the statistical multiplexing model is given by

$$\delta_2 = C_{\text{vbr}}^{\text{new}} - C_{\text{vbr}}^{\text{old}}. \tag{2.31}$$

Combining the lossless multiplexer and statistical multiplexer models, the additional bandwidth Δ_{vbr} is then given by

$$\Delta_{\text{vbr}} = \min\{\delta_1, \delta_2\}. \tag{2.32}$$

The new VBR connection is admitted if a capacity Δ_{vbr} is available in the free pool of bandwidth.

2.3.5   Tagged-Probability-Based CAC

Due to the complexity of directly computing cell loss (i.e., CLR), [70] proposes another approach based on a tight upper bound on the loss probability, called the *tagged probability* [63], which can be computed efficiently. This technique is described below. We then describe three modified CAC algorithms based on NEC's CAC, effective bandwidth, and Lucent's CAC, respectively.

2.3.5.1 Tagged Probability   By definition, the exact loss probability (CLR) in a queuing system with finite buffer B and service rate C, as shown in Figure 2.2(a), is given by

$$\text{CLR} = \frac{E\left[(\lambda - C)^+ | X = B\right]}{\overline{\lambda}} P_{\text{FB}}(X = B), \tag{2.33}$$

where $\overline{\lambda}$ is the average arrival rate of all the sources, λ is the aggregate instantaneous rate, X is the queue length, and $P_{\text{FB}}(X = B)$ is the probability of $X = B$ in the finite-buffer model.

Since the CLR is not readily obtained, many CAC algorithms use the overflow probability, $P_{\text{IB}}(X > B)$, which is the probability that the queue length in an infinite-buffer model exceeds a threshold B, as shown in Figure 2.2(b). This is because $P_{\text{IB}}(X > B)$ is usually an upper bound on CLR.

The tagged probability [63] has been introduced to tighten the bound $P_{\text{IB}}(X > B)$. Consider an infinite-buffer model with threshold B. When the

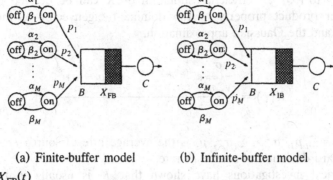

(a) Finite-buffer model   (b) Infinite-buffer model

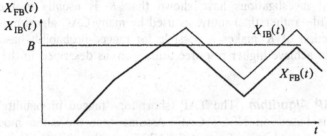

(c) Queue length variation in the different models

Fig. 2.2 Comparison between finite-buffer and infinite-buffer models (subscripts FB and IB used with X are for illustration here).

queue length exceeds the threshold B, the incoming traffic arriving at a rate in excess of C is tagged according to [63]. The tagged probability T_g can be defined by

$$T_g = \frac{E\left[(\lambda - C)^+ | X > B\right]}{\bar{\lambda}} P_{IB}(X > B). \qquad (2.34)$$

Comparing (2.34) with (2.33), [63] shows that the tagged probability T_g is greater than the CLR. Therefore, T_g can be used as an upper bound on the CLR. Define

$$K = \frac{E\left[(\lambda - c)^+ | X > B\right]}{\bar{\lambda}}, \qquad _{ov} = P_{IB}(X > B).$$

We have

$$CLR \leq T_g = KP_{ov}. \qquad (2.35)$$

Thus, we can regard K as the ratio of loss probability to overflow probability by using

$$CLR \approx KP_{ov}. \qquad (2.36)$$

According to [63], an efficient estimation of K can be derived using the Kronecker product property of the dominant eigenvector of the system equation and the Gaussian approximation,

$$K(c) = \frac{\frac{\sigma}{\sqrt{2\pi}}e^{-(c-\mu)^2/2\sigma^2} + (\mu - c)^+}{\bar{\lambda}}, \tag{2.37}$$

where $\mu = \sum_i \mu_i$, $\sigma^2 = \sum_i \sigma_i^2$, μ_i is the average rate of source i, and σ_i is the standard deviation of source i's rate.

Numerical investigations have shown that K is usually in the range 0.001–0.1 [70]—rather than unity, as used by many CAC algorithms. Thus, the introduction of K makes it possible for tagged-probability-based CAC algorithms to achieve higher resource utilization, as described in the following.

2.3.5.2 TAP Algorithm The TAP (short for "tagged probability") algorithm [70] is based on NEC's CAC. Assume that the same model and parameters are used as in Section 2.3.4.2. From the above discussion, we have

$$\varepsilon = P_{ov} = \frac{\text{CLR}}{K}. \tag{2.38}$$

This is different from that of Section 2.3.4.2. Let us define a new threshold T_N and a parameter k_N:

$$T_N = s \cdot \frac{B_{vbr}}{C_{vbr}^{max}}, \tag{2.39}$$

$$k_N = \frac{T_{on}^*}{T_N}, \tag{2.40}$$

where a factor s is introduced to investigate the effectiveness and sensitivity of NEC's modified source model in Section 2.3.4.2. When $s = 1$, T_N is the longest queueing time of a packet in a FIFO queue when the buffer is full.

Define the effective buffer

$$b_E = \frac{B_s}{s}. \tag{2.41}$$

Thus, s is the ratio of the maximum burst size B_s to the effective buffer. Increasing the ratio s can increase the system utilization, as discussed later.

Similarly to Section 2.3.4.2, we can construct another two-state source model with high rate λ_H^* and low rate λ_L^* given by

$$
\lambda_H^* = \begin{cases} k_N \lambda_p^* + (1 - k_N)\lambda_s^* & \text{if } k_N < 1 \text{ or } \dfrac{B_s^*}{sB_{\mathrm{vbr}}} < \dfrac{\lambda_p}{C_{\mathrm{vbr}}^{\max}}, \\[3mm] \lambda_p^* & \text{if } k_N \geq 1 \text{ or } \dfrac{B_s^*}{sB_{\mathrm{vbr}}} \geq \dfrac{\lambda_p}{C_{\mathrm{vbr}}^{\max}}, \end{cases} \tag{2.42}
$$

$$
g\lambda_L^* = \begin{cases} (1 - k_N)\lambda_s^* & \text{if } k_N < 1 \text{ or } \dfrac{B_s^*}{sB_{\mathrm{vbr}}} < \dfrac{\lambda_p}{C_{\mathrm{vbr}}^{\max}}, \\[3mm] 0 & \text{if } k_N \geq 1 \text{ or } \dfrac{B_s^*}{sB_{\mathrm{vbr}}} \geq \dfrac{\lambda_p}{C_{\mathrm{vbr}}^{\max}}. \end{cases} \tag{2.43}
$$

When $s = \frac{1}{2}$, the TAP algorithm becomes NEC's CAC. Henceforth, we choose $s = \frac{1}{2}$ for the sake of comparison.

Next we show some important results from [70], where, unless otherwise stated, all references to proofs are from [70].

Proposition 2.1 $K(c)$ *from (2.37) is monotonic decreasing with increasing* c *when* $c \geq \mu$.

Corollary 2.1 *Given* $\mu < C_{\mathrm{vbr}}^{\mathrm{old}} < C_{\mathrm{vbr}}^{\mathrm{new}}$, *one has* $K(C_{\mathrm{vbr}}^{\mathrm{old}}) > K(C_{\mathrm{vbr}}^{\mathrm{new}})$.

Proposition 2.2 *Define* $g(a) = \mu + a\sigma$, $a(\eta) = 1.8 - 0.46 \log_{10}(\eta)$, $\eta(\varepsilon) = (\mu\sqrt{2\pi}/\sigma)\varepsilon$, $\varepsilon(K) = \mathrm{CLR}/K$. *Then* $g(K)$ *is monotonic increasing with increasing* K $(K > 0)$.

Thus, for a new connection request with UPC parameters $(\lambda_p^*, \lambda_s^*, B_s^*)$, its CLR can be approximated by

$$
\mathrm{CLR} \approx KP_{\mathrm{ov}} = K\varepsilon. \tag{2.44}
$$

According to (2.37), K can be derived as

$$
K = \frac{\left(\sigma_{\mathrm{new}}/\sqrt{2\pi}\right)e^{-(C_{\mathrm{vbr}}^{\mathrm{old}} - \mu_{\mathrm{old}})^2/2\sigma_{\mathrm{new}}^2} + \left(\mu_{\mathrm{old}} - C_{\mathrm{vbr}}^{\mathrm{old}}\right)^+}{\mu_{\mathrm{old}}}, \tag{2.45}
$$

where $C_{\mathrm{vbr}}^{\mathrm{old}}$ is the sum of the effective bandwidths of all existing connections, μ_{old} is the sum of the λ_s's, and

$$
\sigma_{\mathrm{new}}^2 = \sigma_{\mathrm{old}}^2 + \lambda_s^*(\lambda_p^* - \lambda_s^*). \tag{2.46}
$$

Then, we can calculate ε using (2.38), calculate C_{vbr}^{new} through (2.28) to (2.30), and finally get the required additional bandwidth for the new connection from (2.31) and (2.32).

Since $0 < K < 1$, from Proposition 2.2 we have $g(K) < g(1)$, where $g(K)$ is the estimated bandwidth by TAP and $g(1)$ is the estimated bandwidth by NEC's CAC. This means TAP can admit more connections with the same link capacity. Therefore,

Proposition 2.3 *TAP achieves higher utilization than NEC's CAC when* $0 < K < 1$.

2.3.5.3 Modified Effective Bandwidth Assume we have the same model and parameters as those in Section 2.3.2. From the discussion in Section 2.3.5.1, we have $p = P_{ov} = CLR/K$. Thus η is given by

$$\eta = \frac{\log P_{ov}}{B} = \frac{\log(CLR/K)}{B}. \tag{2.47}$$

Then the effective bandwidth of the source, e, is also given by (2.9).

Fact 2.1 *The maximum real eigenvalue $e(\eta)$ of the essentially nonnegative matrix $\Lambda(\eta) = \Lambda - (1/\eta)M$ is monotonic decreasing with increasing η: $e'(\eta) < 0$ ($\eta < 0$).*

Proposition 2.4 $\eta(K)$ *is monotonic decreasing with increasing* K ($K > 0$).

Proposition 2.5 *The effective bandwidth $e(K)$ is monotonic increasing with increasing* K ($K > 0$).

Similarly, for a new connection request with UPC parameters $(\lambda_p^*, \lambda_s^*, B_s^*)$, its CLR can be approximated from (2.44). After obtaining K through (2.45), we can calculate η using (2.47) and the effective bandwidth of the new connection from (2.9). This is called the *modified effective bandwidth* (MEB).

Since $0 < K < 1$, from Proposition 2.5, $e(K) < e(1)$, where $e(K)$ is the MEB and $e(1)$ is the bandwidth estimated by the CAC based on the original effective bandwidth. This means the MEB can admit more connections with the same link capacity. Thus,

Proposition 2.6 *The MEB achieves higher utilization than the CAC based on effective bandwidth when* $0 < K < 1$.

2.3.5.4 Modified Statistical Multiplexing Algorithm The MSM algorithm is based on Lucent's CAC. Again, assume the same model and parameters are used as those in Section 2.3.3. From the discussion in Section 2.3.5.1, we have

$$L = P_{ov} = \frac{CLR}{K}. \tag{2.48}$$

Proposition 2.7 $e_j(K)$ *from* (2.17) *is monotonic increasing with increasing* K ($K > 0$), *where* L *and* $K(c)$ *are from* (2.48) *and* (2.37), *respectively.*

From Corollary 2.1, we know $K(C_{vbr}^{old})$ is an upper bound of $K(C_{vbr}^{new})$ when $\mu < C_{vbr}^{old} < C_{vbr}^{new}$. Furthermore, Proposition 2.7 tells us $e_j(C_{vbr}^{old}) > e_j(C_{vbr}^{new})$. So $e_j(C_{vbr}^{old})$ is a conservative estimate. We can use (2.45) to obtain K. The effective bandwidth e_j of traffic type j can then be derived from (2.15) and (2.16).

Since $0 < K < 1$, from Proposition 2.7, $e_j(K) < e_j(1)$, where $e_j(K)$ is the estimated bandwidth by MSM and $e_j(1)$ is the estimated bandwidth by Lucent's CAC. This means MSM can admit more connections with the same link capacity. Thus,

Proposition 2.8 *MSM can achieve higher utilization than Lucent's CAC when* $0 < K < 1$.

2.3.5.5 Performance Evaluation
To simulate the admission control under realistic traffic conditions, two traces of bursty Motion Picture Expert Group (MPEG) video are used [70].[1] The admissible regions of the CAC schemes of interest will be determined for the traces through numerical methods. The numerical investigations are based on the UPC parameters of the video traces that were obtained by Mark and Ramamurthy's estimation procedure [68]. Since there is no exact analysis for the traces, trace-driven simulations are used as criteria to evaluate the utilization of different CAC schemes given the same UPC parameters of the sources.

In the example in Table 2.1, an MPEG- 1 coded sequence called a *Mobi trace* is used. This empirical data sequence has a peak cell rate of 17.38 Mbit/s and an average cell rate of 8.58 Mbit/s. Using the method developed in [68], we can choose UPC parameters ($\lambda_p = 17.38$ Mbit/s, $\lambda_s = 11.87$ Mbit/s, $B_s = 83$ cells) to characterize the Mobi trace.

Consider an ATM multiplexer with 500-cell buffer and CLR = 10^{-5}. Given the above UPC parameters for the Mobi trace and the different link capacities (i.e., $C_{vbr} = 45, 100, 150$, and 300 Mbit/s), Table 2.1 shows the maximum number of Mobi traces that can be admitted without violating the CLR requirements for various schemes and simulations. The equivalent bandwidth is based on [16]. The private network–network interface (PNNI) CAC is the generic CAC proposed by the ATM Forum [52].

Table 2.1 shows that TAP and NEC's CAC have the same performance for Mobi traces. This is because TAP and NEC's CAC use the same algorithm

---

[1]According to [64–66], the exponential on–off sources often used for evaluating CAC schemes have properties quite different from those observed in measurement studies of many actual traffic streams. In [67], it has been shown that CAC schemes that work well with exponential on–off sources can suffer from considerable inaccuracies when applied to multiple-time-scale sources, such as compressed VBR video.

TABLE 2.1 CAC Performance Comparison Using Mobi Traces[a]

	Link Capacity (Buffer = 500 cells)			
CAC Mechanism	45	100	150	300 Mbit/s
Simulation	4	9	14	28
NEC's CAC	3	8	12	23
TAP	3	8	12	23
Effective bandwidth	3	7	10	21
MEB	3	7	11	22
Lucent's CAC	3	8	12	23
MSM	3	8	12	23
Equivalent bandwidth	3	7	10	21
PNNI	2	5	8	17
Peak Rate	2	5	8	17

[a]The numbers shown in the table are the numbers of MPEG-1 Mobi trace connections that can be admitted by various CAC schemes. ©1999 IEEE.

for lossless multiplexing described in Section 2.3.4.2, and only the lossless equivalent bandwidth is considered, since the lossless equivalent bandwidth (i.e.,the equivalent bandwidth from lossless multiplexing) is less than the lossy equivalent bandwidth (i.e., the equivalent bandwidth from statistical multiplexing) for Mobi traces. Table 2.1 also shows that Lucent's CAC has the same performance as that of NEC's CAC. This is because the two of them use similar assumptions in lossless multiplexing: the ratio between the buffer size and the bandwidth allocated to the individual connections is proportional to the corresponding ratio for total VBR traffic.

In the example in Table 2.2, a more bursty MPEG-1 coded trace called Flower Garden is used. This trace has a peak cell rate of 13.91 Mbit/s and an average cell rate of 3.1 Mbit/s. Based on [68], the UPC parameters

TABLE 2.2 CAC Performance Comparison Using Flower Garden Traces[a]

	Link Capacity (Buffer = 500 cells)			
CAC Mechanisms	45	100	150	300Mbit/s
Simulation	8	17	30	77
NEC's CAC	3	8	16	42
TAP	3	11	20	52
Effective bandwidth	3	7	11	22
MEB	3	7	13	37
Lucent's CAC	3	7	13	26
MSM	3	11	20	50
Equivalent bandwidth	3	7	13	36
PNNI	3	7	13	36
Peak rate	3	7	13	36

[a]The numbers shown in the table are the numbers of MPEG-1 Flower Garden trace connections that can be admitted by various CAC schemes. ©1999 IEEE.

chosen for this trace are $\lambda_p = 13.91$ Mbit/s, $\lambda_s = 3.6$ Mbit/s, $B_s = 1479$ cells. The QoS constraint is CLR $= 10^{-5}$.

Given the above UPC parameters for the Flower Garden trace and the different link capacities (i.e., $C_{\mathrm{vbr}} = 45$ Mbit/s, 100 Mbit/s, 150 Mbit/s, and 300 Mbit/s), Table 2.2 shows the maximum number of Flower Garden traces that can be admitted without violating the CLR requirements for various schemes and simulations.

Table 2.2 shows that TAP is more efficient than any of the other CAC schemes and MSM has comparable performance to that of TAP. Table 2.2 also shows that the larger the link capacity is, the more statistical multiplexing gain TAP achieves. It is noteworthy that Lucent's CAC, which is based on the Chernoff bound, has the same performance as that of equivalent bandwidth, which is based on Gaussian approximation and effective bandwidth. MEB is more efficient than Lucent's CAC and equivalent bandwidth when the link capacity is large.

Even TAP is still very conservative, because TAP still uses an on–off model (λ_p^*, 0) when $T_{\mathrm{on}}^* > T_N$, that is, $B_s^*/(B_{\mathrm{vbr}}/2) > \lambda_p^*/C_{\mathrm{vbr}}$. Frames with a large size like B_s may be very rare. Thus, without more information, such as the rate or frame size distribution, UPC-parameter-based CAC schemes can only choose a conservative estimation.

Since there is more statistical multiplexing gain when the link capacity is large, T_N should change as the link capacity changes. This is the very reason why the ratio s is introduced in the modified source model described in Section 2.3.5.2.

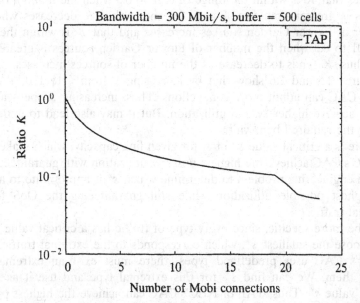

Fig. 2.3   Effect of the number of Mobi connections on ratio K. (©1999 IEEE.)

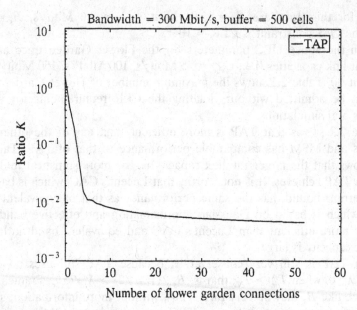

Fig. 2.4   Effect of the number of Flower Garden connections on ratio K. (©1999 IEEE.)

Figure 2.3 shows that the ratio K of loss probability to overflow probability decreases when the number of Mobi sources increases. This figure also indicates that K is within the range of 0.01 to 0.5 when the number of Mobi sources is greater than two. Figure 2.4 shows that K decreases when the number of Flower Garden sources increases and that K is within the range of 0.001 to 0.5 when the number of Flower Garden sources is greater than two. Thus, K tends to decrease as the number of sources increases.

Figures 2.5 and 2.6 show that by increasing s from $\frac{1}{2}$ to 170, TAP and NEC's CAC can admit more connections. Thus, increasing s opens another way to achieve higher system utilization. But it may also lead to under-estimating the required bandwidth.

There is a critical value s^* of s, for given link capacity, which makes TAP or NEC's CAC achieve the highest possible utilization with guaranteed QoS. It is an engineering problem to determine which s^* is appropriate to achieve the highest possible utilization while still guaranteeing the QoS for the extremal traffic.

To be more specific, since every type of traffic has a critical value s^*, we can choose the smallest s^*, which corresponds to the extremal traffic. Since Lucent's CAC uses predefined types, there must exist an extremal type among them. We can find s_j^* for this extremal type and use it as system critical value s^*. Thus, TAP or NEC's CAC can achieve the highest possible utilization with guaranteed QoS.

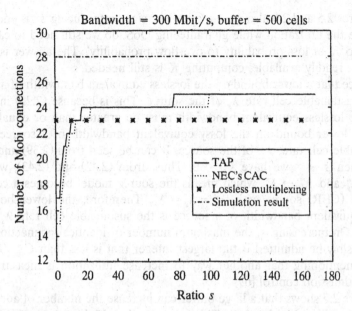

Fig. 2.5 Effect of the ratio s on the number of admissible connections for Mobi traces. (©1999 IEEE.)

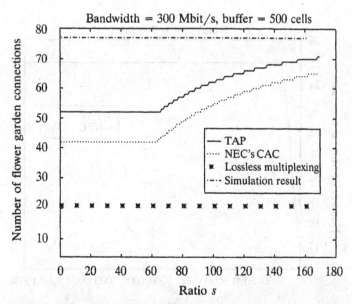

Fig. 2.6 Effect of the ratio s on the number of admissible connections for Flower Garden traces. (©1999 IEEE.)

Figures 2.5 and 2.6 also raise a question: since introducing s^* is enough to increase the utilization while guaranteeing QoS, do we still need to calculate the ratio K of loss probability to overflow probability? The answer is that if s^* is not readily available, computing K is still needed.

Notice that a lower bound on the lossless equivalent bandwidth of a source is the sustainable cell rate λ_s of the source. This is because (2.22) indicates that the lossless equivalent bandwidth must be greater than or equal to λ_s. Also, a lower bound on the lossy equivalent bandwidth of a source is the sustainable cell rate, λ_s, of the source. It can be seen from (2.39) and (2.40) that when $s \to \infty$ we have $k_N \to 0$. Thus, from (2.42) and (2.43), we have $\lambda_H^* \to \lambda_s^*$ and $\lambda_L^* \to \lambda_s^*$, which means the source model becomes a constant bit rate (CBR) source with $\lambda_p = \lambda_s = \lambda_s^*$. Therefore, the lower bound of lossy equivalent bandwidth of a source is the sustainable cell rate λ_s of the source. On increasing s, the maximum number of identical connections that can possibly be admitted is the largest integer that is less than C_{vbr}/λ_s. It is worth mentioning that another way to increase utilization is measurement-based admission control [69].

Figure 2.7 shows that a large buffer can increase the number of admissible connections for Mobi traces. This suggests that Mobi sources (which are NS-VBR) can benefit from a large buffer size. Figure 2.8 shows that the buffer size has no effect on the number of admissible connections for Flower Garden traces, which suggests that Flower Garden sources (which are S-VBR)

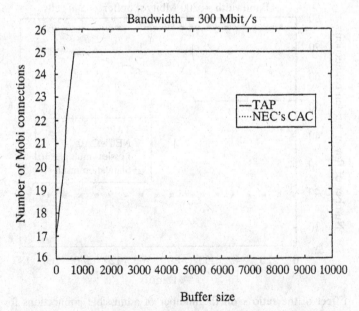

Fig. 2.7 Effect of the buffer size on the number of admissible connections for Mobi traces.

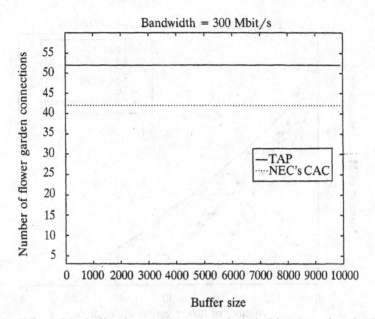

Fig. 2.8 Effect of the buffer size on the number of admissible connections for Flower Garden traces.

cannot gain anything from a large buffer size. Figures 2.7 and 2.8 also shed some light on what traffic types shared-memory ATM switches may have an effect on.

Figure 2.9 shows that

- The effective bandwidth algorithm has the smallest admissible region.
- MEB and Lucent's CAC have comparable performance. When the number of Flower Garden connections is greater than 17, MEB performs somewhat better than Lucent's CAC. This suggests that MEB will perform somewhat better than Lucent's CAC when the number of S-VBR sources is large.
- NEC's CAC performs better than effective bandwidth when the number of Flower Garden connections is greater than 11. It also performs better than Lucent's CAC when the number of Flower Garden connections is greater than 20, and better than MEB when the number of Flower Garden connections is greater than 23. This suggests that NEC's CAC performs better than MEB, Lucent's CAC, and effective bandwidth when the number of S-VBR sources is large. This is because NEC's CAC is based on Gaussian approximation, and Gaussian approximation assumes a large number of connections (i.e., S-VBR sources in this example).
- TAP even performs worse than effective bandwidth when the number of Flower Garden connections is within the range 2 to 6. NEC's CAC

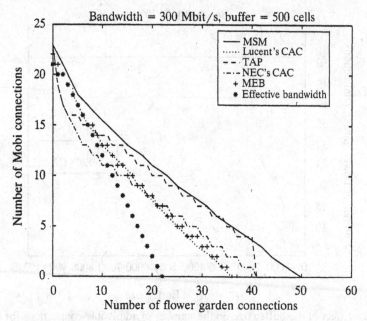

Fig. 2.9   Admissible region for TAP, NEC's CAC, MEB, effective bandwidth, MSM, and Lucent's CAC for CLR $= 10^{-5}$. (©1999 IEEE.)

performs worse than effective bandwidth when the number of Flower Garden connections is within the range 2 to 10. This is simply because there is no statistical multiplexing gain for TAP or NEC's CAC when the number of Flower Garden connections (i.e., S-VBR sources), is small.

- TAP performs better than NEC's CAC, MEB, Lucent's CAC, and effective bandwidth when the number of Flower Garden connections is greater than 9.

- MSM always performs better than NEC's CAC, MEB, Lucent's CAC, and effective bandwidth. It almost always performs better than TAP when the number of Flower Garden connections is within the range 1 to 49. But when there are only Flower Garden connections, the maximum number of connections that MSM can admit (i.e., 50) is 2 less than the maximum admitted by TAP (i.e., 52).

To evaluate the performance of various schemes under a very low CLR requirement, such as CLR $= 10^{-9}$, computations are performed [70] under the same environment as that in Figure 2.9. The results shown in Figure 2.10 are similar to those shown in Figure 2.9, which corroborates the effectiveness of the TAP algorithm under various QoS requirements. Although it has been shown that MSM has the largest admissible region, it also introduces the

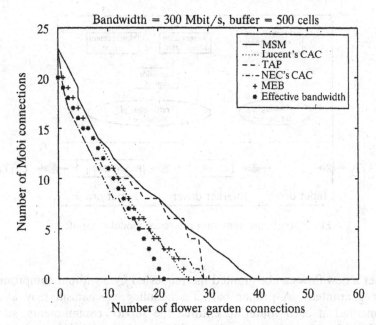

Fig. 2.10 Admissible region for TAP, NEC's CAC, MEB, effective bandwidth, MSM, and Lucent's CAC for CLR = 10^{-9}.

highest complexity (using both the large-deviation approximation and the Gaussian approximation). Therefore, the TAP algorithm seems more promising for CAC on high-speed ATM switches, due to its simplicity and efficiency [70].

2.4 CAC FOR INTEGRATED SERVICES INTERNET

As mentioned in Section 1.2.2, a new service architecture—integrated services Internet (ISI)—is needed to support real-time applications, such as remote video and multimedia conferencing, with various QoS requirements [75, 76]. Two classes of QoS-specific service are currently defined: *guaranteed service* (GS) and *controlled-load service* (CLS) [75–79].

The support of different service classes requires routers to explicitly manage their bandwidth and buffer resources to provide QoS for specific flows. This implies that *resource reservation, admission control, packet scheduling,* and *buffer management* are key building blocks of the ISI.

In particular, a signaling protocol is needed to perform the resource reservation for each flow, such as the RSVP [77]. The RSVP should also be responsible for maintaining the dynamic flow-specific states in the connectionless Internet. Admission control is needed at each router to determine

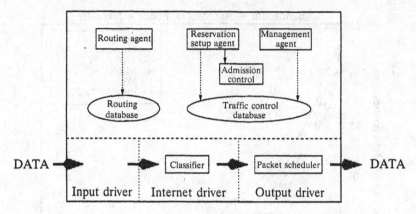

Fig. 2.11 Implementation reference model for routers.

whether a new flow can be granted the requested QoS without compromising earlier guarantees. A proper packet scheduling mechanism may also be implemented at each router to enforce the service commitments, such as delay bound and/or loss rate. In addition, a packet classifier is required to distinguish packets of different flows according to their individual information, such as source and destination IP addresses.

Figure 2.11 shows how these components might fit into an IP router that has been extended to provide integrated services [76]. The router has two broad functional divisions: the forwarding path below the dashed horizontal line, and the background code above the line.

The forwarding path of the router is executed for every packet and is divided into three sections: input driver, Internet forwarder, and output driver. The Internet forwarder interprets the internet working protocol header appropriate to the protocol suite (e.g., the IP header for TCP/IP). For each packet, an Internet forwarder executes a suite-dependent classifier and then passes the packet and its class to the appropriate output driver. The output driver implements the packet scheduler. It has two distinct sections: the packet scheduler, which is largely independent of the detailed mechanics of the interface, and the actual I/O driver.

The background code is simply loaded into router memory and executed by a general-purpose CPU. These background routines create data structures that control the forwarding path. The routing agent implements a particular routing protocol and builds a routing database. The reservation setup agent implements the protocol used to set up resource reservations. If admission control grants a new request, the appropriate changes are made to the classifier and packet scheduler database to implement the desired QoS. Finally, every router supports an agent for network management. This agent must be able to modify the classifier and packet scheduler databases to set up controlled link sharing and to set admission control policies.

The implementation framework for a host is generally similar to that for a router, with the addition of applications. Rather than being forwarded, host data originate and terminate in an application. An application needing a GS or CLS for a flow must somehow invoke a local reservation setup agent. The IP output routine of a host may need no classifier, since the class assignment for a packet can be specified in the local I/O control structure corresponding to the flow.

Below we briefly describe the frameworks of the GS and the CLS and the corresponding admission control. Further details are in [74–79].

2.4.1 Guaranteed Quality of Service

Guaranteed service (GS) is based on the framework of a weighted fair queuing service policy [71–74] combined with token-bucket-constrained access control to achieve firm (mathematically provable) bounds on end-to-end delay. Assume a flow is constrained with a token bucket (r, b), where r is the source average data rate and b is the burst length of source traffic, plus a peak rate p, a minimum policed unit m, and a maximum packet size M. The network (routers) provides a guaranteed bandwidth $R > r$. Below we show how the end-to-end delay can be bounded. The GS is then described in more detail.

2.4.1.1 *Bounds on Delay and Buffer Requirement*   First, we use the notion of an *envelope* process to bound the amount of traffic that arrives at the network ingress in any given time interval. Let $A(t, t + \tau)$ denote the amount of traffic that arrives in the interval $(t, t + \tau]$. We assume that an envelop function, $A(\tau)$, exists such that

$$A(t, t + \tau) \le A(\tau), \qquad t \ge 0, \quad \tau \ge 0. \tag{2.49}$$

By definition, an (r, b, p) flow has an envelop $A(\tau)$, which is given by

$$A(\tau) = \min\{M + p\tau, b + r\tau\}, \qquad \tau \ge 0. \tag{2.50}$$

A network element's (router or host's) implementation of GS is characterized by two error terms, C and D, which represent how the element's implementation of the GS deviates from the fluid model. The error term C is the rate-dependent one. It represents the delay a datagram in the flow might experience due to the rate parameters of the flow. The error term D is the rate-independent, per-element one and represents the worst-case non-rate-based transit time variation through the service element. By definition [74, 79] each network element j must ensure the delay of any packet of the flow be less than

$$\frac{b}{R} + \frac{C_j}{R} + D_j. \tag{2.51}$$

Consider generalized processor sharing (GPS) [72, 73] implementing the bandwidth guarantee mechanism. A newly arriving packet of a flow (which assumably has no backlog at this element currently) may experience a maximum delay

$$\frac{M}{R} + \frac{\text{MTU}}{\text{link capacity}} + \text{link propagation delay} \qquad (2.52)$$

before it reaches the downstream element, where the first term is the worst-cast service time for this packet, the second term is the time for this element to transmit a packet with size equal to the maximum transmission unit (MTU) at its outgoing link capacity, and the third term is simply the physical propagation delay of the outgoing link. In this case, it follows by definition that the C and D terms are given by

$$C = M, \qquad D = \frac{\text{MTU}}{\text{link capacity}} + \text{link propagation delay}. \qquad (2.53)$$

Let $F(t, t + \tau)$ denote the amount of traffic that departs at network element i in $(t, t + \tau]$. There are a tandem of elements $1, 2, \ldots, i$, along the path of the flow, among them each element j introduces a delay term, $C_j/R + D_j$, attributed to its GS implementation. The accumulated sum is given by

$$\theta_i = \frac{C_{\text{tot } i}}{R} + D_{\text{tot } i}, \qquad C_{\text{tot } i} = \sum_{j=1}^{i} C_j, \qquad D_{\text{tot } i} = \sum_{j=1}^{i} D_j. \qquad (2.54)$$

Therefore, by definition the tandem of elements is said to guarantee a reservation level R or a service curve $F(\tau)$ to this flow if there exists

$$F(t, t + \tau) \geq F(\tau) = R(\tau - \theta_i)^+, \qquad (2.55)$$

where $x^+ = \max\{x, 0\}$. Figure 2.12 illustrates these points.

A closed form expression for an upper bound on the delay incurred by a packet, until and including the delay at element i, can be easily computed by calculating the length of the segment EF in Figure 2.12 and is given by

$$\Theta_i = \begin{cases} \dfrac{b - M}{R} \dfrac{p - R}{p - r} + \dfrac{M}{R} + \theta_i & \text{if } p > R, \\[2ex] \dfrac{M}{R} + \theta_i & \text{otherwise}. \end{cases} \qquad (2.56)$$

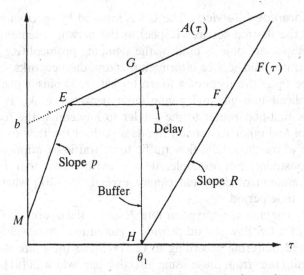

Fig. 2.12   Delay and buffer calculations for an (r, b, p) flow.

The term θ_i is from (2.54). In a similar manner, for the (r, b, p) flow, the buffer requirement at the network element i (segment GH in Figure 2.12) is given by

$$B_i = M + \frac{p - X}{p - r}(b - M) + \theta_i X, \qquad (2.57)$$

where

$$X = \begin{cases} r & \text{if } \dfrac{b - M}{p - r} \le \theta_i, \\[2mm] R & \text{if } \dfrac{b - M}{p - r} > \theta_i \ \text{ and } \ p > R, \\[2mm] p & \text{otherwise}. \end{cases} \qquad (2.58)$$

Equation (2.57) is only an upper bound on the buffer requirement for a given flow, and can be further improved. Notice that the D-term (2.53) actually includes the propagation delay on the link. Since this is a fixed latency, it does not contribute to the buffer requirement at the element, which only depends on the variable portion of the delay bound. Therefore, if the fixed latency from the first element is known, it can be subtracted from the term $C_{\text{tot } i}$ in (2.54) to yield a more accurate bound.

2.4.1.2 *Guaranteed Service*   The GS is invoked by specifying the traffic (TSpec) and the desired service (RSpec) to the network element [79]. Both TSpec and RSpec are objects used by the signaling protocol (e.g., RSVP) to transfer the traffic and service information across the networks.

The TSpec takes the form of a token bucket (r, b) plus a peak rate p, a minimum policed unit m, and a maximum packet size M. It is used to configure the first-hop router to the sender to implement the token bucket access control and other routers on the data path of the flow to enforce the conformance of (re-shape) the flow traffic to its traffic description, such as a burstiness constraint. For example, it is a requirement that for all time periods, the amount of data sent cannot exceed $M + pT$, where T is the length of the time period.

The RSpec contains a reservation rate R and a slack term S. Each router on the path of a GS flow should provide a guaranteed bandwidth R as well as the required buffering according to (2.57). Since these are deterministic guarantees, different from those using effective bandwidth in [81], the admission control becomes simply to check whether there are enough bandwidth and buffer available to meet the requests [79].

The slack term signifies the difference between the delay bound (θ_{max}) obtained by using r instead of R in (2.56) and that obtained by using R. The first bound is always greater than the second bound, because $R > r$. The term S allows some room for a router to adjust the current value of R (surely $R > r$ in the meantime). For instance, if it finds the available bandwidth is not enough for a new flow, it may reduce the reservation level(s) of some existing flow(s) so that this new flow can be granted. Nevertheless, it always guarantees that the worst-case delay is bounded by θ_{max}.

GS ensures that datagrams will arrive within the guaranteed delivery time and will not be discarded due to queue overflows, provided the flow's traffic stays within its specified traffic parameters. The maximum end-to-end queuing delay as shown in (2.56) and the bandwidth (characterized by R) provided along a path will be stable, that is, they will not change as long as the end-to-end path does not change.

The admission control algorithm of a router for GS uses the *a priori* characteristics of sources (so it is usually PBAC) to calculate the worst-case behavior of all the existing flows in addition to the incoming one. And, with the scheduling algorithm, both must ensure that the delay bounds are never violated and packets are not lost when a source's traffic conforms to the TSpec. Network utilization under this model is usually acceptable when flows are smooth. When flows are bursty, however, GS inevitably results in low utilization due to its worst-case service commitments.

Figure 2.13 illustrates an implementation framework for the ISI in a router [75]. As shown, the routers can implement per-flow weighted fair queuing (WFQ) providing rate guarantees, fairness, and isolation [71–74]. Each GS-flow queue is FIFO. Further explanation of this framework, including CLS and best effort, is presented in the following subsection.

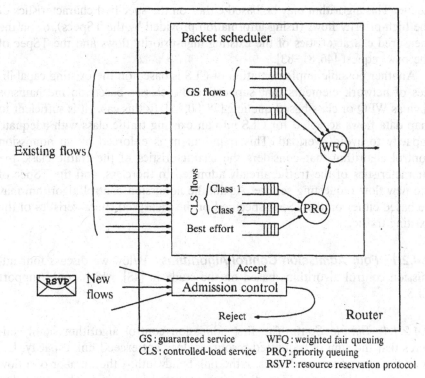

GS : guaranteed service   WFQ : weighted fair queuing
CLS : controlled-load service   PRQ : priority queuing
                                 RSVP : resource reservation protocol

Fig. 2.13   An ISI implementation framework.

2.4.2   Controlled-Load Service

The controlled-load service (CLS) provides the client data flow with a QoS closely approximating what that same flow would receive from a network element that is not heavily loaded or congested. In other words, it is designed for applications that can tolerate variation in packet delays and a minimal loss rate that must closely approximate the basic packet error rate of the transmission medium. The CLS does not accept or use specific target values for control parameters such as delay or loss. Instead, acceptance of a request for CLS is defined to imply a commitment by the network element to provide the requester with service closely equivalent to that provided to best-effort traffic under lightly loaded conditions. Some capacity (admission) control is needed to ensure that this service is received even when the network element is overloaded.

One possible implementation of CLS is to provide a queuing mechanism with two priority levels [75, 78]: a high priority for controlled-load and a lower priority for best-effort service, as shown in Figure 2.13. An admission control algorithm is used to limit the amount of traffic placed in the high-priority queue. This algorithm may be based either on the specified characteristics of the high-priority flows (using information provided by the TSpecs), or on the measured characteristics of the existing high-priority flows and the TSpec of the new request [46, 81–83].

Another possible implementation of CLS is based on the existing capabilities of network elements that support traffic classes based on mechanisms such as WFQ or class-based queuing [78, 80, 83]. In this case, it is sufficient to map data flows accepted for CLS into an existing traffic class with adequate capacity to avoid overload. This requirement is enforced by an admission control algorithm that considers the characteristics of the traffic class, the characteristics of the traffic already admitted to the class, and the TSpec of the new flow requesting service. Again, the admission control algorithm may be based either on the TSpec-specified or the measured characteristics of the existing traffic.

2.4.2.1 Four Admission Control Algorithms
Below we discuss four admission control algorithms based on the study in [48], which could support CLS.

2.4.2.1.1 Simple Sum
The first admission control algorithm simply ensures that the sum of requested resources does not exceed link capacity. Let v be the sum of reserved rates, r the link bandwidth, i the number of a flow requesting admission, and c_i the rate requested by flow i. This algorithm accepts the new flow if the following check succeeds:

$$v + c_i < r. \tag{2.59}$$

This is the simplest admission control algorithm. Hence, it is widely implemented by switch and router vendors. Often, to ensure the low queuing delay called for by CLS, an approximation of the WFQ scheduling discipline is implemented with this admission control algorithm. WFQ assigns each flow its own queue served at its own reserved rate, thereby isolating flows from each other's bursts. However, for the other measurement-based algorithms discussed below, the FIFO scheduling discipline is assumed.

2.4.2.1.2 Measured Sum
Whereas the simple-sum algorithm ensures that the sum of existing reservations plus a newly incoming reservation does not exceed capacity, the *measured-sum* algorithm uses measurement to estimate the load of existing traffic. This algorithm admits the new flow if the following test succeeds:

$$\hat{v} + c_i < \eta r, \tag{2.60}$$

where η is a user-defined utilization target as explained below, and $\hat{\nu}$ the measured load of existing traffic. How load measurement is done is explained later. Upon admission of a new flow j, the load estimate is increased using $\hat{\nu}' = \hat{\nu} + c_j$. As pointed out in [82], in a simple $M/M/1$ queue, the variance of the queue length diverges as the system approaches full utilization. A measurement-based approach is doomed to fail when delay variations are exceedingly large, which will occur at very high utilization. It is thus necessary to identify a utilization target and require that the admission control algorithm strive to keep link utilization below this level.

2.4.2.1.3 Acceptance Region

The second measurement-based algorithm, proposed in [44], computes an acceptance region that maximizes the reward of utilization minus the penalty of packet loss. Given the link bandwidth, the switch buffer space, a flow's token bucket filter parameters and burstiness, and the desired probability of the actual load exceeding the bound, one can compute an acceptance region for a specific set of flow types, beyond which no more flow of those particular types should be accepted. The computation of the acceptance region also assumes Poisson call arrival processes and independent, exponentially distributed call holding times (interested readers are referred to [44] for the computation of the acceptance region).

The measurement-based version of this algorithm ensures that the measured instantaneous load plus the peak rate of a new flow is in the acceptance region. The measured load used in this scheme is not artificially adjusted upon admittance of a new flow. For flows described by a token bucket filter ($\rho; b$) but not peak rate, [46] derives their peak rates $\hat{p}$ from the token bucket parameters using the equation

$$\hat{p} = \rho + b/U, \tag{2.61}$$

where U is a user-defined averaging period. The algorithm adopts the same scheme used by the acceptance region algorithm. If a flow is rejected, the admission control algorithm does not admit another flow until an existing one leaves the network.

2.4.2.1.4 Equivalent Bandwidth

The third measurement-based algorithm computes the equivalent bandwidth for a set of flows using the Hoeffding bounds. The equivalent bandwidth of a set of flows is defined in [16, 46] as the bandwidth $C(\varepsilon)$ such that the stationary bandwidth requirement of the set of flows exceeds this value with the probability at most ε. Here ε is called the loss rate; however, in an environment where a large portion of the traffic is best-effort traffic, real-time traffic with its rate exceeding its equivalent bandwidth is not lost but simply encroaches upon best-effort traffic. In [46], the measurement-based equivalent bandwidth based on Hoeffding bounds

$\hat{C}_H$ assuming peak-rate (p) policing of n flows, is given by

$$\hat{C}_H(\hat{\nu}, \{p_i\}_{1 \leq i \leq n}, \varepsilon) = \hat{\nu} + \sqrt{0.5 \ln(1/\varepsilon) \sum_{i=1}^{n} p_i^2}, \qquad (2.62)$$

where $\hat{\nu}$ is the measured average arrival rate of existing traffic and ε is the probability that the arrival rate exceeds the link capacity. [46] shows that the measured average arrival rate may be approximated by the measured average load.

The admission control check when a new flow i requests admission is

$$\hat{C}_H + p_i \leq r. \qquad (2.63)$$

Upon admission of a new flow, the load estimate is increased using $\hat{\nu}' = \hat{\nu} + p_i$. Again, if a flow's peak rate is unknown, it is derived from its token bucket filter parameters ($\rho; b$) using (2.61). Similarly to the algorithm in [44], if a flow is denied admission, no other flow of a similar type will be admitted until an existing one departs.

Recall that while the admission control algorithms described here are based on meeting QoS constraints on either loss rate or delay bound, the specific values used by the admission control algorithms are not advertised to the users of CLS.

2.4.2.2 Three Measurement Mechanisms
This sub-subsection describes some very simple measurement mechanisms, which may not be the most efficient or the most rigorous, but can help us isolate admission patterns caused by a particular admission control algorithm from those caused by the measurement mechanism. Interested readers are referred to [36, 84] for alternate treatments on measurement mechanisms.

2.4.2.2.1 Time-Window
Following [82], a simple time-window measurement mechanism can be used to measure network load with the measured sum algorithm. As shown in Figure 2.14 [81], an average load is computed every S sampling periods. At the end of a measurement window T, the highest average from the just-ended T is used as the load estimate for the next T-window. When a new flow is admitted to the network, the estimate is increased by the parameters of the new request as explained in Section 2.4.2.1. If a newly computed average is above the estimate, the estimate is immediately raised to the new average. At the end of every T, the estimate is adjusted to the actual load measured in the previous T. A smaller S gives higher maximal averages, resulting in a more conservative admission control algorithm; a larger T keeps a longer measurement history, again resulting in a more conservative algorithm. It would be suitable to keep $T/S \geq 10$ to get a statistically meaningful number of samples.

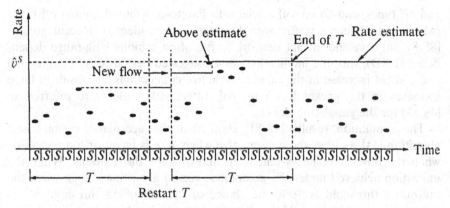

Fig. 2.14   Time-window measurement of network load.

2.4.2.2.2 *Point Samples*   The measurement mechanism used with the acceptance region algorithm takes an average load sample every S' periods [81].

2.4.2.2.3 *Exponential Averaging*   Following [46], an estimate of the average arrival rate can be used instead of instantaneous bandwidth to compute admission decisions with the equivalent bandwidth approach. The overall average arrival rate $\hat{v}^S$ is measured once every S sampling periods. The average arrival rate is then computed using an infinite impulse response function with weight w (e.g., 0.002):

$$\hat{v}' = (1 - w)\,\hat{v} + w\hat{v}^S. \tag{2.64}$$

If the traffic arrival rate changes abruptly from 0 to 1 and then remains at 1, a w of 0.002 allows the estimate to reach 75% of the new rate after 10 sampling periods. A larger w makes the averaging process more adaptive to load changes; a smaller w gives a smoother average by keeping a longer history. Recall that the equivalent bandwidth-based admission control algorithm requires peak-rate policing and derives a flow's peak rate from its token bucket parameters using (2.61) when the peak rate is not explicitly specified. $U = S$ can be set to reflect the peak rate seen by the measurement mechanism. A smaller S not only makes the measurement mechanism more sensitive to bursts; it also makes the peak rate derivation more conservative. A larger S may result in lower averages; however, it also means that the measurement mechanism keeps a longer history because the averaging process (2.64) is invoked less often.

2.4.2.3 *Comparisons*   In [48, 81], simulations have been run on one-link and four-link topologies to compare the above four algorithms. Two kinds of source models are used: an on–off model with exponentially distributed on and off times, and an on–off model with Pareto-distributed on and off times (which can generate traffic with long-range dependence). Recent studies [85, 87, 88] have shown that network traffic often exhibits long-range dependence (LRD), with the implications that congested periods can be quite long and a slight increase in the number of active connections can result in large increases in the packet loss rate [86]. Interested readers are referred to [48, 81] for the simulation details.

The simulation results [48, 81] show that the acceptance-region-based algorithm [44] is either too conservative when p/r is large, or too optimistic when the flows have heavy-tailed on- and off-time distributions. While the utilization achieved under the acceptance region algorithm is high when the utilization threshold is 98%, the choice of the utilization threshold is not from the computations in [44]; rather it is a best-case, but ad hoc, choice for this scenario. Hence, it does not allow the load estimation error to be quantified and assessed any more rigorously than under the measured-sum method. To compute the acceptance region, one must know the source characteristics *a priori*. In environments such as the Internet, where new applications are introduced at a high rate and source characteristics depend not only on the applications but also their use, one cannot make *a priori* characterizations of sources with any degree of certainty.

The equivalent bandwidth-based algorithm [46] is inherently conservative. The measured-sum method seems to work well. In general, while it is clear that an admission control algorithm for CLS should have a utilization target, it is still not clear how to compute this bound from observed traffic characteristics. When computing equivalent bandwidth or acceptance region, taking into account only the sources' peak rate and token bucket filter parameters does not seem sufficient. One must also take into account the sources' burst lengths and idle times distributions.

The simple-sum method, used in conjunction with WFQ scheduling discipline favored by router vendors for its implementations implicitly, gives the worst performance in terms of link utilization. However, the implementation and operational costs of the various admission control algorithms haven't been studied. When these are taken into account, one might not be able to implement anything more complicated than the simple-sum algorithm, given current hardware technology.

REFERENCES

1. D. Anick, D. Mitra, and M. M. Sondhi, "Stochastic theory of a data-handling system with multiple sources," *Bell Syst. Tech. J.*, vol. 61, no. 8, pp. 1871–1895, Oct. 1982.

2. B. Kraimeche and M. Schwartz, "Analysis of traffic access control strategies in integrated service networks," *IEEE Trans. Commun.*, vol. 33, no. 10, pp. 1085–1093, Oct. 1985.

3. J. Y. Hui, "Resource allocation for broadband networks," *IEEE J. Select. Areas Commun.*, vol. 6, no. 9, pp. 1598–1608, Dec. 1988.

4. D. Mitra, "Stochastic theory of a fluid model of producers and consumers coupled by a buffer," *Adv. Appl. Probab.*, vol. 20, pp. 646–676, 1988.

5. S. Ohta, K.-I. Sato, and I. Tokizawa, "A dynamically controllable ATM transport network based on the virtual path concept," *Proc. IEEE GLOBECOM*, pp. 1272–1276, 1988.

6. W. Verbiest, L. Pinnoo, and B. Voeten, "The impact of the ATM concept on video coding," *IEEE J. Select. Areas Commun.*, vol. 6, no. 9, pp. 1623–1632, December 1988.

7. T. Kamitake and T. Suda, "Evaluation of an admission control scheme for an ATM network considering fluctuations in cell loss rate," *Proc. IEEE GLOBECOM*, pp. 1774–1780, 1989.

8. J. Filipiak, "Structured systems analysis methodology for design of an ATM network architecture," *IEEE J. Select. Areas Commun.*, vol. 7, no. 8, pp. 1263–1273, Oct. 1989.

9. M. Decina, T. Toniatti, P. Vaccari, and L. Verri, "Bandwidth assignment and virtual call blocking in ATM networks," *Proc. IEEE INFOCOM*, pp. 881–888, 1990.

10. G. Woodruff, R. Kositpaiboon, G. Fitzpatrick, and P. Richards, "Control of ATM statistical multiplexing performance," *Comput. Netw. ISDN Syst.*, vol. 20, pp. 351–360, 1990.

11. C. Rasmussen and J. Sorensen, "A simple call acceptance procedure in an ATM network," *Comput. Netw. ISDN Syst.*, vol. 20, pp. 197–202, 1990.

12. A. Lombardo, S. Palazzo, and D. Panno, "A framework for sharing bandwidth resources among connectionless and connection-oriented services in B-ISDNs," Presented at 7th Int. Teletraffic Congress, September 1990.

13. N. M. Mitrou and D. E. Pendarakis, "Cell-level statistical multiplexing in ATM networks: analysis, dimensioning and call-acceptance control w.r.t. QoS criteria," Presented at 7th International Teletraffic Congress, September 1990.

14. C. Rasmussen, J. Sorensen, K. S. Kvols, and S. B. Jacobsen, "Source independent call acceptance procedures in ATM networks," *IEEE J. Select. Areas Commun.*, vol. 9, no. 3, pp. 351–358, Apr. 1991.

15. I. Norros, J. W. Roberts, A. Simonian, and J. T. Virtamo, "The superposition of variable bit rate sources in an ATM multiplexer," *IEEE J. Select. Areas Commun.*, vol. 9, no. 3, pp. 378–387, Apr. 1991.

16. R. Guerin, H. Ahmadi, and M. Naghshineh, "Equivalent capacity and Its application to bandwidth allocation in high-speed networks," *IEEE J. Select. Areas Commun.*, vol. 9, no. 7, pp. 968–981, Sept. 1991.

17. H. Saito and K. Shiomoto, "Dynamic call admission control in ATM networks" *IEEE J. Select. Areas Commun.*, vol. 9, no. 7, pp. 982–989, Sept. 1991.

18. G. Ramamurthy and R. S. Dighe, "Distributed source control: a network access control for integrated broadband packet networks," *IEEE J. Select. Areas Commun.*, vol. 9, no. 7, pp. 990–1002, Sept. 1991.

19. S. J. Golestani, "A framing strategy for congestion management," *IEEE J. Select. Areas Commun.*, vol. 9, no. 7, pp. 1064–1077, Sept. 1991.

20. A. Hiramatsu, "Integration of ATM call admission control and link capacity control by distributed neural network," *IEEE J. Select. Areas Commun.*, vol. 9, no. 7, pp. 1131–1138, Sept. 1991.

21. T. Murase, H. Suzuki, S. Sato, and T. Takeuchi, "A call admission control scheme for ATM networks using a simple quality estimate," *IEEE J. Select. Areas Commun.*, vol. 9, no. 9, pp. 1461–1470, Dec. 1991.

22. F. P. Kelly, "Effective bandwidths at multi-class queues" *Queueing Systems*, vol. 9, pp. 5–16, 1991.

23. J. Kurose, "On computing per-session performance bounds in high-speed multi-hop computer Networks," *Proc. ACM SIGMETRICS*, pp. 128–139, Jan. 1992.

24. H. Saito, "Call admission control in an ATM network using upper bound of cell loss probability," *IEEE Trans. Commun.*, vol. 40, no. 9, pp. 1512–1521, Sept. 1992.

25. J. M. Hyman, A. A. Lazar, and G. Pacifici, "A separation principle between scheduling and admission control for broadband switching," *IEEE J. Select. Areas Commun.*, vol. 11, no. 4, pp. 605–616, May 1993.

26. Y.-H. Kim and C.-K. Un, "Analysis of bandwidth allocation strategies with access control restrictions in broadband ISDN," *IEEE Trans. Commun.*, vol. 41, no. 5, pp. 771–781, May 1993.

27. A. I. Elwalid and D. Mitra, "Effective bandwidth of general markovian traffic sources and admission control of high speed networks," *IEEE/ACM Trans. Netwo.*, vol. 1, no. 3, pp. 329–343, Jun. 1993.

28. G. Kesidis, J. Walrand, and C.-S. Chang, "Effective bandwidths for multiclass markov fluids and other ATM sources," *IEEE/ACM Trans. Netw.*, vol. 1, no. 4, pp. 424–428, Aug. 1993.

29. S. H. Low and P. P. Varaiya, "A new approach to service provisioning in ATM networks," *IEEE/ACM Trans. Netw.*, vol. 1, no. 5, pp. 547–553, Oct. 1993.

30. J. W. Roberts, "Traffic control in B-ISDN," *Comput. Netw. ISDN Syst.*, vol. 25, pp. 1055–1064, 1993.

31. L. Gun and R. Guerin, "Bandwidth management and congestion control framework of the broadband network architecture," *Comput. Netw. ISDN Syst.*, vol. 26, pp. 61–78, 1993.

32. Z. Dziong, K-Q. Liao, and L. Mason, "Effective bandwidth allocation and buffer dimensioning in ATM based networks with priorities," *Comput. Netw. ISDN Syst.*, vol. 25, pp. 1065–1078, 1993.

33. C. Shim, I. Ryoo, J. Lee, and S-B. Lee, "Modeling and call admission control algorithm of variable bit rate video in ATM networks," *IEEE J. Select. Areas Commun.*, vol. 12, no. 2, pp. 332–344, Feb. 1994.

34. S. Abe and T. Soumiya, "A traffic control method for service quality assurance in an atm network," *IEEE J. Select. Areas Commun.*, vol. 12, no. 2, pp. 322–331, Feb. 1994.

35. H. Zhang and E. W. Knightly, "Providing end-to-end statistical performance guarantee with bounding interval dependent stochastic models," *Proc. ACM SIGMETRICS*, pp. 211–220, May 1994.

36. R. Warfield, S. Chan, A. Konheim, and A. Guillaume, "Real-time traffic estimation in ATM networks," presented at Int. Teletraffic Congress, Jun. 1994.

37. H. Zhang and D. Ferrari, "Improving utilization for deterministic service in multimedia communication," presented at IEEE Int. Conf. Multimedia Computing and Systems, 1994.

38. S. Chong, S.Q. Li, and J. Ghosh, "Predictive dynamic bandwidth allocation for efficient transport of real-time VBR video over ATM," *IEEE J. Select. Areas Commun.*, vol. 13, no. 1, pp. 12–23, Jan. 1995.

39. S.-Q. Li, S. Chong, and C.-L. Hwang, "Link capacity allocation and network control by filtered input rate in high-speed networks, *IEEE/ACM Trans. Net.*, vol. 3, no. 1, pp. 10–25, Feb. 1995.

40. M. Grossglauser, S. Keshav, and D. Tse, "RCBR: a simple and efficient service for multiple time-scale traffic," *Proc. ACM SIGCOMM*, pp. 219–230, 1995.

41. G. de Veciana, G. Kesidis, and J. Walrand, "Resource management in wide-area ATM networks using effective bandwidths," *IEEE J. Select. Areas Commun.*, vol. 13, no. 6, pp. 1081–1090, Aug. 1995.

42. A. Elwalid, D. Mitra, and R. H. Wentworth, "A new approach for allocating buffers and bandwidth to heterogeneous regulated traffic in an ATM node," *IEEE J. Select. Areas Commun.*, vol. 13, no. 6, pp. 1115–1127, Aug. 1995.

43. C.-S. Chang and J. A. Thomas, "Effective bandwidth in high-speed digital networks," *IEEE J. Select. Areas Commun.*, vol. 13, no. 6, pp. 1091–1100, Aug. 1995.

44. R. Gibbens, F. Kelly, and P. Key, "A decision-theoretic approach to call admission control in ATM networks," *IEEE J. Select. Areas Commun.*, vol. 13, no. 6, pp. 1101–1114, Aug. 1995.

45. D. E. Wrege, E. W. Knightly, H. Zhang, and J. Liebeherr, "Deterministic delay bounds for VBR video in packet-switching networks: fundamental limits and trade-offs," *IEEE/ACM Trans. Netw.*, vol. 4, no. 3, pp. 352–363, Jun. 1996.

46. S. Floyd, "Comments on measurement-based admissions control for controlled-load service," ftp://ftp.ee.lbl.gov/papers/admit.ps.Z.

47. Z. Dziong, M. Juda, and L.G. Mason, "A framework for bandwidth management in ATM networks–aggregate equivalent bandwidth estimation approach," *IEEE/ACM Trans. Net.*, vol. 5, no. 1, pp. 134–147, Feb. 1997.

48. S. Jamin, "A Measurement-based admission control algorithm for integrated services packet networks," Ph.D. dissertation, Dept. of Computer Science, University of Southern California, Aug. 1996.

49. ATM Forum, *User-Network Interface Specification Version 3.0*, 1993.

50. A. Elwalid and D. Mitra, "Effective bandwidth of general Markovian traffic sources and admission control of high speed networks," *IEEE/ACM Trans. Netw.*, vol. 1, no. 3, pp. 329–343, June 1993.

51. J. Suh, "Call admission and source characterization in high speed networks," Ph.D. Dissertation, Dept. of Electrical Engineering, Polytechnic University, Dec. 1995.

52. ATM Forum, *Private Network-Network Interface Specification Version 1.0* (*PNNI 1.0*), 1996.

53. A. Elwalid, D. Mitra, and R. Wentworth, "A new approach for allocating buffers and bandwidth to heterogeneous, regulated traffic in an ATM node," *IEEE J. Select. Areas Commun.*, vol. 13, no. 6, pp. 1115–1127, Aug. 1995.

54. L. He and A. Wong, "Connection admission control design for Globeview-2000 ATM core switches," *Bell Labs Tech. J.*, pp. 94–111, Jan.–Mar. 1998.

55. F. Lo Presti, Z.-L. Zhang, J. Kurose, and D. Towsley, "Source time scale and optimal buffer/bandwidth trade-off for regulated traffic in an ATM node," *IEEE Infocom'97*, pp. 676–683, Mar. 1997.

56. B. Jamoussi, S. Rabie, and O. Aboul-Magd, "Performance evaluation of connection admission control techniques in ATM networks," *IEEE Globecom'96*, pp. 659–664, Nov. 1996.

57. G. Ramamurthy and Qiang Ren, "Multi-class connection admission control policy for high speed ATM switches," *IEEE Infocom'97*, pp. 965–974, Mar. 1997.

58. B. T. Doshi, "Deterministic rule based traffic descriptors for broadband ISDN: worst case behavior and connection acceptance control," *IEEE Globecom'93*, pp. 1759–1764, Nov. 1993.

59. D. Mitra and J. A. Morrison, "Independent regulated processes to a shared unbuffered resource which maximize the loss probability," unpublished manuscript, 1994.

60. T. Worster, "Modeling deterministic queues: the leaky bucket as an arrival process," *Proc. ITC-14*, J. Labetoulle and J. W. Roberts, Eds., Elsevier, New York, pp. 581–590, 1994.

61. N. Yamanaka, Y. Sato, and K. I. Sato, "Performance limitations of leaky bucket algorithm for usage parameter control of bandwidth allocation methods," *IEICE Trans. Commun.*, vol. E75-B, no. 2, pp. 82–86, 1992.

62. S. Rajagopal, M. Reisslein, and K. W. Ross, "Packet multiplexers with adversarial regulated traffic," *IEEE Infocom'98*, pp. 347–355, Mar. 1998.

63. J. Song and R. Boorstyn, "Efficient loss estimation in high speed networks," *IEEE ATM Workshop'98*, pp. 360–367, May 1998.

64. M. Garret and W. Willinger, "Analysis, modeling and generation of self-similar VBR video traffic," *ACM SIGCOMM'94*, pp. 269–280, Aug. 1994.

65. A. Lazar, G. Pacifici, and D. Pendarakis, "Modeling video sources for real time scheduling," *ACM Multimedia Syst. J.*, vol. 1, no. 6, pp. 253–266, Apr. 1994.

66. H. Zhang and E. Knightly, "RED-VBR: a renegotiation-based approach to support delay-sensitive VBR video," *ACM Multimedia Syst. J.*, vol. 5, no. 3, pp. 164–176, May 1997.

67. E. Knightly, "On the accuracy of admission control tests," presented at IEEE ICNP'97, Oct. 1997.

68. B. Mark and G. Ramamurthy, "Real-time estimation of UPC parameters for arbitrary traffic sources in ATM networks," *IEEE Infocom'96*, pp. 384–390, Mar. 1996.

69. Q. Ren and G. Ramamurthy, "A hybrid model and measurement based connection admission control and bandwidth allocation scheme for multi-class ATM networks," presented at IEEE Globecom'98, Nov. 1998.

70. D. Wu and H. J. Chao, "Efficient bandwidth allocation and call admission control for VBR service using UPC parameters," *Int. J. Commun. Syst.*, vol. 13, no. 1, pp. 29–50, Feb. 2000.

71. A. Demers, S. Keshav, and S. Shenker, "Analysis and simulation of a fair queueing algorithm," *Proc. ACM SIGCOMM*, pp. 1–12, Sept. 1989.

72. A. K. Parekh and R. G. Gallager, "A generalized processor sharing approach to flow control in integrated services networks: the single node case," *IEEE/ACM Trans. Netw.*, vol. 1, no. 3, pp. 344–357, Jun. 1993.

73. A. K. Parekh and R. G. Gallager, "A generalized processor sharing approach to flow control in integrated services networks: the multiple node case," *IEEE/ACM Trans. Netw.*, vol. 2, no. 2, pp. 137–150, Apr. 1994.

74. L. Georgiadis, R. Guerin, V. Peris, and R. Rajan, "Efficient support of delay and rate guarantees in an internet," *Proc. ACM SIGCOMM*, pp. 106–116, Aug. 1996.

75. D. D. Clark, S. Shenker, and L. Zhang, "Supporting real-time applications in an integrated services packet networks: architecture and mechanism," *Proc. ACM SIGCOMM*, pp. 14–26, Aug. 1992.

76. R. Braden, D. Clark, and S. Shenker, "Integrated services in the internet architecture: an overview," RFC 1633, Internet Engineering Task Force (IETF), Jun. 1994.

77. R. Braden, L. Zhang, S. Berson, S. Herzog, and S. Jamin, "Resource ReSerVation Protocol (RSVP)—Version 1 functional specification," RFC 2205, Internet Engineering Task Force (IETF), Sept. 1997.

78. J. Wroclawski, "Specification of the controlled-load network element service," RFC 2211, Internet Engineering Task Force (IETF), Sept. 1997.

79. S. Shenker, C. Partridge, and R. Guerin, "Specification of guaranteed quality of service," RFC 2212, Internet Engineering Task Force (IETF), Sept. 1997.

80. S. Floyd and V. Jacobson, "Link-sharing and resource management models for packet networks," *IEEE/ACM Trans. Netw.*, vol. 3, no. 4, pp. 365–386, Aug. 1995.

81. S. Jamin, S. J. Shenker, and P. B. Danzig, "Comparison of measurement-based admission control algorithms for controlled-load service," *Proc. IEEE INFOCOM*, Mar. 1997.

82. S. Jamin, P. B. Danzig, S. J. Shenker, and L. Zhang, "A measurement-based admission control algorithm for integrated services packet networks," *IEEE/ACM Trans. Netw.*, vol. 5, no. 1, pp. 56–70, Feb. 1997.

83. D. Wu, Y. T. Hou, Z.-L. Zhang, and H. J. Chao, "A framework of architecture and traffic management algorithms for achieving QoS provisioning in integrated services networks," *Int. J. Parallel & Distributed Sys. Netw.*, vol. 3, no. 2, pp. 64–81, May 2000.

84. C. Casetti, J. Kurose, and D. Towsley, "A new algorithm for measurement-based admission control in integrated services packet networks," *Proc. Protocols for High Speed Networks Workshop*, Oct. 1996.

85. W. E. Leland, M. S. Taqqu, W. Willinger, and D.V. Wilson, "On the self-similar nature of ethernet traffic (extended version)," *IEEE/ACM Trans. Netw.*, vol. 2, no. 1, pp. 1–15, Feb. 1994.

86. V. Paxson and S. Floyd, "Wide-area traffic: the failure of poisson modeling," *Proc. ACM SIGCOMM*, pp. 257–268, Aug. 1994. An extended version of this paper is available at ftp://ftp.ee.lbl.gov/papers/poisson.ps.Z.

87. J. Beran, R. Sherman, M. S. Taqqu, and W. Willinger, "Long-range dependence in variable-bit-rate video traffic," *IEEE Trans. Commun.*, vol. 4, pp. 1566–1579, 1995.

88. M. Garrett and W. Willinger, "Analysis, modeling and generation of self-similar VBR video traffic," *Proc. ACM SIGCOMM*, pp. 269–279, Sep. 1994.

CHAPTER 3

TRAFFIC ACCESS CONTROL

Traffic access control consists of a collection of specification techniques and mechanisms to (1) specify the expected traffic characteristics and service requirements (e.g., peak rate, required delay bound, loss tolerance) of a data stream, (2) shape data streams (e.g., reducing their rates and/or burstiness) at the edges and selected points within the network, and (3) police data streams and take corrective actions (e.g., discard, delay, or mark packets) when traffic deviates from its specification. For example, ATM has usage parameter control (UPC). Similar mechanisms are required at the edge of autonomous networks for Internet integrated services (Intserv) and Internet differentiated services (Diffserv).

The components of traffic access control are directly related to the mechanisms of admission control and scheduling that implement QoS-controlled services. End-to-end performance cannot be guaranteed for completely arbitrary traffic streams. Most real-time queuing and scheduling mechanisms require some control of the rate and burstiness of data moving through the system.

Policing functions monitor traffic flows and take corrective actions when the observed characteristics deviate from those specified. The actions taken by policing functions are determined by a service level agreement between the user and the network providers. The location of policing functions (e.g., at the network edge and at stream merge points) are usually determined by the network providers.

The traffic contract (or flow specification) function of traffic access control provides the common language by which applications and network elements communicate service requirements. The semantics of the service interface between applications and the network is inherently embodied in the flow specifications. Section 3.1 describes ATM traffic contract and control algorithms. Then we present two examples of traffic shapers, one for ATM networks in Section 3.2, and the other for packet networks in Section 3.3, to show how the traffic access control works in different networking environments.

3.1   ATM TRAFFIC CONTRACT AND CONTROL ALGORITHMS

3.1.1   Traffic Contract

ATM connections are monitored and enforced according to their traffic contracts by network operators at the user–network interface (UNI) and network–node interface (NNI) (Fig. 3.1). If a connection does not conform to the traffic contracts, the UPC (i.e., policing) or network parameter control (NPC) function will take appropriate actions with respect to violating cells (i.e., cells not conforming to their traffic contracts), such as discarding them or tagging them to a lower priority. To reduce the probability that cells are tagged or discarded, a shaping multiplexer (a multiplexer with traffic-shaping function) is used to delay violating cells in a buffer and transmit them to the next node only when they become conforming.

A traffic parameter describes an inherent characteristic of a traffic source. It may be quantitative or qualitative. Traffic parameters include peak cell rate (PCR), sustainable cell rate (SCR), maximum burst size (MBS), minimum cell rate (MCR), and maximum frame size (MFS).

The *source traffic descriptor* is the set of traffic parameters of the source. It is used during the connection establishment to capture the intrinsic traffic characteristics of the connection requested by a particular source.

The *connection traffic descriptor* consists of all parameters and the conformance definition used to specify unambiguously the conforming cells of the

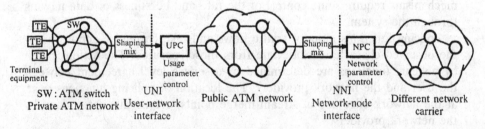

Fig. 3.1 ATM shaping multiplexers and usage and network parameter control (policing) functions at the UNI and NNI.

connection, namely:

- the source traffic descriptor (i.e., PCR, SCR, MBS, MFS, and MCR),
- the CDVT,
- the conformance definition.

Call admission control (CAC) procedures may use the connection traffic descriptor to allocate resources and to derive parameter values for the operation of the UPC.

The values of the traffic contract parameters can be specified either explicitly or implicitly. A parameter value is specified *explicitly* when its value is assigned by the end-system using signaling for switched virtual channels (SVCs), or when it is specified by the network management system (NMS) for permanent virtual channels (PVCs). A parameter value specified at subscription time is also considered to be explicitly specified. A parameter value is specified *implicitly* when its value is assigned by the network using default rules, which in turn depend on the information explicitly specified by the end-system.

3.1.2 PCR Conformance, SCR, and BT

The PCR traffic parameter specifies an upper bound on the rate at which traffic can be submitted on an ATM connection. Enforcement of this bound by the UPC allows the network to allocate sufficient resources to ensure that the network performance objectives [e.g., for the cell loss ratio (CLR)] can be achieved.

The *sustainable cell rate* (SCR) is an upper bound on the average rate of the conforming cells of an ATM connection, over time scales that are long relative to those for which the PCR is defined. Enforcement of this bound by the UPC can allow the network to allocate sufficient resources, but less than those based on the PCR, and still ensure that the performance objectives (e.g., for the CLR) can be achieved.

The *burst tolerance* (BT) is the duration of the period in which the source is allowed to submit traffic at its peak rate (PCR). Note that ITU-T Recommendation I.371 refers to BT as "intrinsic burst tolerance" (IBT).

3.1.3 Cell Delay Variation Tolerance

ATM layer functions (e.g., cell multiplexing) may alter the traffic characteristics of connections by introducing cell delay variation (CDV). When cells from two or more connections are multiplexed, cells of a given connection may be delayed while cells of another connection are being inserted at the output of the multiplexer. Similarly, some cells may be delayed while physical layer overhead or operations, administration, and maintenance (OAM) cells are inserted. Consequently with reference to the peak emission interval T (i.e., the reciprocal of the contracted PCR), some randomness may affect the interarrival time between consecutive cells of a connection (i.e., T) as monitored at the UNI (public or private). The upper bound on the clumping measure is the CDVT.

3.1.4   Generic Cell Rate Algorithm

The GCRA, also known as the leaky bucket algorithm, is used to define conformance with respect to the traffic contract [1–5]. For each cell arrival, the GCRA determines whether the cell conforms to the traffic contract of the connection. The UPC function may implement the GCRA, or one or more equivalent algorithms, to enforce conformance. Although traffic conformance is defined in terms of the GCRA, the network is not required to use this algorithm (or the same parameter values) for the UPC. Rather, the network may use any UPC as long as its operation supports the QoS objectives of a compliant connection.

The GCRA is a virtual scheduling algorithm or a continuous-state leaky bucket algorithm as defined in the flowchart in Figure 3.2. The GCRA is used to define, in an operational manner, the relationship between the PCR and the CDVT, and the relationship between the SCR and the BT. The BT can be derived from the PCR, SCR, and MBS (explained in the next section). In addition, the GCRA is used to specify the conformance, at the public or private UNI, of the declared values of the above two tolerances, as well as the declared values of the traffic parameters PCR, SCR, and MBS.

The GCRA is defined with two parameters: the *increment* (I) and the *limit* (L). The notation GCRA(I, L) means the generic cell rate algorithm with the value of the increment parameter set equal to I and the value of the limit parameter set equal to L, where I and L are not restricted to integer values. The GCRA is formally defined in Figure 3.2. The two algorithms in the figure are equivalent in the sense that for any sequence of cell arrival times, $\{t_a(k), \ k \geq 1\}$, the two algorithms determine the same cells to be conforming and thus the same cells to be nonconforming. The two algorithms are easily compared by noticing that at each arrival epoch $t_a(k)$, after the algorithms have been executed, we have TAT = X + LCT, as in Figure 3.2.

Multiple instances of the GCRA with possibly different values of I and L may be applied to multiple flows (CLP = 0 and CLP = 0 + 1) of the same connection, or to the same flow. A cell is then conforming only if it conforms to all instances of the GCRA against which cells with its CLP state are tested. For example, if one instance of the GCRA tests the CLP = 0 flow and one instance tests the CLP = 0 + 1 flow, then a CLP = 0 cell is conforming only if it conforms to both instances of the GCRA. In this same

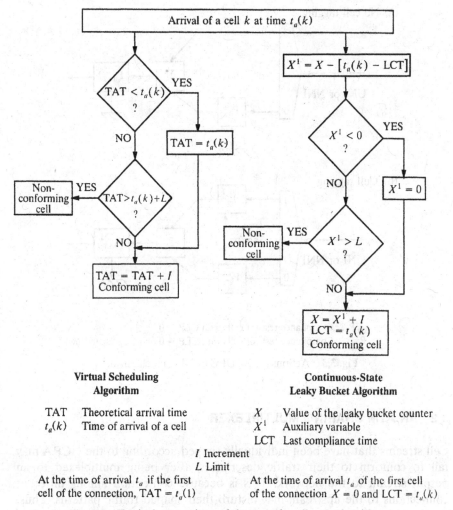

Fig. 3.2 Equivalent versions of the generic cell rate algorithm.

configuration, a CLP = 1 cell is conforming only if it conforms to the instance of the GCRA that tests the CLP = 0 + 1 flow.

If tagging is used, a tagged cell is conforming only if it conforms as a CLP = 1 cell. The state of a particular instance of the GCRA is updated only by the cells that conform as part of a flow tested by that instance of the GCRA. For example, a conforming tagged cell will not update the state of an instance of the GCRA that tests the CLP = 0 flow, since the tagged cell conforms as a CLP = 1 cell. Figure 3.3 shows the actions of the UPC/NPC mechanism (Figure 1.2 gives the ATM service category attributes.)

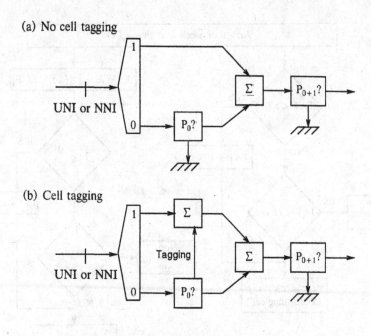

$P_0?$: Compliance test for cells with CLP = 0
$P_{0+1}?$: Compliance test for cells with CLP = 0 or 1 (aggregate flow)

Fig. 3.3   Actions of the UPC or NPC mechanism.

3.2   AN ATM SHAPING MULTIPLEXER

Cell streams that have been individually shaped according to the GCRA may fail to conform to their traffic descriptors after being multiplexed to an output stream to the network. This is because contention among different connections to the upstream may disturb their shaped traffic patterns. Thus, it is important to perform traffic shaping after multiplexing, in what is called an *ATM shaping multiplexer*. This section points out several challenging issues in designing an ATM traffic shaper that can support a large number of virtual connections (VCs) and ensures that every VC strictly complies with the traffic descriptor [17, 18].

The rest of this section is organized as follows. Section 3.2.1 presents the regularity condition—dual leaky bucket. Section 3.2.1 describes a so-called departure-event-driven traffic-shaping algorithm (DEDTS) to strictly enforce each connection to comply with its traffic descriptor at the multiplexed output stream. Section 3.2.3 shows an implementation architecture for the ATM shaping multiplexer that performs the DEDTS algorithm and can be implemented to accommodate any number of VCs by using off-the-shelf components. Section 3.2.4 describes the finite bits overflow problem. Section 3.2.5 presents the performance study for two different shaping algorithms (arrival-event-driven and departure-event-driven) through computer simulations. Section 3.2.6 presents conclusions. Interested readers are referred to [17, 18] for further details.

3.2.1 Regularity Condition—Dual Leaky Bucket

Throughout this section, we use the so-called dual leaky bucket, denoted GCRA $(T_p, \tau_p; T_s, \tau_s)$, as a regularity condition, which defines conformance to the traffic parameters T_p and T_s with associated CDVT (τ_p) and BT (τ_s) at a given interface. The *sustained emission interval* (T_s) and *peak emission interval* (T_p) are obtained as the reciprocals of the SCR and PCR normalized by the line bandwidth, respectively. The parameter BT (τ_s) includes the IBT (τ_s^*) and the CDVT at a given interface. The IBT characterizes the burstiness of a traffic source in conjunction with the SCR at the physical service access point (PHY-SAP) of an equivalent terminal.

Figure 3.4 shows a dual leaky bucket, composed of two leaky buckets (a sustainable-rate bucket and a peak-rate bucket) for each connection, which adopts the *continuous-state* representation [6] rather than the discrete credit token representation to eliminate division operations.

If an incoming cell of a connection arrives at the dual leaky bucket and finds its bucket levels below the limits τ_s and τ_p, it is a conforming cell. If either one of the buckets exceeds the limit, it is considered a violating cell. The two buckets each drain out at a continuous rate of 1 unit of content per cell time, and, after transmitting one cell, increase by the amount of T_s and T_p, respectively. The sizes of the sustainable- and peak-rate buckets are $T_s + \tau_s$ and $T_p + \tau_p$, respectively.

The CDVT (τ_p) and BT (τ_s) have different values at different points in the network. For example, at the PHY-SAP of an equivalent terminal, τ_s and τ_p

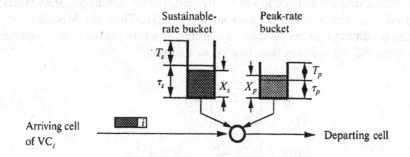

T_s : Sustained emission interval T_p : Peak emission interval
τ_s : Burst tolerance τ_p : Cell delay variation tolerance
X_s : Sustainable-rate bucket level X_p : Peak-rate bucket level

Fig. 3.4 A continuous-state representation of the dual leaky bucket.

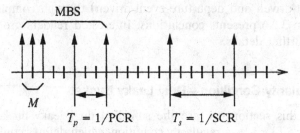

M : Maximum number of back-to-back cells
MBS : Maximum burst size
PCR : Peak cell rate
SCR : Sustainable cell rate

Fig. 3.5   A typical traffic pattern conforming to the dual-leaky-bucket algorithm.

are defined to accommodate only the source traffic characteristics. On the other hand, at the public UNI, they need to include the accumulated CDV that occurs at the private ATM network. In this section, we denote τ_s^* and τ_p^* the intrinsic tolerances for the traffic descriptors at the PHY-SAP. Determination of the CDVT is beyond the scope of this section and is not discussed.

In general, there may be multiple traffic patterns that simultaneously conform to GCRA $(T_p, \tau_p; T_s, \tau_s)$. Figure 3.5 shows a typical traffic pattern with the associated traffic parameters. Here, we formulate the maximum number of back-to-back cells (M) and the maximum burst size (MBS) that can be seen at the output of the dual leaky bucket. Note that these parameters are connection traffic descriptors rather than source traffic descriptors. Although in the ITU-T and ATM Forum the term MBS is defined in conjunction with IBT to represent the source traffic characteristics, we use it here as a general term to represent the MBS at a given interface along the connection path. Instead, the term *intrinsic MBS* (IMBS) is used to represent the MBS associated with IBT. Thus, the MBS is defined as the maximum number of cells transmitted at a rate *greater than or equal to*[1] the peak cell rate, and is thus limited as follows:

$$M = \left\lfloor \frac{\tau_p}{T_p - 1} \right\rfloor + 1 \tag{3.1}$$

$$\text{MBS} = \left\lfloor \frac{\tau_s + \tau_p}{T_s - T_p} - (M - 1) \right\rfloor + 1.$$

[1]On the contrary, the IMBS is defined as the maximum number of cells that are transmitted at the PCR. That some cells may be transmitted at a rate greater than the PCR is due to the nonzero CDVT.

X_s	(0, 5)	(4, 9)	(8, *)	(7, *)	(6, *)	(5, *)	(4, 9)	(8, *)	(7, *)	(6, *)	(5, *)	(4, 9)	(8, *)	(7, *)	(6, *)	(5, *)	(4, 9)	(8, *)	(7, *)	(6, *)	(5, *)	(4, 9)
X_p	(0, 3)	(2, 5)	(4, *)	(3, *)	(2, *)	(1, *)	(3, *)	(2, *)	(1, *)	(0, *)	(0, *)	(0, 3)	(2, *)	(1, *)	(0, *)	(0, *)	(0, 3)	(2, *)	(1, *)	(0, *)	(0, *)	(0, 3)
	0	1	2	3	4	5	6	7	8	9	10	11	12	13	14	15	16	17	18	19	20	21
	CELL	CELL	NO CELL	NO CELL	NO CELL	NO CELL	CELL	NO CELL	NO CELL	NO CELL	NO CELL	CELL	NO CELL	NO CELL	NO CELL	NO CELL	CELL	NO CELL	NO CELL	NO CELL	NO CELL	CELL

X_p : Peak-rate bucket level X_s : Sustainable-rate bucket level (A, B) : A = previous value, B = updated value (* = unchanged)

Fig. 3.6 A back-to-back cell stream shaped by GCRA(3, 2; 5, 4).

For a case where $\tau_p = 0$ (i.e., at the PHY-SAP), the minimum cell interval will not be smaller than the T_p, and thus $M = 1$ and MBS = $1 + \lfloor \tau_s/(T_s - T_p) \rfloor$ [6] (i.e., it equals to IMBS).

Furthermore, to support any real numbers for the PCR, the CDVT (τ_p) needs to be set to a nonzero value in some cases. For example, if $T_p = 1.5$ (allowing two cells transmitted in three time slots) there may be two cells transmitted back to back, due to the slotted system in ATM networks. Thus, a minimum of $\tau_p = 0.5$ needs to be assigned to the case of $T_p = 1.5$. If τ_p is always set to zero, the GCRA $(T_p, 0; T_s, \tau_s)$ will only allow certain rates of PCR that let T_p be integers.

For illustration, consider an example where a back-to-back cell stream is shaped by the dual-leaky-bucket algorithm with the following parameters: line rate = 150 Mbit/s, peak rate = 50 Mbit/s, sustainable rate = 30 Mbit/s, CDVT (τ_p) = 2 cells, and BT (τ_s) = 4 cells. It is easy to compute T_s = (line rate)/(sustainable rate) = 5, T_p = (line rate)/(peak rate) = 3, and $M = 2$ cells from (3.1). Figure 3.6 shows the output cell sequence. Suppose the first cell of this stream arrives at the dual leaky bucket at the beginning of time slot 0 with both buckets being empty, i.e., $X_s(0^+) = X_p(0^+) = 0$, where t^+ and $(t + 1)^-$ are used to indicate the beginning and end of slot t, respectively. As mentioned before, since $X_s(0^+) < \tau_s$ and $X_p(0^+) < \tau_p$, this cell is conforming and it is forwarded. After transmitting this cell, the two buckets are increased by the amounts $T_s = 5$ and $T_p = 3$, respectively, making $X_s(1^-)$ = 5 and $X_p(1^-) = 3$, as indicated with $(0, 5)$ and $(0, 3)$ in Figure 3.6.

At the beginning of each time slot, each bucket drains out 1 unit of content unless it is empty. Thus, $X_s(1^+) = X_s(1^-) - 1 = 4$ and $X_p(1^+) = X_p(1^-) - 1 = 2$. Since $\tau_s = 4$ and $\tau_p = 2$, the second cell is also conforming and it is forwarded. Again, after sending the second cell, $X_s(2^-) = X_s(1^+) + T_s = 9$ and $X_p(2^-) = X_p(1^+) + T_p = 5$, as shown in Figure 3.6.

Recall that at the beginning of each time slot, an input cell is allowed to pass through the dual leaky bucket only if both X_s and X_p are less than or equal to τ_s and τ_p, respectively. Since this condition is not valid at the beginning of slot 2, the third cell is considered a violating cell and is buffered (i.e.,shaped in this particular case). The bucket sizes keep unchanged at the end of this slot, as indicated by * in Figure 3.6. So are the cases at time slots 3, 4, and 5. Only at slot 6, when each bucket has drained out 4 units of content (i.e., 1 per slot), so that $X_s(6^+) = 4 = \tau_s$ and $X_p(6^+) = 0 < \tau_p = 2$, will the next cell become conforming and be forwarded, as shown in Figure 3.6. In conformance to $M = 2$, there are at most two back-to-back cells in the output sequence as illustrated in Figure 3.6.

3.2.2   ATM Shaping Multiplexer Algorithm

When the dual-leaky-bucket algorithm is used for shaping users' traffic by delaying violation cells, a cell buffer is required to temporarily store those violating cells. As pointed out earlier, as multiple shaped connections are multiplexed to a single transmission line, each cell stream may no longer comply with the associated GCRA(T_p, τ_p; T_s, τ_s) due to contention of the multiple connections at the output stream. Here, we present an ATM shaping multiplexer that strictly shapes each individual connection to conform to its GCRA(T_p, τ_p; T_s, τ_s).

3.2.2.1 *Shaping and Scheduling*   The conventional method of designing the traffic shaper for multiple connections is to separate the traffic-shaping function and the scheduling function [10, 11]. Figure 3.7 shows an abstract model of conventional traffic shaping and scheduling. When a cell of a virtual connection VC arrives at the traffic shaper, it is stored in its associated VC buffer. If both bucket levels are below the limits, the cell is eligible to be transmitted and is moved to a scheduler (e.g., a FIFO server). The bucket levels are increased when the cell passes through the dual leaky bucket. If two or more cells from different connections are eligible to transmit at the same time slot, only one cell will be chosen by the scheduler for transmission and the rest of them will wait in a buffer. This approach is called *arrival-event-driven traffic shaping* (AEDTS) because the bucket levels are increased at the cells' arrival time. Although the output cell stream of each buffered dual leaky bucket complies with the GCRA(T_p, τ_p; T_s, τ_s), as multiples of the mare multiplexed into a single transmission line, each shaped cell stream may violate GCRA(T_p, τ_p; T_s, τ_s) at the output of the multiplexer. This is due to cell contention among eligible cells for transmission. For example, if a properly shaped cell stream is delayed due to the loss of contention, it may clump with the following cells, introducing a larger CDV. Computer simulations have shown that this CDV is not negligible. Some intelligent scheduling schemes other than the FIFO discipline may overcome this problem by rearranging cells' departure sequences. However, that may be too complex to be practical.

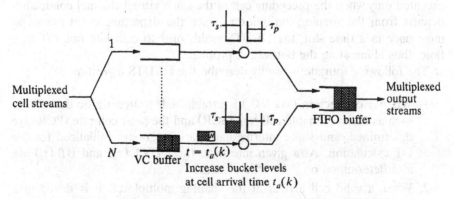

Fig. 3.7   An abstract model of the arrival-event-driven traffic shaping.

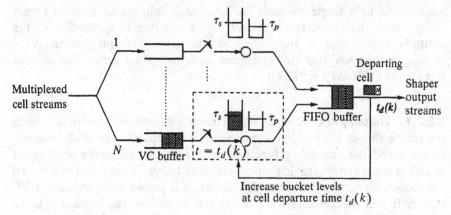

Fig. 3.8   An abstract model of departure-event-driven traffic shaping.

To strictly limit the CDVT and BT at the output of the shaping multi-plexer, the actual departure time of the cell plays a key role in the shaping function. Unlike the AEDTS, *departure-event-driven traffic shaping* (DEDTS) increases the bucket levels when a cell departs from the shaping multiplexer (as shown in Fig. 3.8).

3.2.2.2 *Departure-Event-Driven Traffic Shaping Algorithm*   To facilitate the implementation of the shaping multiplexer, we timestamp cells according to dual-leaky-bucket levels. The DEDTS keeps a real time (RT) clock that ticks and increases by one in every cell time slot. Cells in the shaping multiplexer cannot be transmitted until their departure times (DTs) are due (equal to RT) or overdue (greater than RT). However, using only the time-stamp method may cause a bottleneck problem when it is required to update multiple connections' bucket levels and calculate multiple cells' DTs in the same cell time slot.

In our DEDTS algorithm, bucket level update and DT calculation are executed only when the preceding cell of the same virtual channel connection departs from the shaping multiplexer. Since the departure event occurs at most once in a time slot, the DEDTS needs only to calculate one DT at a time, thus eliminating the bottleneck problem.

The following four steps briefly describe the DEDTS algorithm:

1. When a connection (say VC_i) is established, source traffic descriptors such as the sustainable cell rate (SCR) and the peak cell rate (PCR) are determined, and some intermediate parameters are initialized for the DT calculation. At a given interface, the CDVT (τ_p) and BT (τ_s) are also determined on the traffic contracts.

2. When a valid cell arrives at the shaping multiplexer, it is stored in a logical queue [a linked list of cells that have the same virtual channel

identifier (VCI) or virtual path identifier (VPI)], and the number of backlogged cells (NB_i) is increased by one. For each logical queue, only the head-of-line (HOL) cell is assigned a DT. The DT calculation algorithm will be discussed in Section 3.2.2.3. When a cell arrives and finds the logical queue empty ($NB_i = 0$), the cell will be assigned a DT with either the arrival time (AT) or a precalculated DT, whichever is larger.

3. Cells that have the same DT are linked together in a timing queue. As the RT clock ticks and reaches the DT value, all cells in the timing queue become eligible for transmission and join a departure queue, which links all the cells that are eligible for transmission. The detailed operations are explained in Section 3.2.3:.

4. As soon as a cell departs from the shaping multiplexer, a new DT is calculated and assigned to the new HOL cell in the same logical queue. If there is no cell in the same logical queue, the newly calculated DT will be stored in a lookup table for the future arriving cell.

3.2.2.3 *DT Calculation and Bucket Level Update*   Here, we describe how the DEDTS algorithm calculates the departure time and updates the bucket levels. The required variables for calculating the DT are shown in Figure 3.9. Some variables such as DT, X_s, and X_p change dynamically, while others such as T_s, T_p, τ_s, and τ_p are constant and remain unchanged

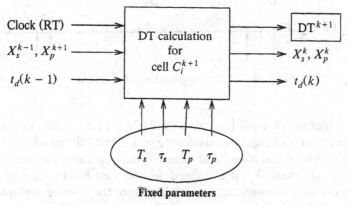

Fixed parameters

C_i^{k+1} : the $(k + 1)$th cell of the ith virtual connection
$t_d(k)$: the actual departure time of C_i^k
X_s^k : the sustainable-rate bucket level after transmitting C_i^k at time $t_d(k)$
X_p^k : the peak-rate bucket level after transmitting C_i^k at time $t_d(k)$
DT^{k+1} : the calculated departure time for C_i^{k+1}.

Fig. 3.9 Variables for calculating cells' departure time.

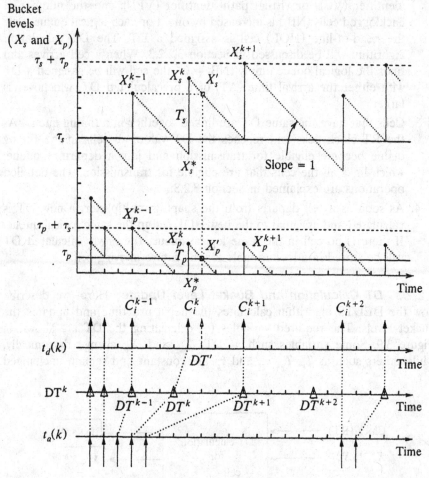

Fig. 3.10 Timing diagram for DT calculation.

during the lifetime of a call [except in available bit rate (ABR) flow control; its parameters may change]. Let us denote by C_i^k the kth cell of VC_i, and by DT^k and $t_d(k)$ the *calculated* and *actual* departure times for the kth cell, respectively. X_s^k and X_p^k are defined as the sustainable- and peak-rate bucket levels after transmitting the kth cell from the shaping multiplexer at time $t_d(k)$. Figure 3.10 shows these variables in a timing diagram to assist in the explanation of the algorithm. The horizontal line represents the time axis. The solid arrows approaching the lower horizontal line are actual cells arriving at the shaping multiplexer. The solid arrows emanating from the upper horizontal line are actual cells departing from the shaping multiplexer. The small triangles in the middle line represent the calculated DT. The dashed arrows on the time axis represent the tentative DT values.

The DEDTS calculates the DT based on both bucket levels (X_s and X_p), which are updated at the same time. When calculating the DT for the next

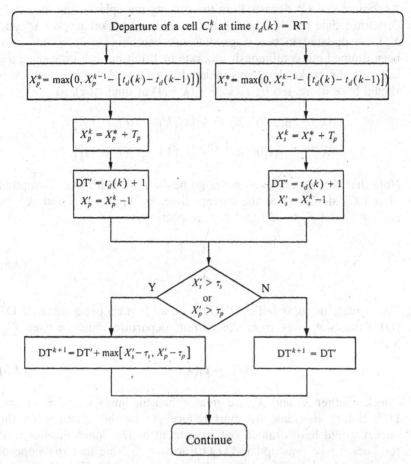

Fig. 3.11 The flowchart of the DEDTS algorithm for DT calculation and bucket level update.

cell, we use the stored values in a lookup table such as DT, X_s, and X_p. Note that DT^2, X_s, and X_p are real numbers, allowing us to support any sustainable or peak cell rates.

For better explanation of the algorithm (Figure 3.11), let us assume that the current real time RT is exactly the departure time of cell C_i^k on the time axis, i.e., $RT = t_d(k)$. Note that the actual departure time $t_d(k)$ and the corresponding calculated departure time DT^k of cell C_i^k may not be the same, because of cell contention of multiple connections, as shown in Figure 3.10. Note also that the cell arrival time, $t_a(k)$, is shown for illustration purposes only and is not used in the DT calculation.

---

[2] In a slotted environment such as ATM, only integer values of the DT are allowed. Although we keep the real numbers of the DT in the lookup table, we assign integers by rounding up the DT.

1. As soon as cell C_i^k departs from the shaping multiplexer, the next cell's departure time (DT^{k+1}) is calculated, and the bucket levels (X_s^k and X_p^k) are updated in two steps. Because each bucket's contents have been drained out continuously at a rate of 1 unit per cell time since the last time the bucket level was updated at $t_d(k-1)$, both bucket levels should have decreased by $t_d(k) - t_d(k-1)$ at time $t_d(k)$, so

$$X_s^* = \max\{0, \; X_s^{k-1} - [t_d(k) - t_d(k-1)]\}$$
$$X_p^* = \max\{0, \; X_p^{k-1} - [t_d(k) - t_d(k-1)]\}. \tag{3.3}$$

Note that the bucket levels never go below zero. With the assumption of cell C_i^k departing at the current time, we update X_s^k and X_p^k by adding T_s and T_p to X_s^* and X_p^*, respectively:

$$X_s^k = X_s^* + T_s,$$
$$X_p^k = X_p^* + T_p. \tag{3.4}$$

2. To calculate the next cell's DT (DT^{k+1}), we first choose a tentative DT (DT') one slot away from the current departure time, i.e. (see Fig. 3.10),

$$DT' = t_d(k) + 1. \tag{3.5}$$

3. Check whether X_s and X_p are greater than the limits τ_s and τ_p at time DT'. Before checking, we must reduce X_s by the amount that the bucket would have drained out its content at DT' since the last time the bucket level was updated $[t_d(k)]$ in step 1. Note that the slope of the bucket level decrease is 1, as shown in Figure 3.10. Thus, the bucket levels (X_s and X_p) at time DT' become

$$X_s' = X_s^k - 1,$$
$$X_p' = X_p^k - 1. \tag{3.6}$$

4. If neither bucket level X_s' or X_p' is greater than its limit (i.e., $X_s' \le \tau_s$ and $X_p' \le \tau_p$), then DT^{k+1} is assigned the value DT'. On the other hand, if either bucket level is greater than its limit, DT^{k+1} is assigned a time when both bucket levels will reduce to the limits, as shown in Figure 3.10. Let C_i^{k+1} be the earliest possible DT among the many choices that will pass the GCRA($T_p, \tau_p; T_s, \tau_s$) test. By delaying the cell's DT by $\max\{(X_s' - \tau_s), (X_p' - \tau_p)\}$, both bucket levels will reduce to their limits at DT^{k+1}. Thus, DT^{k+1} will be assigned as follows:

$$DT^{k+1} = DT' + \max\{(X_s' - \tau_s), (X_p' - \tau_p)\}. \tag{3.7}$$

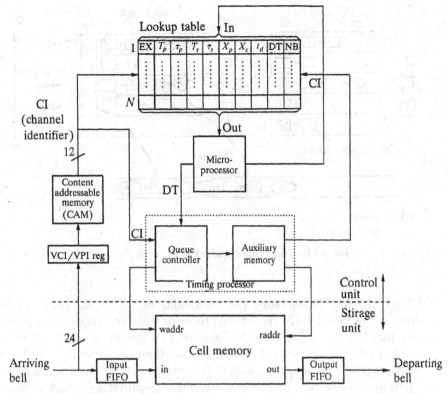

Fig. 3.12 A memory-based architecture for the ATM shaping multiplexer.

3.2.3 Implementation Architecture

This section presents a memory-based implementation architecture for the departure-event-driven traffic shaping. [10] proposed an architecture that was implemented with the existing VLSI chip called the Sequencer [15]. However, as the number of connections becomes large, the Sequencer-based architecture requires many Sequencer chips and is not cost-effective. On the other hand, the memory-based architecture can be implemented using off-the-shelf parts and can handle any number of VCs.

Figure 3.12 shows a memory-based architecture, which is divided into two units. The storage unit is composed of a cell memory, an input FIFO, and an output FIFO. The control unit, composed of a VCI or VPI register, a content-addressable memory (CAM), a microprocessor, a lookup table, and a timing processor, generates appropriate addresses to read (write) cells from (to) the cell memory based on the DEDTS algorithm. The timing processor includes a queue controller and an auxiliary memory, which is a key component in the memory-based architecture and is detailed in the following section. Here, we briefly explain the operations of the memory-based shaping multiplexer. Its detailed design in a hardware implementation is in [13].

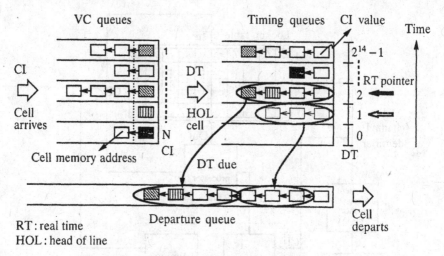

Fig. 3.13 Three major queues in the timing processors.

3.2.3.1 Timing Processor

The operations of the timing processor are illustrated in Figure 3.13. The timing processor has three major queues. First, cells that belong to the same VC are linked together in a logical queue, called the *VC queue*. Second, cells that have the same timestamp (DT) are linked together in a queue, called the *timing queue*. Third, cells whose departure time is due or overdue are linked together in a logical queue, called the *departure queue*. The contents in the VC queue are addresses of cells stored in cell memory. The contents of both the timing queue and the departure queue are channel identifier (CI) values. There is also an *idle-address linked list* (IALL) that keeps the idle (available) spaces of the cell memory.

Newly arriving cells are appended to corresponding VC queues according to their CI values. When a cell becomes the HOL cell of a VC queue, it is assigned a DT and appended to the corresponding timing queue. As the RT clock ticks and as the RT pointer passes the DT, the timing queue whose DT is identical with RT will be appended to the departure queue as shown in Figure 3.13. As the RT pointer moves from 1 to 2, cells at DT = 2 are appended to the departure queue. The HOL cell of the departure queue is read out one at a time slot. Its content, CI, is then used to access the HOL cell of the corresponding VC queue, where the cell address is obtained to transmit the cell in the cell memory.

3.2.3.2 Write-In Procedure

When a cell arrives at the shaping multiplexer, it is first stored in the input FIFO in Figure 3.12. Its VCI or VPI is extracted from the cell header to access the corresponding CI from the content addressable memory (CAM). The cell address of the cell memory is fetched from the IALL. The address is then used as an index to store the cell in the Cell Memory.

Meanwhile, the CI is used to access the corresponding VC's NB (number of backlogged cells) from the lookup table, which has N entries for N VCs. If the NB is equal to zero, its precalculated DT is accessed from the lookup table and compared with RT. The value of $\max\{RT, \lceil DT \rceil\}$ is used as an index to link the cell to a timing queue. On the other hand, if NB is not equal to zero, the newly arrived cell will simply be appended to the VC queue. Note that only the HOL cell from each VC will be able to join the timing queue, while all other cells are stored in the VC queue and wait until they become the HOL cells.

3.2.3.3 Readout Procedure
As the RT clock ticks, the value of RT is used by the queue controller as an index to access a timing queue, whose DT is equal to RT. This timing queue is then appended to the tail of the departure queue.

Cells in the departure queue are the cells that are eligible to be transmitted. In every cell time slot, the HOL cell in the departure queue is transmitted and the next cell becomes the new HOL cell.

Whenever there is a cell to be sent out, the first step is to fetch its CI from the departure queue. The CI is used as an index to find a VC queue, where cells' addresses are linked together. The cell body of the HOL cell in the VC queue is read out from the cell memory, placed in the output FIFO, and transmitted to the network. Meanwhile, the CI is also used to read out the corresponding contents of the look-up table for the microprocessor to calculate the DT of the next cell. If the NB is not zero (i.e., the VC queue is not empty), a newly calculated DT is used as an index to join the new HOL cell to an associated timing queue. On the other hand, if the NB is zero, the newly calculated DT is stored in the lookup table for the next arriving cell. The available cell location is linked to the IALL.

3.2.4 Finite Bits Overflow Problem

When a cell departs from the shaping multiplexer, there is a chance that the next cell from the same virtual connection is not available. In this situation, the calculated DT is stored in the lookup table for later use. Later, when a new HOL cell of the virtual connection arrives at the shaping multiplexer, it compares the DT in the lookup table with the RT. If the precalculated DT is not expired (DT is greater than or equal to RT), then it is simply assigned to the arriving cell. But if the precalculated DT is expired (DT is smaller than RT), then there is no reason to assign it to the newly arriving cell, since the cell is already eligible to transmit. Thus, the precalculated DT is obsolete, and RT is assigned to the newly arriving cell.

However, by simply comparing the values of precalculated DT and RT, it is not possible to distinguish whether the precalculated DT is *retarded* or *advanced*. This is because the register keeping track of the RT and the memory storing a DT have finite bits (e.g., $B = 14$ bits) and overflows in every 2^B cell time slots. As an example, Figure 3.14(a) shows a snapshot of

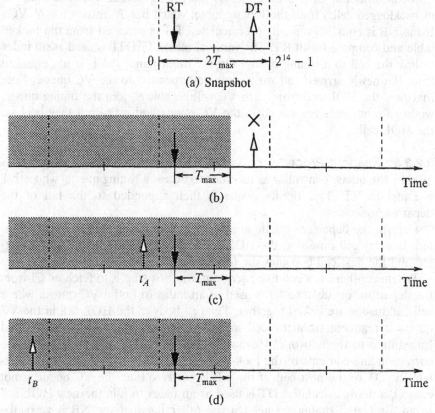

Fig. 3.14 Case I: DT retarded; DT ≥ RT and DT − RT > T_{max}.

DT and RT values at the time of a new HOL cell's arrival. Without any previous history or certain constraints, it is impossible to know which value is actually greater. Fortunately, this problem can be solved by extending the architecture a little, in view of the following very important property:

The precalculated DT for the new HOL cell is always, at most, T_{max} cell times ahead in time at the time of the cell arrival.

Here, T_{max} is the sustained emission interval of the smallest SCR it supports.

The following four cases cover all the possible combinations of DT and RT. First of all, the range of DT and RT is determined from the smallest sustainable cell rate that it supports. The number of bits (B) allocated to the RT and DT must be twice greater than the largest average emission interval T_{max} (e.g., the minimum SCR). For example, if min SCR is 32 Kbit/s running on a 155-Mbit/s transmission line, then

$$B = \left\lfloor \log_2 \frac{2 \times 155 \times 10^6}{32 \times 10^3} t \right\rfloor + 1 = 14 \text{ bits.}$$

Case I *The shaded area in Figure 3.14 shows the possible times where the precalculated DT can be located for a given time (RT) at which the new cell arrives at the shaping multiplexer. Suppose that the previous cell departed at the time right before the new cell arrives. Then the latest possible precalculated DT for the connection is* $T_{max} = 1/SCR$ *time units away from that departure time. From the snapshot, if the precalculated DT were more than* T_{max} *time units away from the RT (which is of course impossible), then we would immediately notice that the precalculated DT had expired.Thus, in this case, the DTs in Figure 3.14(c) and (d) are the valid choices, and they are all DT-retarded.*

Case II *Case II is when the new cell arrives at the shaping multiplexer and finds that the precalculated DT is ahead of it but the difference is within* T_{max}, *as shown in Figure 3.15(a). In this case, Figure 3.15(b), (c), and (d) are all possible choices. Thus, it cannot be distinguished whether the DT is retarded or advanced.*

Here, the periodic purging mechanism is introduced to overcome this situation. The periodic purging mechanism purges the expired precalculated DT in round-robin fashion from the lookup table by marking the EX bit in the table (see Figure 3.16). Note that the periodic purging mechanism only applies to the virtual connection of which the number of backlogged cells is zero (i.e., $NB = 0$), to ensure that the arriving cell is the new HOL cell of the connection. The ideal purging mechanism is to purge all the expired DTs in every time slot. But N memory accesses in one time slot is impossible if N is too large. Thus, each virtual connection is visited in every 2^{B-1} time slots. If the precalculated DT entry is expired, then the EX bit is marked. This guarantees that the precalculated DT cannot last more than 2^{B-1} time slots in the lookup table. Figure 3.17 shows an example when $N = 7$ and $B = 14$. The shaded areas represent the possible unmarked, precalculated DTs of each virtual connection. In this figure, the small upward arrows represent the time that the periodic purging mechanism checks the specified VC. Every VC is visited once in every $T_{max} = 2^{14-1} = 8192$ time slots by the periodic purging mechanism. In general, for any given VC, the precalculated DT, which is retarded more than T_{max} time units, is guaranteed to be marked with the periodic purging mechanism.

Then, the number of memory accesses in one cell time is $N/2^{B-1}$. For example, if $B = 14$ and $N = 4096$, then we have $4096/8192 = 0.5$ memory accesses. Thus, in every two cell times, one DT entry must be checked with the periodic purging mechanism. In Figure 3.15(c) and (d), by the time the new cell

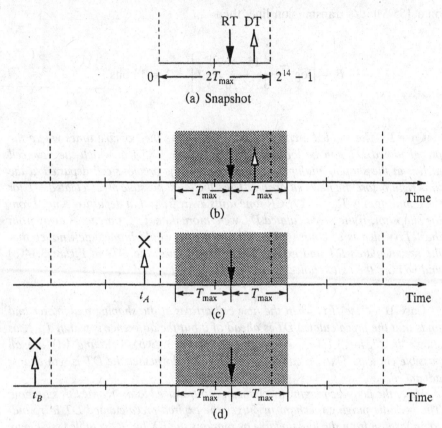

Fig. 3.15 Case II: DT advanced; DT $\geq$ RT and DT $-$ RT $\leq T_{max}$.

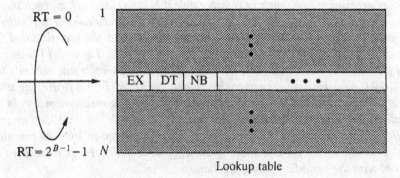

Fig. 3.16 The periodic purging mechanism.

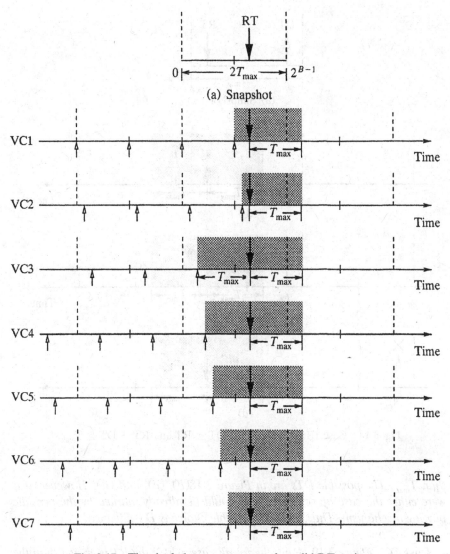

Fig. 3.17 The shaded areas represent the valid DT regions.

arrives (RT), the precalculated DTs t_A and t_B are already marked by the periodic purging mechanism. Thus, if the arriving cell sees the DT unmarked, then it is definitely DT-advanced.

Case III *Cases III and IV are similar to Cases I and II except that the RT is greater than the precalculated DT in the snapshot. Case III is shown in Figure 3.18. In Figure 3.18(a), the RT is greater than DT and the difference is greater*

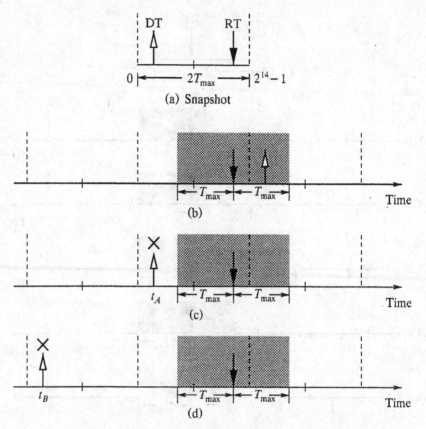

Fig. 3.18   Case III: DT advanced; DT $\leq$ RT and RT $-$ DT $\geq$ T_{max}.

than T_{max}. *The possible DTs are in Figure* 3.18(b), (c), *and* (d). *However, if it were either the case of* (c) *or* (d), *it would be already marked by the periodic purging mechanism. Thus, the only possible choice is DT-advanced.*

Case IV . *Figure* 3.19 *shows the fourth case, where the RT is greater than the precalculated DT and the difference is less than* T_{max}. *If the precalculated DT is advanced, as shown in Figure* 3.19(b), *then it is not in the eligible range, meaning that it is too advanced. Thus, it is considered either as it is in the snapshot or as an expired DT. Both Figure* 3.19(c) *and* (d) *are DT-retarded.*

In summary, this DT comparison algorithm is shown in a flowchart in Figure 3.20. The same algorithm is applied when the periodic purging mechanism is checking whether the DT entry is expired or not. The DT expiration checking is done in two steps. The first step is done by the periodic purging mechanism. The second step is done when the new HOL

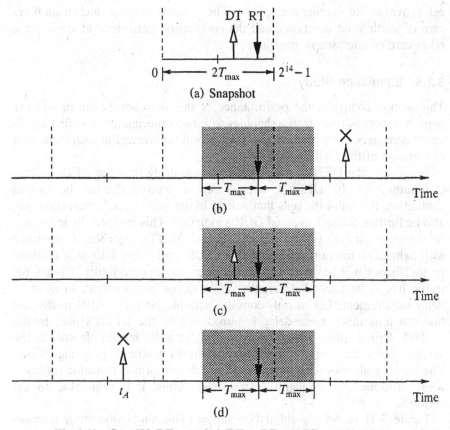

Fig. 3.19 Case IV: DT retarded; DT $\leq$ RT and RT $-$ DT $\leq T_{max}$.

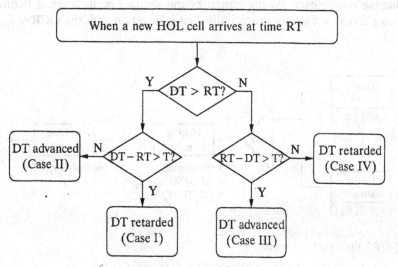

Fig. 3.20 DT comparison algorithm.

cell arrives at the shaping multiplexer. The periodic purging mechanism takes care of multiple bit overflows, and the comparison performed at arrival time takes care of microscopic overflow.

3.2.5 Simulation Study

This section compares the performance of the arrival-event-driven and the departure-event-driven traffic shapings with two experiments. The first experiment compares the average delay. The second experiment measures the rate of violation of the $GCRA(T_p, \tau_p; T_s, \tau_s)$.

Intuitively, the DEDTS will introduce more delay than the AEDTS. That is because, for the DEDTS, once a cell is delayed due to the loss of contention, the following cells that belong to the same virtual connection may also be further delayed to avoid GCRA violation. This is especially important for constant bit rate (CBR) services. If the DEDTS algorithm is operated with tight CDV tolerances when shaping CBR traffic, the delay that is added by the shaper will be accumulated and increase monotonically. Thus, CBR traffic has to be treated differently, as in queue management, to meet its delay requirement. Let us only consider variable bit rate (VBR) traffic and find out how much more delay is introduced by the DEDTS than by the AEDTS. Our simulation study shows that there is not much difference in the average delay between these two event-driven traffic-shaping algorithms. The DEDTS allows each connection to strictly conform to its traffic contracts while introducing a negligible extra delay. Thus, it is preferable to the AEDTS.

Figure 3.21 shows the simulation model. This simulation study assumes that N heterogeneous traffic sources generate cells and they are multiplexed into a FIFO buffer of infinite size. The multiplexed output stream is fed into a shaping multiplexer. At the output of the shaping multiplexer, a fictitious policing function monitors each individual connection with the $GCRA(T_p, \tau_p; T_s, \tau_s)$.

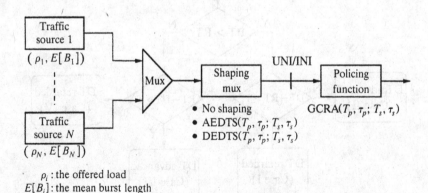

ρ_i: the offered load
$E[B_i]$: the mean burst length

Fig. 3.21 The simulation model.

An on–off source model is used for the traffic sources in a discrete-time environment where time is slotted. Assume that the traffic source model alternates between active and silent modes. During the active mode, cells are generated back to back; during the silent mode, no cells are generated. The lengths of the active and silent periods are geometrically distributed with averages $E[B]$ and $E[I]$, respectively. Let us define p = Prob{starting a new burst per time slot} and q = Prob{the arrived cell is the last cell in the burst}. Then the probability that the burst has i cells is

$$P[B = i] = (1 - q)^{i-1}q, \quad i \geq 1.$$

The probability that an idle period lasts for j time slots is

$$P[I = j] = (1 - p)^{j}p, \quad j \geq 0.$$

Thus,

$$E[B] = \frac{1}{q} \quad \text{and} \quad E[I] = \frac{1 - p}{p}.$$

The offered load ρ is equal to $E[B]/(E[B] + E[I])$.

The source traffic descriptors, the SCR and the PCR, are set as follows. The SCR is set to the offered load ρ of each connection. The PCR is chosen arbitrarily to see the effect of limiting the peak cell rate. The sources generate a peak cell rate of 1 cell/slot. The IBT (τ_s^*) is set to accommodate the IMBS, which is assigned the mean burst length $(E[B])$ of the traffic source. The intrinsic CDVT (τ_p^*) is assigned the value one, as discussed in Section 3.2.1. The source traffic types and their source traffic descriptors are listed in Table 3.1.

The set of experiments investigate an average delay of the shaping multiplexer and the violation rate of GCRA($T_p, \tau_p; T_s, \tau_s$) by varying the CDV tolerances. The results shown in Figure 3.22 are obtained by generating 50×10^6 cells with the same traffic patterns for the AEDTS and DEDTS. The CDVT (τ_s) and BT (τ_p) vary proportionally to the IBT (τ_s^*) and intrinsic

TABLE 3.1   Traffic Source Types and Parameter

Type	No.	SCR ($= 1/T_s$)	PCR ($= 1/T_p$)	τ_s^*	τ_p^*
S1	28	0.005	0.010	900	1
S2	12	0.010	0.040	525	1
S3	10	0.010	0.020	550	1
S4	8	0.025	0.050	80	1
S5	5	s0.050	0.500	252	1
S6	1	0.100	0.500	152	1

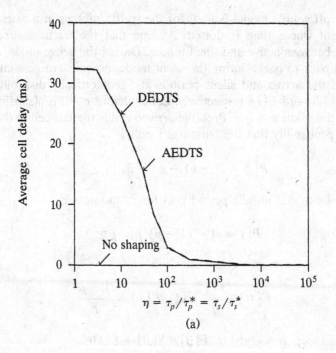

(a)

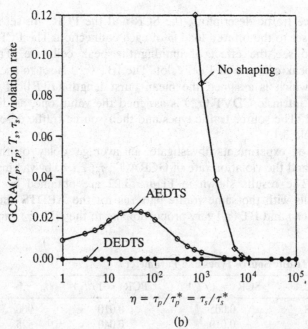

(b)

Fig. 3.22 The average delay and GCRA($T_p, \tau_p; T_s, \tau_s$) violation rate.

CDVT (τ_p^*):

$$\tau_s = \eta\tau_s^*, \qquad \tau_p = \eta\tau_p^*.$$

where $\eta \geq 1$. If η is less than 1, the DEDTS may not be able to clear out the buffer.

As η increases, the average delays of AEDTS and DEDTS decrease. When η becomes 10^4, the average delay is almost zero, representing no shaping effect. In other words, if the CDV tolerances are large enough, an arriving cell is transmitted immediately and experiences no delay. As shown in Figure 3.22(a), the average delays of the two algorithms are almost identical. The difference is about 2 cell slots in this experiment.

Figure 3.22(b) shows the GCRA(T_p, τ_p; T_s, τ_s) violation rates at the output of the shaping multiplexer. This result is obtained by the same simulation setup as the first experiment. The output stream shaped by the DEDTS algorithm always conforms to the GCRA(T_p, τ_p; T_s, τ_s), while the one shaped by the AEDTS algorithm violates it noticeably. Although the AEDTS reduces the CDV tolerances to some degree as compared to the no-shaping case, the violation rate is still quite high. Again, the simulation experiment shows that when η becomes 10^4, all cells are considered as conforming even without the shaping function. For our simulation setup, when $\eta = 30$, the effect of cell delay variation on the regularity condition GCRA is the worst.

The simulation experiments show that the AEDTS and DEDTS algorithms have similar performance in average delay. The AEDTS requires larger CDV tolerances at the UPC and NPC functions to achieve low cell violation rate and does not provide a guarantee on burstiness. The DEDTS guarantees each connection strictly conforming to any given CDVT values. Thus, the shaping multiplexer eliminates the difficulty of determining the CDVT values. A network operator may just set the CDVT based on the delay requirement.

3.2.6 Summary

To meet various QoS requirements, users' traffic must be forced to conform to a set of traffic descriptors, sustainable and peak cell rates (SCR and PCR), and the associated cell delay variation tolerance (CDVT) and burst tolerance (BT). If a connection does not comply with the traffic contracts, some action has to be taken against the violating traffic.

It is very challenging to implement an ATM shaping multiplexer capable of shaping traffic from thousands of virtual connections on an input line using existing hardware technology. Even though the traffic of each individual virtual connection may have been shaped properly, the statistical multiplexing may disturb each connection's traffic flow pattern, and that may violate the negotiated traffic parameters. Moreover, it is difficult to determine the CDVT values, due to the multiplexing of random traffic from many virtual connections.

The departure-event-driven traffic-shaping (DEDTS) algorithm presented above, along with its implementation architecture for an ATM shaping multiplexer, can guarantee that each virtual connection's traffic flow will strictly conform to the negotiated parameters and that no cell will be enforced by the UPC and NPC functions. The algorithm combines both shaping and scheduling functions by timestamping cells using the DEDTS algorithm, which eliminates the bottleneck problem when shaping thousands of virtual connections at the input line. The DEDTS eliminates the difficulty of choosing proper CDVT and BT values. The simulation study shows that any extra delay that may be introduced in the DEDTS is negligible.

3.3 AN INTEGRATED PACKET SHAPER

It has been suggested to use a traffic shaper at the network edge to facilitate QoS provision by delaying submitted traffic to conform to its traffic descriptor. Most proposed traffic shapers [14–17] use leaky bucket control to buffer nonconforming packets and schedule them for later transmission. The traffic shaper is usually followed by a scheduler that schedules conforming packets according to their allocated bandwidth. Several scheduling schemes [19–25] have been proposed in the packet network to guarantee an individual connection's bandwidth requirement and to provide an optimal end-to-end delay bound. Traffic shapers have been proposed to handle only fixed-length packets (e.g., ATM cells). Here we present a mechanism to shape variable-length packets using a similar traffic descriptor.

This section introduces a new implementation architecture that integrates a traffic shaper and a worst-case fairness index (WFI) scheduler.[3] It is called the WFI packet shaper. This new shaper can provide an optimal WFI comparable to other scheduling policies.

3.3.1 Basics of a Packet Traffic Shaper

3.3.1.1 Strict Conformance The packet-based traffic shaper studied in this section is based on the departure-event-driven shaping algorithm [17]. It shapes users' traffic in such a way that the outgoing packet (with variable length) stream of each connection strictly conforms to its traffic descriptor. As shown in Figure 3.23, each connection has a dedicated logical buffer in which its arriving packets queue before they are eligible. Each connection also has its own bucket with size B and bucket level indicator X. It should be noted that the value of B is predetermined in accordance with the parameters of the corresponding traffic descriptor, such as average rate (R_a), peak rate (R_p), and maximum burst size (MBS). This will be discussed later.

---

[3] It suffices to say that a packet scheduler achieving minimum WFI performs best in approximating the ideal generalized processor sharing (GPS) scheduler. More details are in Chapter 4.

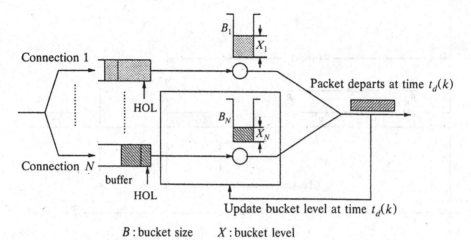

B : bucket size   X : bucket level

Fig. 3.23   An abstract model of a traffic shaper.

When the kth packet (with length L^k) of the investigated connection departs from the shaper (here "departs" means the last bit of the packet departs) at time $t_d(k)$, the bucket level X will increase by an amount of $(L^k/R_a)R$, where R is the outgoing link rate of the shaper. On the other hand, the bucket level always decreases at a rate R if the bucket is not empty. Note that the physical meaning of L^k/R_a is the time needed for this packet to be transmitted with its average rate R_a. Thus, the bucket level X reflects the difference between the actual service and the theoretical service that the connection receives. Although R is rather irrelevant here, it makes the scheme consistent with the packet fair queueing (PFQ) algorithms. We will leave the details of the PFQ algorithms to Chapter 4.

A packet is said to be *conforming* if and only if the departure of the packet will not cause the bucket to overflow. In other words, we have to keep $X \leq B$ at each checking point in order to make the packet stream conforming. For this purpose, upon the departure of a packet, the next HOL packet (if any) of the same connection is timestamped with a conforming time according to its bucket size B and the current bucket level X. However, even if a packet is delayed for transmission due to loss of contention to other packets that have the same timestamp (conforming time) value, when it departs from the shaper, the amount of delay will be taken into consideration when calculating the next packet's timestamp. Thus, it pushes back the transmission time of the next packet of the same connection and achieves strict conformance. Although the departure-event-driven shaping introduces extra delay, it is essential when traffic enforcement is performed at the network edge according to, for example, a traffic conditioning agreement (TCA) in the differentiated services.

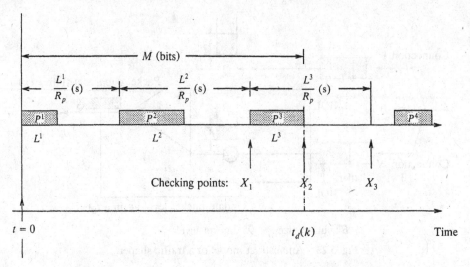

Fig. 3.24   Bucket size calculation for given R_a, R_p, and MBS.

3.3.1.2 *Bucket Size Determination*   The bucket size B of a connection is a crucial parameter for the shaper to shape its packet stream. For a given set of traffic descriptors (R_a, R_p, and MBS), B can be determined as follows. With reference to Figure 3.24, assume that the bucket is empty when $t \leq 0$. At $t = 0$, the connection starts a busy period. Packets of the connection can then be transmitted with its peak rate until the sum of the packet lengths equals the maximum burst size M of the connection ($L^1 + L^2 + L^3 = M$ in the example). After that we will assume that the bucket is full and the subsequent packets can only be transmitted with its average rate. Note that the time needed to transmit these packets with peak rate, theoretically, is M/R_p [$(L^1 + L^2 + L^3)/R_p$ in the example]. During this time interval, the bucket level has increased by $(M/Ra)R$, but also decreased by $(M/R_p)R$ according to the leaky bucket algorithm. It is then easy to show that the bucket size B should be

$$B = \frac{M}{R_a} \cdot R - \frac{M}{R_p} \cdot R. \tag{3.8}$$

Let $T_a = R/R_a$ and $T_p = R/R_p$. The above equation can be rewritten as

$$B = M \cdot (T_a - T_p). \tag{3.9}$$

Referring to Figure 3.24 again, let P^x (P^3 in the example) be the last packet that can be transmitted with peak rate, that is, $\sum_{y=1}^{x} L^y = M$. The derivation of the bucket size can be performed at different checking points: the start point of the transmission of P^x (X_1), the end point (X_2), or the

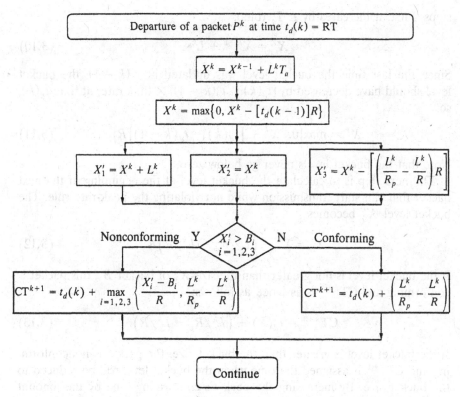

Fig. 3.25 Conforming time calculation of a new head-of-line packet.

point where the next packet can be transmitted without violating the peak rate rule (X_3). The above derivation is based on X_3. Nevertheless, Section 3.4 shows that it doesn't matter at which checking point the bucket size is set; the calculated conforming time is the same. In other words, as long as the checking point where the bucket level and the bucket size are compared is the place where the bucket size is set, the results of all three checking points are identical.

3.3.1.3 *Conforming Time Calculation*
Depending on the checking point, there are different ways of calculating the conforming time (CT) of the packet, as shown in Figure 3.25. Section 3.4 shows that the CT is independent of the checking point. If the packet is conforming (the time that the packet starts transmission is not earlier than the CT), it will not be discarded or marked.

Assume the current real time, RT, is the departure time of packet P^k (when the last bit of the packet departs), i.e., RT $= t_d(k)$. When packet P^k with packet length L^k departs from the shaper and the next packet (P^{k+1}) becomes the HOL packet, the bucket levels X^k, in bits, are updated in two steps. First, it increases by $L^k T_a$ (bits):

$$X^k = X^{k-1} + L^k a. \tag{3.10}$$

Since the last time the bucket level was updated at $t_d(k-1)$, the bucket level should have decreased by $[t_d(k) - t_d(k-1)] \times$ (link rate) at time $t_d(k)$, so

$$X^k = \max\{0, \ X^k - [t_d(k) - t_d(k-1)]R\}. \tag{3.11}$$

Note that the bucket levels never go below zero.

The next step is to calculate the bucket level at the beginning of the next packet that can start transmission while not violating the peak-rate rule. The bucket level X'_3 becomes

$$X'_3 = X^k - \left(L^k/R_p - L^k/R\right)R. \tag{3.12}$$

If the bucket level is not greater than the size X_3 of bucket B_3, this packet is conforming and CT^{k+1} is assigned as follows:

$$CT^{k+1} = t_d(k) + \left(L^k/R_p - L^k/R\right). \tag{3.13}$$

If the bucket level is greater than the bucket size, the packet is nonconforming and CT^{k+1} is assigned the time when the bucket level will be reduced to the bucket size. By increasing the packet's conforming time by the amount beyond the bucket size, the bucket level will be reduced to the bucket size. CT^{k+1} will be assigned as follows. Assuming that the checking point is at the beginning of the next packet that can start transmission while not violating the peak-rate rule and that the current packet is nonconforming, then

$$CT^{k+1} = t_d(k) + \max\left\{\frac{X'_3 - B_3}{R}, \frac{L^k}{R_p} - \frac{L^k}{R}\right\}. \tag{3.14}$$

From (3.9), we have

$$B_3 = M \cdot (T_a - T_p), \tag{3.15}$$

$$X'_3 - B_3 = X^k - (L^k/Rp - L^k/R) \cdot R - M \cdot (T_a - T_p). \tag{3.16}$$

$$X'_3 - X_3 = X^k - \left[M \cdot (T_a - T_p) + L^k \cdot (T_p - 1)\right]. \tag{3.17}$$

From (3.14) and (3.17),

$$CT^{k+1} = t_d(k) + \max\left\{\frac{X^k - M \cdot (T_a - T_p) - L^k \cdot (T_p - 1)}{R}, \frac{L^k}{R_p} - \frac{L^k}{R}\right\}. \tag{3.18}$$

3.3.2 Integrating Traffic Shaping and WFI Scheduling

Although a traffic shaper can shape users' traffic to conform strictly to its traffic descriptor, it fails to provide end-to-end delay bound and fairness for individual connections when multiple packets from different connections conform for transmission at the same time. Therefore, a scheduler is needed to follow the traffic shaper to guarantee each connection's delay bound and bandwidth requirement.

Packet fair queuing (PFQ) algorithms [19, 23, 26, 27, 28] based on generalized processor sharing (GPS) [19] are widely accepted as an effective scheduling scheme in the packet network. When the server is ready to transmit the next packet at time τ, the PFQ picks, among all the packets queued in the system at τ, the packet that has the smallest virtual finishing time F,

$$F_i^k = S_i^k + L_i^k/r_i, \tag{3.19}$$

$$S_i^k = \max\{V(t), F_i^{k-1}\}, \tag{3.20}$$

where L_i^k is the length of the packet of connection i, r_i is the allocated average rate of the connection i, and $V(t)$ is the system virtual time of the server, or the normalized work of each backlogged connection. Packets in the PFQ system with minimum WFI need to pass an eligibility test. A packet is eligible when

$$S_i^k \leq V(t). \tag{3.21}$$

In [26], the virtual clock mechanism [20] is used for the scheduler. It is known that the virtual clock algorithm fails to provide fairness. However, once it is combined with the eligibility test (called the *shaped virtual clock algorithm*), it can achieve the optimal WFI. As proven below, the conforming shaping function can be seamlessly integrated with the shaped virtual clock algorithm, which also achieves the optimal WFI.

Theorem *With the shaped virtual clock scheduling scheme, packets that are eligible are also conforming.*

Proof Assume an HOL packet P_i^k with CT_i^k is conforming and transmits at infinite rate from the conforming shaper to the eligible shaper at real time RT_1:

$$CT_i^k = RT_1. \tag{3.22}$$

Since the system virtual time $V(t)$ in the virtual clock server is the real time [i.e., $V(t) = RT_1$], and according to (3.20) packet P_i^k arrives at the eligible shaper at real time RT_1, its S-value is equal to

$$S_i^k = \max\{V(t), F_i^{k-1}\} = \max\{RT_1, F_i^{k-1}\}. \tag{3.23}$$

Therefore,

$$S_i^k = \max\{CT_i^k, F_i^{k-1}\}. \tag{3.24}$$

Then the conforming shaper can be combined with the eligible shaper to form a single shaper, with its S-value equal to $\max\{CT_i^k, F_i^{k-1}\}$. When packet P_i^k becomes eligible at time RT_2 ($RT_2 \geq RT_1$),

$$V(t) = RT_2 \tag{3.25}$$

and

$$V(t) \geq S_i^k. \tag{3.26}$$

From (3.24), (3.25), and (3.26),

$$RT_2 \geq CT_i^k. \tag{3.27}$$

Therefore, when packet P_i^k becomes eligible at time RT_2, it is also conforming.

3.3.3 A Logical Structure of the WFI Packet Shaper

Figure 3.26 shows a logical structure of the WFI packet shaper. The shaper that performs conformity and eligibility functions is implemented with a calendar queue based on the S-values, i.e., $S_i^k = \max\{CT_i^k, F_i^{k-1}\}$. Each slot in the calendar queue represents a distinct value of the virtual starting time. Associated with each slot j is a list of packets whose S-value is j. The packets in each S-queue are maintained in a sorted order of their F-values. The packet with the smallest F-value appears at the head of the S-queue.

A separate priority queue of packets is maintained by the scheduler. Packets are transmitted in increasing order of their F-values. Packets in the scheduler are all conforming and eligible. When a new packet arrives from a

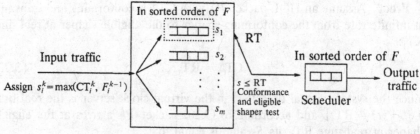

Conforming and eligible shaper

Fig. 3.26 A logical structure of the WFI packet shaper.

connection that has no backlogged packets in the system, the packet must be added either to the shaper or to the scheduler. Since $V(t)$ in the virtual clock scheduler is the real time, if the computed virtual starting time (i.e., $S_i^k = \max\{CT_i^k, F_i^{k-1}\}$) of the arriving packet is not greater than the RT, then, according to the theory described above, it is a conforming and eligible packet. This packet is sent directly to the scheduler, assigned $F_i^k = S_i^k + L_i^k/r_i$, and sorted by the scheduler according to its F-value. If the computed S-value of the arriving packet is greater than the RT, it is either a nonconforming or a noneligible packet. The packet joins the shaper calendar queue at the location of its S-value, and is sorted according to its F-value. When the RT is equal to or greater than the S-value, this packet is both conforming and eligible, and it is released to the scheduler.

When a packet departs from the WFI packet shaper (i.e., the last bit leaves the system) at time t, a new eligible packet with the smallest time-stamp is chosen to serve. At the end of its service, the packet departs at, say, $t + \tau$, and $V(t)$ is updated to $t + \tau$. All the HOL packets in the shaper queues whose S-values are within the time interval $(t, t + \tau]$ have now become conforming and eligible. Figure 3.26 describes the operations of arrival and departure events. However, transferring all such HOL packets to the scheduler queue will cause a bottleneck. A slotted mechanism [29] has been introduced to solve this problem. It is known that for efficient hardware handling, packets are usually divided into fixed-length segments in routers and switches before they are stored in the buffer and forwarded to the next node(s). As a result, a packet scheduler can be viewed as a slotted system. A time slot T corresponds to the time that is needed to transmit a segment at the link capacity R. Assuming a fixed-rate server and $\tau = T$, the work that the packet server performs in a time slot is a constant $W(t, t + \tau) = R \times T$, which can be normalized to 1, like that in ATM switches. When a virtual clock is used as the scheduling policy and the slotted mechanism is employed, $V(t)$ (i.e., RT) is incremented by one at every time slot. Then the same goal can be achieved as that in [26]. That is, only the following two packets at most need to be transferred to the scheduler in each time slot:

- The packet at the head of the shaper queue associated with $S = RT + 1$, where $RT + 1$ is the updated value of the real time.
- If the packet transmitted by the scheduler in the previous slot had a virtual starting time value of S_c, the packet at the head of the shaper queue associated with S_c, if any, is also moved to the scheduler queue.

3.3.4 Implementation of the WFI Packet Shaper

This section describes an implementation architecture for the WFI packet shaper. The Sequencer chip [15] can be used to facilitate the traffic shaper and queue management. Figure 3.27 shows the operation of the Sequencer.

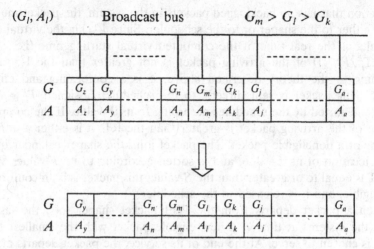

Fig. 3.27 Parallel sorting in the Sequencer.

Each module contains two fields: the priority value G and the address A. Note that the smaller the priority value, the higher the priority. When a new packet with priority value G_l arrives, G_l is broadcast to every module and is compared with each module's G-value simultaneously. Let us assume that G_l is less than G_m but is greater than G_k. The addresses of these two packets are A_k and A_m. All pairs on the right of A_k, including A_k itself, remain at their positions, while the others are shifted to the left. The vacant position is replaced with the pair composed of the new packet's priority value field (G_l) and address (A_l).

3.3.4.1 *Operations upon Packet Arrival*

To facilitate the implementation of the combined shaper and scheduler (the WFI packet shaper) in a single chip, a new Sequencer chip called the Sequencer II is proposed [30]. When a new HOL packet is to join the WFI packet shaper, the Sequencer II first combines its S and F to form a new priority value G (i.e., $G = \langle lS, F \rangle$) and then stores it in one of two regions, as shown in Figure 3.28. The right region corresponds to the *scheduler* in the logical structure of the WFI packet shaper (Fig. 3.26), and the left region corresponds to the *shaper*. If $S \leq V$, it is a conforming and eligible packet. The packet's G-value is set to $\langle 0, F \rangle$, which is then stored in the right region (with the smallest F in the rightmost position). If S is greater than V, it is either a nonconforming or a noneligible packet and is stored in the left region. The same S-values are grouped together in a subregion, corresponding to the priority queue in the shaper. Subregions are stored in the left side of the Sequencer II in increasing order of S. Within each subregion, the packets are sorted in increasing order of F.

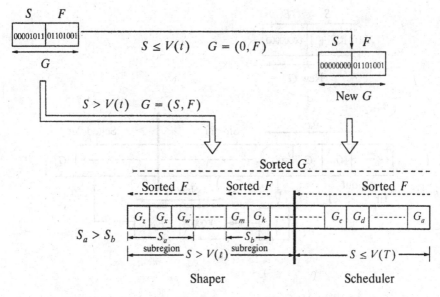

Fig. 3.28 Operations of the Sequencer II upon packet arrival.

3.3.4.2 *Operations upon Packet Departure* Figure 3.29 shows the operations of the Sequencer II when a packet departs from the scheduler at RT. As described before, at most two HOL packets with $S = RT + 1$ and $S = S_c$ need to be moved from the shaper to the scheduler, i.e., from the left region to the right region of the Sequencer II. To find the packet at the HOL of the shaper queue associated with $S = RT + 1$, a value x_1 is sent to the Sequencer II to perform a *pseudo-insertion*, as shown in Figure 3.29(a). The x_1 value is $\langle RT + 1, 0 \rangle$. The operation of the pseudo-insertion is not to insert the value, but rather to use it to find others in the Sequencer II. By pseudo-inserting x_1 in the Sequencer, we can find the packet with the smallest F-value in the smallest-S subregion. For instance, the packet with a priority value of $G_d = \langle S_d, F_d \rangle$ is found by inserting x_1. If its S-value is equal to or smaller than the $RT + 1$, this packet is eligible. It is pushed out from the left region and inserted in the Sequencer II with a priority value $G = \langle 0, F_d \rangle$. It is then stored in the right region of the Sequencer II.

Meanwhile, if a packet with $S = S_c$ departs from the scheduler, an x_2-value is sent to do a pseudo-insertion in the Sequencer II, as shown in Figure 3.29(b). The x_2-value is $\langle S_c, 0 \rangle$. By inserting the x_2-value in the Sequencer II, we can find the packet with the smallest F-value in a subregion whose S is equal to S_c. In Figure 3.29(b), the packet with priority value $G_k = \langle S_c, F_k \rangle$ is found. Since the packet is always eligible, it is pushed out from the left region and inserted in the Sequencer II with a priority value $G = \langle 0, F_k \rangle$.

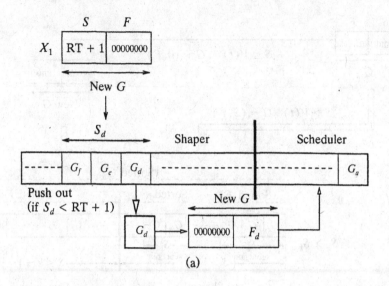

(a)

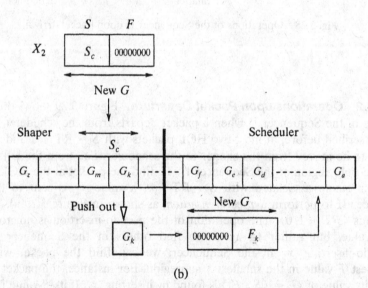

(b)

Fig. 3.29 Operations of the Sequencer II upon packet departure.

As described above, when a packet in the left region (shaper) is eligible, it is pushed out from the left region and joins the right region (scheduler). The operation of the pushout function in the Sequencer II is shown in Figure 3.30. Assume that a packet with a priority value G_x arrives, and G_x is less than G_m but greater than G_k. All pairs on the right of A_k, including A_k itself, remain at their positions while the others are shifted to the right with the pair G_m/A_m to be pushed out.

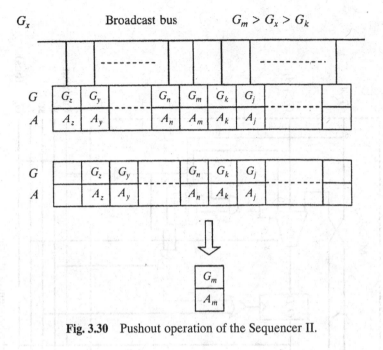

G_x Broadcast bus $G_m > G_x > G_k$

Fig. 3.30 Pushout operation of the Sequencer II.

3.3.4.3 Sequencer II Chip

Figure 3.31 shows the block diagram of the Sequencer II chip. Each module is represented by a dashed box. The Sequencer II chip is the same as the original Sequencer chip [15], except it adds some extra circuits to perform the pushout function. When a pushout signal is not asserted, it works like the original Sequencer chip. When the pushout signal is asserted, it performs the pushout function. As shown in Figure 3.31, $G_{i+1}A_{i+1}$ is shifted to register i when the shift-right (sr) and the shift-right-clock (srck) signals are both asserted.

Table 3.2 shows the truth table for generating the s_n and s_l, where $X = G_{i-1}$, $Y = G_i$, and $Z = G_n$. For case (a) in Table 3.2, where a new pair $G_n A_n$ is to be latched to register i, both s_n and s_l signals are asserted to select the $G_n A_n$ for register i's input (D) and to pass the shift-left-clock (slck) signal to register i's clock input. For case (b) in Table 3.2, sr, srck, and pushout signals are all asserted with $X \leq Z < Y$. Register i's output (Q) will be pushed out to the bus, and $G_{i+1}A_{i+1}$ is latched to register i with the clock signal srck. For case (c) in Table 3.2, only the s_l signal is asserted, which results in $G_{i-1}A_{i-1}$ being selected and latched to register i with the clock signal slck. For case (d) in Table 3.2, sr, srck, and pushout signals are all asserted with $Z < X \leq Y$, and $G_{i+1}A_{i+1}$ is to be latched in the register with the clock signal srck. For case (e) in Table 3.2, the s_l signal is deserted while the s_n signal is "don't care," and thus register i maintains its original value, $G_i A_i$.

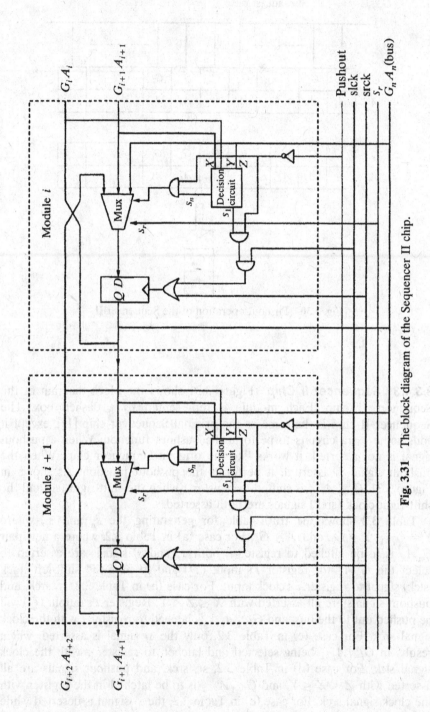

Fig. 3.31 The block diagram of the Sequencer II chip.

TABLE 3.2   Truth Table for the s_n and s_l Signals

Cases	s_n	s_l	Action
(a) pushout = 0, $X \leq Z < Y$	1	1	Latch $G_n A_n$ in register i
(b) pushout = 1, $X \leq Z < Y$	1	1	Push out $G_i A_i$ to the bus, and latch $G_{i+1} A_{i+1}$ in register i
(c) pushout = 0, $Z < X \leq Y$	0	1	Latch $G_{i-1} A_{i-1}$ in register i
(d) pushout = 1, $Z < X \leq Y$	0	1	Latch $G_{i+1} A_{i+1}$ in register i
(e) $X \leq Y \leq Z$	d	0	Retain $G_i A_i$ in register i (d = don't care)

When a packet arrives at the WFI packet shaper (Sequencer II), the priority value and the address pair, $G_n A_n$, are broadcast to every module in the Sequencer II chip. Based on the priority values of G_i, G_{i-1}, and G_n, the decision circuit generates proper signals, s_n and s_l, to shift the new pair $G_n A_n$ to register i [case (a) in the truth table], shift the pair $G_{i-1} A_{i-1}$ from the right to register i [case (c) in the truth table], or retain the original value $G_i A_i$ [case (e) in the truth table]. When a packet departs from the Sequencer II, x_1, x_2, or $G_n A_n$ is broadcast to every module. Based on the value of G_i, G_{i-1}, and G_n, the decision circuit generates proper signals, s_n and s_l, to push out $G_i A_i$ to the bus and latch $G_{i+1} A_{i+1}$ to register i [case (b) in the truth table], shift the pair $G_{i+1} A_{i+1}$ from the left to register i [case (d) in the truth table], or retain the original value $G_i A_i$ [case (e) in the truth table].

3.3.4.4 Summary   This section describes integrating a traffic shaper with a shaped virtual clock scheduler. The integrated WFI packet shaper can not only shape users' traffic to strictly conform to its traffic descriptor, but also provide the minimum WFI. Also presented is an implementation architecture for the WFI packet shaper with a new Sequencer II chip.

3.4 APPENDIX: BUCKET SIZE DETERMINATION

This section shows that it doesn't matter at which checking point the bucket size is set: the calculated conforming time is the same. This statement can be proven by calculating the bucket level and bucket size at different checking points.

First, suppose the beginning of packet P_3 is the checking point (i.e., X_1):

$$M = L^1 + L^2 + L^3, \tag{3.28}$$

where L^1, L^2, and L^3 are the packet lengths of packets P^1, P^2, and P^3, respectively. Assume the bucket is empty at time $t = 0$. Since the elapsed time from the first bit of the maximum burst to this checking point X_1 is $L^1/R_p + L^2/R_p$,

$$X_1 = MT_a - \left(L^1/R_p + L^2/R_p \right) R. \tag{3.29}$$

By denoting $T_p = R/R_p$ and using (3.28),

$$X_1 = M \cdot (T_a - T_p) + L^3 T_p. \tag{3.30}$$

If the bucket level at this checking point is chosen to be the bucket size, then $B_1 = X_1$.

Second, let the end of packet P^3 be the checking point (i.e., X_2). Since the elapsed time from the first bit of the maximum burst to this checking point is $L^1/R_p + L^2/R_p + L^3/R$,

$$X_2 = MT_a - \left[\left(L^1/R_p + L^2/R_p \right) + L^3/R \right] R. \tag{3.31}$$

From (3.28) and $T_p = R/R_p$, we have

$$X_2 = M \cdot (T_a - T_p) + L^3 \cdot (T_p - 1). \tag{3.32}$$

If the bucket level at this checking point is chosen to be the bucket size, then $B_2 = X_2$.

Third, let the beginning of the next packet that can start transmission while not violating the peak-rate rule be the checking point (i.e., X_3). Since the elapsed time from the first bit of the maximum burst to this checking point is $L^1/R_p + L^2/R_p + L^3/R_p$,

$$X_3 = MT_a - \left(L^1/R_p + L^2/R_p + L^3/R_p \right) R. \tag{3.33}$$

This can be simplified to

$$X_3 = M \cdot (T_a - T_p). \tag{3.34}$$

If the bucket level at this checking point is chosen to be the bucket size , then $B_3 = X_3$.

Based on different checking points, there are different ways of calculating the conforming time (CT) value of the packet. As shown in Figure 3.25, the CT is independent of the checking point. If the packet is conforming (the time that the packet starts transmission is not earlier than the CT), this packet will not be discarded or marked.

Assume the current real time (RT) is the departure time of packet P_i^k (the last bit of the packet departs), i.e., $\text{RT} = t_d(k)$.

1. When packet P^k with packet length L^k departs from the shaper and its next packet becomes the HOL packet, the bucket level X^k, in terms of bits, is updated in two steps. First, it increases by $L^k T_a$ (bits),

$$X^k = X^{k-1} + L^k T_a. \tag{3.35}$$

Second, since the last time the bucket level was updated at $t_d(k - 1)$, the bucket level should have decreased by $[t_d(k) - t_d(k - 1)] \times$ (link rate) at time $t_d(k)$, or

$$X^k = \max\{0,\ X^k - [t_d(k) - t_d(k-1)]R\}. \tag{3.36}$$

Note that the bucket levels never go below zero.

2. The next step is to calculate the bucket level X^k at the different checking points. If the checking point is at the beginning of the current packet, the bucket level X_1' becomes

$$X_1' = X^k + L^k. \tag{3.37}$$

If the checking point is at the end of the current packet, the bucket level X_2' becomes

$$X_2' = X^k. \tag{3.38}$$

If the checking point is at the beginning of next packet that can start transmission while not violating the peak rate rule, the bucket level X_3' becomes

$$X_3' = X^k - \left(L^k/R_p - L^k/R\right)R. \tag{3.39}$$

3. If the bucket level is not greater than the corresponding bucket size B, this packet is conforming and CT^{k+1} is assigned as the following:

$$CT^{k+1} = t_d(k) + \left(L^k/R_p - L^k/R\right). \tag{3.40}$$

If the bucket level is greater than the bucket size, the packet is nonconforming and CT^{k+1} is assigned the time when the bucket level will be reduced to the bucket size. By increasing the packet's conforming time by the amount beyond the bucket size, the bucket level will be reduced to the bucket size. CT^{k+1} will be assigned as follows: Assuming that the checking point is at the beginning of the current packet and that this packet is nonconforming, then

$$CT^{k+1} = t_d(k) + \max\left\{\frac{X_1' - B_1}{R}, \frac{L^k}{R_p} - \frac{L^k}{R}\right\}. \tag{3.41}$$

From (3.30), we have

$$B_1 = M \cdot (T_a - T_p) + L^k T_p. \tag{3.42}$$

$$X_1' - B_1 = (X^k + L^k) - \left[M \cdot (T_a - T_p) + L^k T_p\right]. \tag{3.43}$$

$$X_1' - B_1 = X^k - \left[M \cdot (T_a - T_p) + L^k \cdot (T_p - 1)\right]. \tag{3.44}$$

From (3.41) and (3.44),

$$CT^{k+1} = t_d(k) + \max\left\{\frac{X^k - M \cdot (T_a - T_p) - L^k \cdot (T_p - 1)}{R}, \frac{L^k}{R_p} - \frac{L^k}{R}\right\}. \tag{3.45}$$

In the same way, assuming that the checking point is at the end of the current packet and that this packet is nonconforming, then

$$CT^{k+1} = t_d(k) + \max\left\{\frac{X_2' - B_2}{R}, \frac{L^k}{R_p} - \frac{L^k}{R}\right\}. \tag{3.46}$$

From (3.32), we have

$$B_2 = M \cdot (T_a - T_p) + L^k \cdot (T_p - 1), \tag{3.47}$$

$$X_2' - B_2 = X^k - \left[M \cdot (T_a - T_p) + L^k \cdot (T_p - 1)\right]. \tag{3.48}$$

From (3.46) and (3.48),

$$CT^{k+1} = t_d(k) + \max\left\{\frac{X^k - M \cdot (T_a - T_p) - L^k \cdot (T_p - 1)}{R}, \frac{L^k}{R_p} - \frac{L^k}{R}\right\}. \tag{3.49}$$

Assuming that the checking point is at the beginning of the next packet that can start transmission while not violating the peak-rate rule and that the current packet is nonconforming, then

$$CT^{k+1} = t_d(k) + \max\left\{\frac{X_3' - B_3}{R}, \frac{L^k}{R_p} - \frac{L^k}{R}\right\}. \tag{3.50}$$

From (3.34), we have

$$B_3 = M \cdot (T_a - T_p), \tag{3.51}$$

$$X_3' - B_3 = X^k - \left(L^k/R_p - L^k/R\right)R - M \cdot (T_a - T_p), \tag{3.52}$$

$$X_3' - X_3 = X^k - \left[M \cdot (T_a - T_p) + L^k \cdot (T_p - 1)\right]. \tag{3.53}$$

From (3.50) and (3.53),

$$CT^{k+1} = t_d(k) + \max\left\{\frac{X^k - M \cdot (T_a - T_p) - L^k \cdot (T_p - 1)}{R}, \frac{L^k}{R_p} - \frac{L^k}{R}\right\}. \tag{3.54}$$

Note that from the above equations, the difference between the bucket level and the bucket size is the same for all three checking points. In other words, as long as the checking point where the bucket level and bucket size are compared is the place where the bucket size is set, the results of all three checking points are the same.

REFERENCES

1. J. S. Turner, "New directions in communications (or which way to the information age?)," *IEEE Commun. Mag.*, vol. 24, no. 10, pp. 8–15, Oct. 1986.

2. E. P. Rathgeb, "Modeling and performance comparison of policing mechanisms for ATM networks," *IEEE J. Select. Areas Commun.*, vol. SAC-9, no. 1, pp. 325–334, Apr. 1991.

3. M. Sidi, W.-Z. Liu, I. Cidon, and I. Gopal, "Congestion control through input rate regulation," *IEEE Trans. Commun.*, vol. 41, no. 3, Mar. 1993.

4. P. Boyer, F. M. Guillemin, M. J. Servel, and J.-P. Coudreuse, "Spacing cells protects and enhances utilization of ATM network links," *IEEE Netw.*, vol. 6, no. 5, pp. 38–49, Sep. 1992.

5. F. Guillemin, P. Boyer, A. Dupuis, and L. Romoeuf, "Peak rate enforcement in ATM networks," *IEEE INFOCOM*, Florence, Italy, vol. 2, pp. 753–758 (6A.1), May 1992.

6. The ATM Forum, *Traffic Management Specification Version* 4.1, Mar. 1999.

7. ITU-T, Recommendation I.371: "Traffic control and congestion control in B-ISDN," Perth, Nov. 1995.

8. R. Braden, D. Clark, and S. Shenker, "Integrated services in the Internet architecture: an overview," Informational, RFC 1633, Internet Engineering Task Force (IETF), Jun. 1994.

9. S. Blake, D. Black, M. Carlson, E. Davies, and Z. Wang, "An Architecture for Differentiated Services," RFC 2475, Internet Engineering Task Force (IETF), Dec. 1998.

10. H. J. Chao, "Architecture design for regulating and scheduling user's traffic in ATM networks," *SIGCOMM Symp. Communications Architectures and Protocols*, Baltimore, pp. 77–87, Aug. 1992.

11. H. Zhang and D. Ferrari, "Rate-controlled static-priority queueing," Technical Report TR-92-003, Computer Science Division, University of California, Berkeley, Nov. 1992.

12. L. Zhang, "Virtual clock: a new traffic control algorithm for packet switching networks," *SIGCOMM Symp. Communications Architectures and Protocols*, pp. 19–29, Sep. 1990.

13. P.-C. Chen, "The design of a timing processor for ATM traffic shaper," Master's thesis, Polytechnic University, Brooklyn, NY, Jul. 1995.

14. H. J. Chao, "Design of leaky bucket access control schemes in ATM networks," *IEEE Proc. Int. Conf. on Communications*, pp. 180–187, Jun. 1991.

15. H. J. Chao and N. Uzun, "A VLSI Sequencer chip of ATM traffic shaper and queue manager," *IEEE J. Solid-State Circuits*, vol. 27, no. 11, Nov. 1992.

16. T. Moors, N. Clarke, and G. Mercankosk, "Implementing traffic shaping," *IEEE Proc. Conf. on Local Computer Networks*, pp. 307–314, Oct. 1994.

17. H. J. Chao and J. S. Hong, "Design of an ATM shaping multiplexer with guaranteed output burstiness," *Comput. Syst. Sci. Eng.*, vol. 12, no. 2, Mar 1997.

18. J. S. Hong, "Design of an ATM shaping multiplexer algorithm and architecture," Ph.D. dissertation, Electrical Engineering Department, Polytechnic University, Brooklyn, NY, Jan. 1997.

19. A. K. Parekh and R. G. Gallager, "A generalized processor sharing approach to flow control—the single node case," *Proc. IEEE INFOCOM'92*, vol. 2, pp. 915–924, May 1992.

20. L. Zhang, "Virtual clock: a new traffic control algorithm for packet switch networks," *Proc. ACM SIGCOMM'90*, pp. 19–29, Sep. 1990.

21. H. Zhang and S. Keshav, "Comparison of ratebased service disciplines," *Proc. ACM SIGCOMM*, pp. 113–121, Sep. 1991.

22. D. Stiliadis and A. Varma, "Efficient fair-queueing algorithms for ATM and packet networks," Technical Report UCSC-CRL-95-59, University of California, Santa Cruz, CA, Dec. 1995.

23. J. C. R. Bennett and H. Zhang, "WF^{2}Q: worst-case fair weighted fair queueing," *Proc. IEEE INFOCOM'96*, pp. 120–128, Mar. 1996.

24. D. Stiliadis and A. Varma, "Rate proportional servers: a design methodology for fair queueing algorithms," *IEEE/ACM Trans. Netw.*, vol. 6, no. 2, pp. 164–174, Apr. 1998.

25. D. Stiliadis and A. Varma, "Latency-rate servers: a general model for analysis of traffic scheduling algorithms," *IEEE/ACM Trans. Netw.*, vol. 6, no. 5, pp. 611–624, Oct. 1998.

26. D. Stiliadis and A. Varma, "A general methodology for designing efficient traffic scheduling and shaping algorithms," *Proc. IEEE INFOCOM'97*, 1997.

27. L. Georgiadis, R. Guerin, V. Peris, and K. N. Sivarajan, "Efficient network QoS provisioning based on per node traffic shaping," *IEEE/ACM Trans. Netw.*, vol. 4, no. 4, pp. 482–501, Aug. 1996.

28. J. Rexford, F. Bonomi, A. Greenberg, and A. Wong, "Scalable architectures for integrated traffic shaping and link scheduling in high-speed ATM switches," *IEEE J. Select. Areas in Commun.*, vol. 15, no. 5, pp. 938–950, Jun. 1997.

29. H. J. Chao, Y. Jeng, X. Guo, and C. H. Lam, "Design of packet fair queuing schedulers using a RAM-based searching engine," *IEEE J. Select. Areas Commun.*, vol. 17, no. 6, pp. 1105–1126, Jun. 1999.

30. L. S. Chen, "Architecture design of packet scheduling for delay bound guarantee," Ph.D. Dissertation, Polytechnic University, Brooklyn, NY, Aug. 2000.

CHAPTER 4

PACKET SCHEDULING

Packet networks allow users to share resources such as buffer and link bandwidth. However, the problem of contention for the shared resources arises necessarily. Given a number of users (flows or connections) multiplexed at the same link, a packet scheduling discipline is needed to determine which packets to serve (or transmit) next. In other words, sophisticated scheduling algorithms are required to prioritize users' traffic to meet various QoS requirements while fully utilizing network resources. For example, real-time traffic is delay-sensitive, while data traffic is loss-sensitive. This chapter presents a historical overview of different packet scheduling algorithms. Section 4.1 briefly reviews the classification of such algorithms. Sections 4.2 to 4.13 are each dedicated to a particular scheme, from simple FIFO to round-robin, stop-and-go, hierarchical round-robin, earliest-due-date, rate-controlled static priority, generalized processor sharing (GPS), weighted fair queuing (WFQ), virtual clock, self-clocked fair queuing, worst-case fair weighted fair queuing (WF^{2}Q), and WF^{2}Q + . Section 4.4 discusses the delay bound of packet fair queuing algorithms in the multiple-node case. A comparison is given in Section 4.15. In Section 4.16, a core-stateless scheduling algorithm is described.

4.1 OVERVIEW

Numerous packet scheduling algorithms have been developed for support of packet-switched network services. In general, the different network services that they provide can be classified as follows:

- *Best Effort*   Services make no commitments about QoS. The scheduling algorithms of this kind do not take into account any requested or realized QoS properties of packet flows. An example of such a scheme is first in, first out (FIFO), or first come, first served (FCFS).
- *Better Than Best Effort*   Services make no deterministic guarantees about delay, but make a best effort to try to support requested QoS requirements. While schemes that offer strict guarantees on delay must enforce isolation of packet flows, this kind of scheme is able to achieve a higher level of sharing of network switching resources. An example of such schemes is FIFO+ .
- *Guaranteed Throughput*   Services ensure each flow a negotiated bandwidth regardless of the behavior of all other traffic. When strict admission control and traffic access controls are used to limit the arrival rate of packets at the system, per-flow upper delay bounds can be achieved. Examples of such schemes include weighted fair queuing (WFQ), virtual clock, and WF^2Q.
- *Bounded Delay Jitter*   Services guarantee upper and lower bounds on observed packet delays. These services are implemented by non-work-conserving scheduling disciplines. Strict admission control based on *a priori* traffic characterizations and traffic shapers are typically required. An example of such schemes is jitter-earliest-due-date (jitter-EDD).

Scheduling schemes are typically evaluated along several dimensions, such as tightness of delay bounds, achievable network utilization, fairness and protection, protocol overhead, computational cost, and robustness. Recently the implementation of packet fair queuing schedulers, which aim at approximating the generalized processor sharing (GPS) policy [15, 17], is becoming a central issue for providing multimedia services with various QoS requirements in asynchronous transfer mode (ATM) switches and next-generation Internet protocol (IP) routers [2, 3]. The objective is to design an efficient and scalable architecture that can support hundreds of thousands of sessions (virtual channels in ATM or flows in IP) in a cost-effective manner.

The GPS is an ideal WFQ service policy based on a fluid-flow model. It can provide network delay bound for leaky-bucket-constrained traffic and has been intensively studied [4–16, 18]. However, because the fluid GPS is not practical, a class of packet fair queuing (PFQ) algorithms has been proposed to emulate the fluid GPS to achieve the desired performance [14, 20–28, 33]. All of them are based on maintaining a global function, referred to as either *system virtual time* or *system potential*, which tracks the progress of GPS. This global function is used to compute a *virtual finish time* (or *timestamp*) for each packet or the head-of-line (HOL) packet of each session in the system. The timestamp of a packet is the sum of its *virtual start time* and the time needed

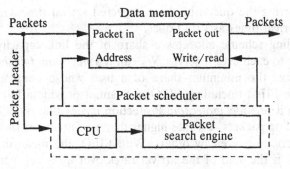

Fig. 4.1 Packet scheduler.

to transmit the packet at its reserved bandwidth. Packets are served in increasing order of their timestamps.

The implementation cost of a PFQ algorithm is determined by two components: (1) computing the system virtual time function, and (2) maintaining the relative ordering of the packets via their timestamps in a priority queue mechanism. Figure 4.1 illustrates a packet scheduler, which, for instance, can be located at the output of a switch or router. The CPU is responsible for computing timestamps and other system control. The packet search engine is responsible for selecting the next packet for transmission according to the timestamp values.

4.2 FIRST IN, FIRST OUT

The simplest possible scheduling algorithm is *first in, first out* (FIFO). In this algorithm, the scheduler transmits incoming packets in the order they arrive at the output queue, and drops packets that arrive at a full queue. However, the scheduler cannot differentiate among users. Thus, it cannot explicitly allocate some users lower mean delays than others. Every user in the same service class suffers the same delay jitter. Besides, the jitter tends to increase dramatically with the number of hops, since the packet's queuing delays at different hops are uncorrelated [35].

FIFO + is an attempt to induce FIFO-style sharing (equal jitter for all users in the aggregate class) across all the hops along the path to minimize jitter [35]. It is done as follows. For each hop, we measure the average delay seen by packets in each class at that node. We then compute for each packet the difference between its particular delay and the class average. We add (or subtract) this difference to (or from) a field in the header of the packet, which thus accumulates the total offset for this packet from the average for its class. This field allows each node to compute when the packet *should* have arrived if it had indeed been given average service. The node then schedules the packet as if it had arrived at the expected average time. This is done by ordering the queue by these expected arrival times, rather than by the actual arrival times as FIFO does.

A scheduling scheme allocates a share of the link capacity and output queue buffer to each user it serves. We call it a *max–min fair share* allocation if it maximizes the minimum share of a user whose demand is not fully satisfied. The FIFO (including FIFO +) cannot provide a fair share allocation; nor can it provide *protection*. Protection means that misbehavior by one user (by sending packets at a rate higher than its fair share) should not affect the performance received by others. With FIFO, the mean delay of a user may increase if the sum of the arrival rates over all users increases. In the following, we introduce a number of scheduling disciplines that can provide both fairness and protection.

4.3 ROUND-ROBIN

In a round-robin scheduler, newly arriving packets queue up by users. The server polls each queue in a cyclic order and serves a packet from any nonempty queue encountered. A misbehaving user overflows its own queue, and the others are unaffected. Thus, round-robin can provide protection. Round-robin is an attempt to treat all users equally and provide each of them an equal share of the link capacity. It performs reasonably well when all users have equal weights and all packets have the same size (like cells in ATM networks).

If users have different weights, then weighted round-robin (WRR) serves a user in proportional to its weight. Consider a round-robin cell system in which there are two users, A and B with weights $w_A = 3$ and $w_B = 7$ cells, respectively. The scheme attempts to give a 30% $[= w_A/(w_A + w_B)]$ share of the link capacity to user A and 70% $[= w_B/(w_A + w_B)]$ share to user B. One outgoing cell sequence possible in a round is *AAABBBBBBB*. A better implementation of WRR operates according to a frame structure and gives *ABBABBABBBB*. That is, user A doesn't need to wait for 7 cell time slots before having its cell sent.

Deficit round-robin (DRR) [32] modifies WRR to allow it to handle variable packet sizes in a fair manner. The basic idea is to use round-robin with a quantum of service assigned to each queue. The only difference from traditional round-robin is that if a queue was not able to send a packet in the previous round because its packet size was too large, the remainder from the previous quantum is added to the quantum for the next round. Thus, deficits are recorded. Queues that were shortchanged in a round are compensated in the next round. The detailed algorithm is as follows.

Assume that each flow i is allocated Q_i bits in each round; there is an associated state variable DC_i recording the deficits. Since the algorithm works in rounds, we can measure time in terms of rounds. A round is one round-robin iteration over the queues that are backlogged. To avoid examining empty queues, an auxiliary list *ActiveList* is kept and consists of a list of indices of queues that contain at least one packet. Packets coming in on different flows are stored in different queues. Let the number of bytes of the head-of-line (HOL) packet for queue i in round k be bytes$_i(k)$. Whenever a packet arrives at a previously empty queue i, i is added to the end of *ActiveList*. Whenever index i is at the head of *ActiveList*, say, in round k, the algorithm computes $DC_i \leftarrow DC_i + Q_i$, sends out queue i's HOL packet subject to the condition that bytes$_i(k) \leq DC_i$ and (if the condition is true) updates $DC_i \leftarrow DC_i -$ bytes$_i(k)$. If, at the end of this service opportunity, queue i still has packets to send, the index i is moved to the end of *ActiveList*; otherwise, DC_i is reset to zero and i is removed from *ActiveList*. The DRR, however, is fair only over time scales longer than a round time. At a shorter time scale, some users may get more service (in terms of bits sent) than others.

4.4 STOP-AND-GO

Stop-and-go uses a framing strategy [11]. In such a strategy, the time axis is divided into frames, which are periods of some constant length T. Stop-and-go defines departing and arriving frames for each link. At each switch, the arriving frame of each incoming link is mapped to the departing frame of the output link by introducing a constant delay θ, where $0 \leq \theta < T$. The transmission of a packet that has arrived on any link l during a frame f should always be postponed until the beginning of the next frame.

Stop-and-go ensures that packets on the same frame at the source stay in the same frame throughout the network. If the traffic at the source is $(r; T)$-smooth (i.e., no more than rT bits are transmitted during any frame of size T), it satisfies the same characterization throughout the network. By maintaining traffic characteristics throughout the network, end-to-end delay bounds can be guaranteed in a network of arbitrary topology as long as each local server can ensure local delay bounds for traffic characterized by the $(r; T)$ specification.

The framing strategy introduces the problem of coupling between delay bound and bandwidth allocation granularity. The delay of any packet at a single switch is bounded by two frame times. To reduce the delay, a smaller T is desired. However, since T is also used to specify traffic, it is tied to bandwidth allocation granularity. Assuming a fixed packet size P, the minimum granularity of bandwidth allocation is P/T. To have more flexibility in allocating bandwidth, or a smaller bandwidth allocation granularity, a larger T is preferred. It is clear that low delay bound and fine granularity of bandwidth allocation cannot be achieved simultaneously in a framing strategy like stop-and-go.

To get around this coupling problem, a generalized version of stop-and-go with multiple frame sizes is proposed [11]. In the generalized stop-and-go, the time axis is divided into a hierarchical framing structure. For G-level framing with frame sizes $T_1, \ldots, T_G$, we have $T_g = f_g T_{g+1}$ for $g = 1, \ldots,$

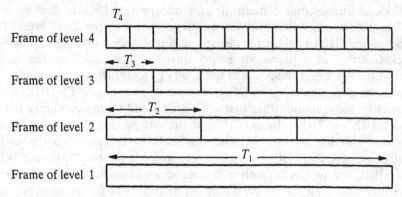

Fig. 4.2 G-level framing with $G = 4$, $f_1 = 3$, $f_2 = 2$, $f_3 = 2$.

$G - 1$, as illustrated in Figure 4.2. Packets on a level-g connection need to observe the stop-and-go rule with frame size T_g, i.e., level-g packets that arrived at an output link during a T_g frame will not become eligible for transmission until the start of the next T_g frame. Also, for two packets with different frame sizes, the packet with a smaller frame size has a nonpreemptive priority over the packet with a larger frame size.

In particular, at each node, any level-g arriving frame on input link i is mapped to a level-g departing frame on output link j by the introduction of a constant delay $\Theta_{i,j}^g$ as shown in Figure 4.3, where $0 \le \Theta_{i,j}^g < T_g$. According to the stop-and-go service discipline, the transmission of a level-g packet arriving during a level-g frame on link i should always be postponed until the beginning of the mapped departing frame.

Stop-and-go ensures that packets in the same frame at the source stay in the same frame throughout the network. As a result, if the traffic stream of a

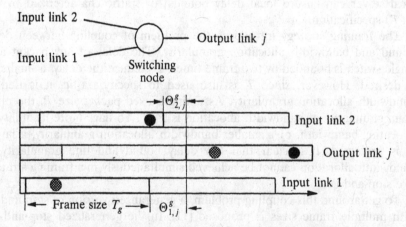

Fig. 4.3   Frame mismatch delay at a switching node.

connection is characterized at the source as (r, T_g)-smooth, it satisfies the same characterization throughout the network. The delay of a level-g frame at a switching node can be bounded by

$$T_g \leq \text{frame delay} < 2T_g.$$

Moreover, the intraframe delay of each packet in the switching node can be bounded by $\pm T_g$. Thus, the delay of a packet can be bounded by

$$0 < \text{cell delay} < 3T_g.$$

As a result, the buffer needed to allocate to an (r, T_g)-smooth session can be given by Little's law, which is $3T_g \times r$.

With the multiframe stop-and-go, it is possible to provide low delay bounds to some channels by putting them in frames with a smaller frame time, and to allocate bandwidth with fine granularity to other channels by putting them in levels with a larger frame time. However, the coupling between delay and bandwidth allocation granularity still exists within each frame.

4.5 HIERARCHICAL ROUND-ROBIN

HRR [7] is similar to stop-and-go in that it also uses a multilevel framing strategy. A slot in one level can be allocated either to a connection or to a lower-level frame. The server cycles through the frame and serves packets according to the assignment of slots. If the server cycles through a slot assigned to a connection, one packet from that connection is transmitted; if it cycles through a slot assigned to a lower level frame, it serves one slot from the lower level frame in the same fashion.

Similar to stop-and-go, HRR also maintains traffic smoothness inside the network. However, there are also important differences between HRR and stop-and-go. As an example, assume that three packet transmission times are allocated to a connection in each frame. In stop-and-go, packets that are transmitted in the same frame at the entrance to the network will be transmitted in the same frame on all the links traversed by the connection. The difference between delays experienced by any two packets from the source to any server is bounded by T, the frame size. In HRR, packets that are transmitted in the same frame at the entrance to the network do not necessarily stay in the same frame inside the network; however, the property that no more than three packets from the connection are transmitted during one frame time holds throughout the network.

Since HRR uses the framing strategy, it also has the problem of coupling between delay and bandwidth allocation granularity.

4.6   EARLIEST DUE DATE

In classic earliest due date (EDD) [8, 9] scheduling, we assign each packet a deadline, and the scheduler serves packets in the order of their deadlines. If the scheduler is overcommitted, then some packets miss their deadlines. Obviously, with EDD, packets assigned deadlines closer to their arrival times receive a smaller delay than packets assigned deadlines farther away from their arrival times.

Delay EDD [9] is an extension of EDD that specifies the process by which the scheduler assigns deadlines to packets. A delay-EDD scheduler reserves bandwidth at a session's peak rate. The scheduler sets a packet's deadline to the time at which it should be sent had it been received no faster than its peak rate. Thus, every packet from a session obeying the peak-rate constraint receives a hard delay bound, which is independent of its bandwidth reservation. Therefore, delay EDD separates the bandwidth and delay bounds, but at the cost of using peak-rate allocation, giving up temporal statistical multiplexing gain.

Jitter EDD [10] extends delay EDD to provide a delay-jitter bound (i.e., a bound on the maximum delay difference between two packets). Jitter EDD incorporates a delay-EDD scheduler preceded by a delay-jitter regulator. After a packet has been served at each server, a field in its header is stamped with the difference between its deadline and the actual finishing time. A regulator at the entrance of the next server holds the packet for this period before it is made eligible to be scheduled.

A scheduler implementing delay-jitter regulation can remove the effect of queuing delay variability in the previous node, thus avoiding the burstiness built up within the network. More precisely, if a_k^n and e_k^n are the arrival and eligibility times for the kth packet at the nth node, respectively, then

$$e_k^0 = a_k^0,$$
$$e_k^{n+1} = e_k^n + d_n + l_{n, n+1}, \tag{4.1}$$

where d_n is the delay bound at the previous node, and $l_{n, n+1}$ is the link propagation delay between nodes n and $n + 1$. The kth packet is eligible for service at the first node the moment it arrives. However, at subsequent nodes it becomes eligible for service only after a fixed time interval of length $d_n + l_{n, n+1}$, which is the longest possible delay in the previous node and in the previous link.

So, if a packet is served before its delay bound at the previous node, the delay-jitter regulator at the downstream node adds sufficient delay to convert this to the longest possible delay. Thus, a network of jitter-EDD schedulers can give sessions end-to-end bandwidth, delay, and delay-jitter bounds. However, the delay-jitter regulator is hard to implement. Not only does it require the network operator to know a bound on the propagation delay on each link; it also requires the network to maintain clock synchrony at adjacent nodes at all times. Since, in the real world, clocks drift out of synchrony unless corrected, delay-jitter regulation implicitly assumes the presence of a mechanism for maintaining clock synchrony, which is not feasible.

4.7 RATE-CONTROLLED STATIC PRIORITY

While the EDD algorithm can provide flexible delay bounds and bandwidth allocation, it is based on a sorted priority mechanism, which, without a hardware-based mechanism, is difficult to implement in high-speed networks. *Rate-controlled static priority* (RCSP) is designed to achieve flexibility in the allocation of delay and bandwidth, as well as simplicity of implementation.

As shown in Figure 4.4, an RCSP server consists of two components: a rate controller and a static priority scheduler [12, 13]. Logically, a rate controller consists of a set of regulators corresponding to each of the sessions traversing the server. When a packet arrives at the server, an eligibility time is calculated and assigned to the packet by the regulator. The packet is held in the regulator until its eligibility time, and then it is handed to the scheduler for transmission. Different ways of calculating the eligibility time of a cell result in different types of regulators.

For example, a traffic stream can be characterized by $(X_{\min}, X_{\text{ave}}, I)$ if the interarrival time between any two packets in the stream is more than $X_{\min}$, and the average packet interarrival time during any interval of length I is more than X_{ave}. It is obvious that $X_{\min} \leq X_{\text{ave}} < I$ must hold.

For a $(X_{\min}, X_{\text{ave}}, I)$ regulator, the eligibility time of the kth packet from session i, $e_{i,k}$, is defined with reference to the eligibility times of packets

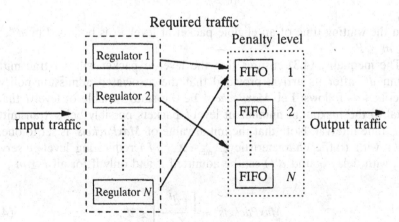

Fig. 4.4 One regulator for each of the N sessions.

arriving earlier at the server from the same session, where

$$e_{i,k} = -I \quad \text{for} \quad k \le 0,$$

$$e_{i,1} = a_{i,1},$$

$$e_{i,k} = \max\left(e_{i,k-1} + X_{\min}, e_{i,k-\lfloor I/X_{\text{ave}}\rfloor+1} + I, a_{i,k}\right), \quad \text{for} \quad k > 1,$$

(4.2)

where $a_{i,k}$ is the time the kth packet from session i arrived at the server.

From this definition, we can see that $e_{i,k} \ge a_{i,k}$ always holds, i.e., a packet is never eligible before its arrival. Also, if we consider the sequence of packet eligibility times at the server, $\{e_{i,k}\}_{k=1,2,\ldots}$, it always satisfies the $(X_{\min}, X_{\text{ave}}, I)$ *traffic characterization*.

The scheduler in the RCSP server uses a nonpreemptive static priority (SP) policy: it always selects the packet at the head of the highest-priority queue that is not empty. The SP scheduler consists of a number of priority levels with each priority level corresponding to a delay bound. Each session is assigned to a priority level during the call admission time. Multiple sessions can be assigned to the same priority level, and all packets from the sessions associated with the same priority level are appended to the end of the queue for that priority level.

Let the number of the priority levels be P, and $d^1, d^2, \ldots, d^P$ ($d^1 < d^2 < d^3 < \cdots < d^P$) be the delay bounds associated with the P priority levels, respectively, in an RCSP switch. Assume that the jth connection among i_l traversing the switch at priority level l has the traffic specification $(X^l_{\min_j}, X^l_{\text{ave}_j}, I^l_j)$. Also assume that the link speed is one. Zhang and Ferrari [12] have proved that if

$$\sum_{l=1}^{m} \sum_{j=1}^{i_l} \left\lceil \frac{d^m}{X^l_{\min_j}} \right\rceil + 1 \le d^m,$$

(4.3)

then the waiting time of an eligible packet at level m is bounded by d^m, for $1 \le m \le P$.

The inequality (4.3) ensures that the level-m packet will get transmitted within d^m after its arrival provided that the associated admission policy is described as follows. Let *MaxPackets^l* be the state variable of level l that is equal to the maximum number of level-l packets possibly being transmitted in d^l time interval. Note that the initial value of *MaxPackets^l* is set to one. A call i with traffic characteristic $(X_{\min_i}, X_{\text{ave}_i}, I_i)$ requesting level-m service (i.e., with delay bound d^m) can be admitted if and only if for all $l \ge m$,

$$MaxPackets^l + \left\lceil \frac{d^l}{X_{\min_i}} \right\rceil \le d^l.$$

(4.4)

After session i has been admitted as a level-m call, we set

$$MaxPackets^m \leftarrow MaxPackets^m + \left\lceil \frac{d^m}{X_{\min_i}} \right\rceil.$$

If we require that the output traffic stream satisfy $(X_{\min}, X_{ave}, I)$, the SP scheduler must feed the departure time of the kth packet from session i, $d_{i,k}$, back to the corresponding regulator, so that the eligibility time setup algorithm given in (4.2) can be modified as follows:

$$e_{i,k} = -I, \quad \text{for} \quad k \leq 0,$$

$$e_{i,1} = a_{i,1}, \tag{4.5}$$

$$e_{i,k} = \max\{d_{i,k-1} + X_{\min}, d_{i,k-\lfloor I/X_{ave}\rfloor+1} + I, a_{i,k}\}, \quad \text{for} \quad k > 1.$$

4.8 GENERALIZED PROCESSOR SHARING

Generalized processor sharing (GPS) [15, 16] is an ideal scheduling policy in that it provides an exact max–min fair share allocation. GPS is fair in the sense that it allocates the whole outgoing capacity to all backlogged sessions in proportion to their minimum rate (bandwidth) requirements. Basically, the algorithm is based on an idealized fluid-flow model. That is, we assume that a GPS scheduler is able to serve all backlogged sessions instantaneously and that the capacity of the outgoing link can be split infinitesimally and allocated to these sessions. However, in real systems only one session can be served at each time and packets cannot be split into smaller units. An important class of so-called packet fair queuing (PFQ) algorithms can then be defined in which the schedulers try to schedule the backlogged packets by approximating the GPS scheduler, such as weighted fair queuing (WFQ), virtual clock, and self-clock fair queuing (SCFQ), which are discussed in the following sections. We first introduce the ideal GPS algorithm.

Assume that a set of N sessions (connections), labeled $1, 2, \ldots, N$, share the common outgoing link of a GPS server. For $i \in \{1, 2, \ldots, N\}$, let r_i denote the minimum allocated rate for session i. The associated admission policy should guarantee that

$$\sum_{i=1}^{N} r_i \leq r \tag{4.6}$$

where r is the capacity of the outgoing link.

Let $B(t)$ denote the set of backlogged sessions at time t. According to the GPS [15], the backlogged session i will be allocated a service rate $g_i(t)$ at time t such that

$$g_i(t) = \frac{r_i}{\sum_{j \in B(t)} r_j} \times r. \tag{4.7}$$

We will use an example to illustrate the service rate allocation principle of GPS servers. Let $A_i(\tau, t)$ be the amount of arrivals of session i during the interval $(\tau, t]$, $W_i(\tau, t)$ the amount of service received by session i during the same interval, and $Q_i(t)$ the amount of session i traffic queued in the server at time t, i.e.,

$$Q_i(t) = A_i(\tau, t) - W_i(\tau, t).$$

Note that, whenever the system becomes idle, all parameters can be reset to zero.

Definition 4.1 A **system busy period** is a maximal interval of time during which the server is always busy with transmitting packets.

Definition 4.2 A **backlogged period for session** i is any period of time during which packets of session i are continuously queued in the system.

Definition 4.3 A **session** i **busy period** is a maximal interval of time $(\tau_1, \tau_2]$ such that for any $t \in (\tau_1, \tau_2]$, packets of session i arrive at rate greater than or equal to r_i, i.e.,

$$A_i(\tau_1, t) \geq r_i(t - \tau_1), \quad \text{for} \quad t \in (\tau_1, \tau_2].$$

With reference to Figure 4.5, assume that the capacity of the server is $r = 1$, and three connections, labeled 1, 2, and 3, share the same outgoing link of the server, where $r_1 = \frac{1}{6}$, $r_2 = \frac{1}{3}$, and $r_3 = \frac{1}{2}$.

Suppose that each packet has a fixed length that needs exactly one unit of time to transmit. At time $t = 0$, session 1 starts a session busy period in which packets from session 1 arrive at the server at a rate of one packet per unit time. At $t = 1$, packets from session 2 also start to arrive at the server at

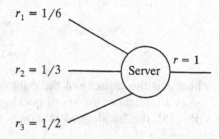

Figure 4.5 A GPS server with three in-coming sessions.

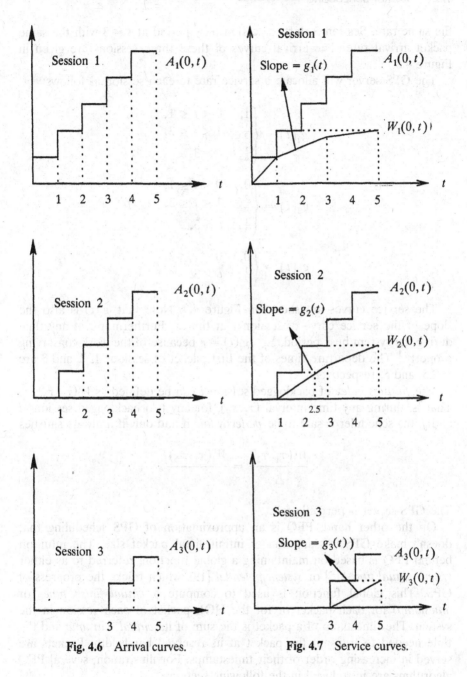

Fig. 4.6 Arrival curves.   **Fig. 4.7** Service curves.

the same rate. Session 3 starts a session busy period at $t = 3$ with the same packet arrival rate. The arrival curves of these three sessions are given in Figure 4.6.

The GPS server will allocate a service rate to each session as follows:

$$g_1(t) = \begin{cases} 1, & 0 < t \le 1, \\ \frac{1}{3}, & 1 < t \le 3, \\ \frac{1}{6}, & t > 3, \end{cases}$$

$$g_2(t) = \begin{cases} 0, & 0 < t \le 1, \\ \frac{2}{3}, & 1 < t \le 3, \\ \frac{1}{3}, & t > 3, \end{cases}$$

$$g_3(t) = \begin{cases} 0, & 0 < t \le 3, \\ \frac{1}{2}, & t > 3. \end{cases}$$

The service curves are shown in Figure 4.7. Note that $g_i(t)$ is also the slope of the service curve of session i at time t. Furthermore, at any time during the system busy period, $\sum_{i=1}^{N} g_i(t) = r$ because of the work-conserving property.[1] The departure times of the first packet of sessions 1, 2, and 3 are 1, 2.5, and 5, respectively.

The *fairness index* of backlogged session i can be defined as $W_i(\tau_1, \tau_2)/r_i$. That is, during any time interval $(\tau_1, \tau_2]$, for any two backlogged sessions i and j, the scheduler is said to be *perfectly fair* if and only if it always satisfies

$$\frac{W_i(\tau_1, \tau_2)}{r_i} = \frac{W_j(\tau_1, \tau_2)}{r_j}.$$

The GPS server is perfectly fair.

On the other hand, PFQ is an approximation of GPS scheduling that doesn't make GPS's assumption of infinitesimal packet size. The intuition behind PFQ is based on maintaining a global function, referred to as either *system virtual time* [15] or *system potential* [19], which tracks the progress of GPS. This global function is used to compute a *virtual finish time* (or *timestamp*) for each packet or for the HOL packet of each session in the system. The timestamp of a packet is the sum of its *virtual start time* and the time needed to transmit the packet at its reserved bandwidth. Packets are served in increasing order of their timestamps. For illustration, several PFQ algorithms are introduced in the following sections.

---

[1]A server is *work-conserving* if it is never idle whenever there are packets to be transmitted. Otherwise, it is *non-work-conserving*.

4.9   WEIGHTED FAIR QUEUING

Although the GPS service principle is perfectly fair, the idealized fluid-flow model is not practical to implement. However, we can simulate the GPS server and then schedule the backlogged packets in accordance with the packet behavior of the simulated GPS server. A WFQ (also called packetized GPS) system is defined with respect to its corresponding GPS system. Let d_p^{GPS} be the time at which packet p will depart (finish service) under GPS. A good approximation of GPS would be a scheme that serves packets in increasing order of d_p^{GPS}. However, this is not always possible without causing the discipline to be non-work-conserving. This is because when the packet system is ready to choose the next packet to transmit, the next packet to depart under GPS may not have arrived at the packet system yet. Waiting for it requires knowledge of the future and also causes the system to be non-work-conserving. *In WFQ, the server simply assigns the departure time of a packet in the simulated GPS server as the timestamp of that packet, and then the server transmits packets in increasing order of these timestamps.* When the server is ready to transmit the next packet at time τ, it picks the first packet that would complete service in the corresponding GPS system if no additional packets were to arrive after time τ.

WFQ [15] uses the concept of *virtual time* to track the progress of GPS that will lead to a practical implementation of packet-by-packet GPS. Define an *event* as any arrival or departure of a session from the GPS server, and let t_j be the time at which the jth event occurs (simultaneous events are ordered arbitrarily). Let the time of the first arrival of a busy period be denoted as $t_1 = 0$. Now observe that, for each $j = 2, 3, \ldots$, the set of sessions that are busy in the interval (t_{j-1}, t_j) is fixed. We denote this set as B_j. The virtual time $V(t)$ is defined to be zero for all times when the server is idle. Consider any busy period, and let the time that it begins be time zero. Then, $V(t)$ evolves as follows:

$$V(0) = 0,$$

$$V(t_{j-1} + \tau) = V(t_{j-1}) + \frac{r\tau}{\sum_{i \in B_j} r_i} \quad \text{for } \tau \le t_j - t_{j-1}, j = 2, 3 \ldots . \quad (4.8)$$

The rate of change of V, namely $dV(t_j + \tau)/d\tau$, is $r/\sum_{i \in B_j} r_i$, and each backlogged session i receives service at rate $r_i \, dV(t_j + \tau)/d\tau$, i.e., $g_i(t_j + \tau)$ according to (4.7). Thus, V can be interpreted as increasing at the marginal rate at which backlogged sessions receive service.

Now suppose that the kth packet from session i arrives at time $a_{i,k}$ and has length $L_{i,k}$. Then denote the virtual times at which this packet begins and completes service as $S_{i,k}$ (also called the *virtual start time* [15] or *start potential* [19]) and $F_{i,k}$ (the *virtual finish time* [15] or *finish potential* [19]), respectively. Defining $F_{i,0} = 0$ for all i, we have

$$S_{i,k} = \max\{F_{i,k-1}, V(a_{i,k})\},$$

$$F_{i,k} = S_{i,k} + \frac{L_{i,k}}{r_i}. \qquad (4.9)$$

The role of $V(a_{i,k})$ is to reset the value of $S_{i,k}$ when queue i becomes active (i.e., receives one packet after being empty for a while) to allow for the service it missed [19, 29]. Therefore, the start times of backlogged queue can stay close to each other (they are the same in a GPS server).

For the above example, let the kth packet from backlogged session i be labeled $(i, d_{i,k}^{\mathrm{GPS}})$, where $d_{i,k}^{\mathrm{GPS}}$ is the departure time of this packet in the simulated GPS server. Figure 4.8 shows the service curves and the departure

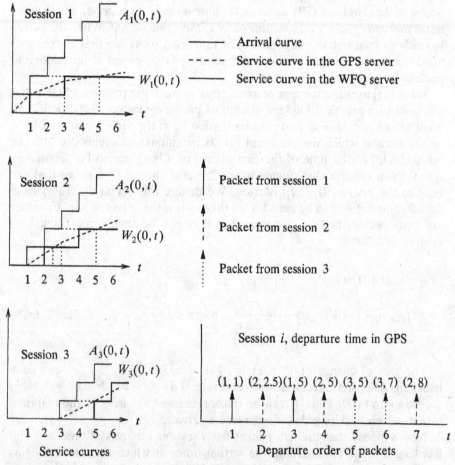

Fig. 4.8 Service curves and the departure order of packets in the WFQ server. Packets depart according to the departure times as if the packets were served by a GPS server.

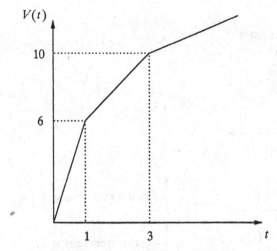

Fig. 4.9 Virtual time function $V(t)$.

order of packets in the WFQ server for the previous example. As shown on the right-hand side of Figure 4.8, packets depart according to the departure time in the GPS system. Packets that have the same departure time are served arbitrarily.

On the other hand, let each packet be labeled $(i, F_{i,k})$. Figure 4.9 shows the virtual time $V(t)$ according to (4.8). Figure 4.10 shows the virtual finish time curves and the departure order of packets for the previous example. $F_i(t)$ denotes a staircase function of time t with $F_i(a_{i,k}) = F_{i,k}$; $a_{i,k}$ is the arrival time of the kth packet of session i and is calculated according to (4.9). Note that the departure order is the same as that in Figure 4.8.

In [15], Parekh and Gallager establish the following relationships between the GPS system and its corresponding packet WFQ system:

$$d_{i,k}^{\text{WFQ}} - d_{i,k}^{\text{GPS}} \leq \frac{L_{\max}}{r} \qquad \forall i, k, \qquad (4.10)$$

$$W_i^{\text{GPS}}(0, \tau) - W_i^{\text{WFQ}}(0, \tau) \leq L_{\max} \qquad \forall i, \tau, \qquad (4.11)$$

where $d_{i,k}^{\text{WFQ}}$ and $d_{i,k}^{\text{GPS}}$ are the times at which the kth packet in session i departs under WFQ and GPS, respectively; $W_i^{\text{WFQ}}(0, \tau)$ and $W_i^{\text{GPS}}(0, \tau)$ are the total amounts of service received by session i (the number of session i bits transmitted) by time τ under WFQ and GPS, respectively; and $L_{\max}$ is the maximum packet length among all the sessions.

Another parameter, called *latency* [19], can be defined and used to compare the performance of the WFQ and the GPS servers.

Definition 4.4 The **latency** of a server $\mathcal{S}$, $\Theta_i^{\mathcal{S}}$, is the minimum nonnegative number that satisfies

$$W_{i,j}^{\mathcal{S}}(\tau, t) \geq \max\{0, r_i(t - \tau - \Theta_i^{\mathcal{S}})\}. \qquad (4.12)$$

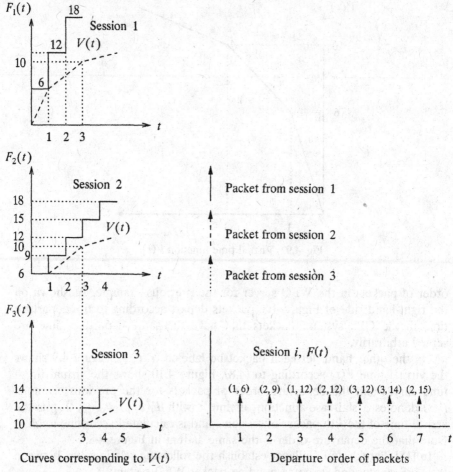

Fig. 4.10   Virtual finish timestamps, $F(t)$, and the departure order of packets. $F(t)$ is calculated with knowledge of the virtual time $V(t)$. Packets depart according to their $F(t)$-values.

for any time t after time τ when the jth busy period started and until the packets that arrived during this period are served.

With reference to Figure 4.11, the inequality (4.12) defines an envelope to bound the minimum service offered to session i during a busy period. It is easy to show that the latency $\Theta_i^{\mathscr{S}}$ represents the worst-case delay seen by the first packet of a busy period of session i.

In the GPS server, the newly backlogged session can get served immediately with a rate equal to or greater than its required transmission rate. As a result, the latency is zero.

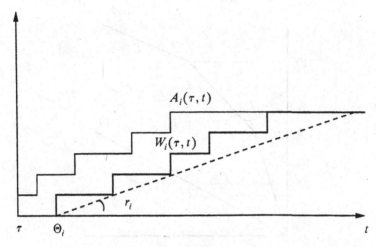

Fig. 4.11 Latency.

In the WFQ server, however, the worst-case delay of the first packet of a backlogged period for session i is $d_{i,1}^{\text{WFQ}} - a_{i,1}$, where $a_{i,1}$ is the arrival time of that packet. From the inequality (4.10) we have

$$d_{i,1}^{\text{WFQ}} - a_{i,1} \le d_{i,1}^{\text{GPS}} + \frac{L_{\max}}{r} - a_{i,1} \le \frac{L_i}{r_i} + \frac{L_{\max}}{r},$$

where L_i is the maximum packet size of session i. Thus, we can conclude that the latency of session i in the WFQ server is bounded by $L_i/r_i + L_{\max}/r$.

The WFQ algorithm has a time complexity of $O(N)$ because of the overhead in keeping track of sets B_j (which is essential in the updating of virtual time), where N is the maximum number of backlogged sessions in the server. Nevertheless, we can find another function of time to approximate the virtual time function such that the computation complexity of the scheduling algorithm can be further reduced. As shown in the following sections, all PFQ algorithms use a similar priority queue mechanism that schedules packet transmission in increasing order of their timestamps, but they differ in choices of system virtual time function and packet selection policies [19, 29].

4.10 VIRTUAL CLOCK

The virtual clock (VC) scheduler uses a real time function to approach the virtual time function. That is, the scheduler assigns

$$V^{\text{VC}}(t) = t \quad \text{for} \quad t \ge 0. \tag{4.13}$$

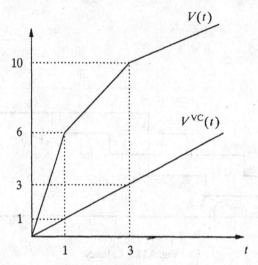

Fig. 4.12 The VC scheduler uses real time to approximate the virtual time.

The kth packet from session i will be assigned a timestamp $F_{i,k}$ from (4.9) and (4.13), namely,

$$F_{i,k} = \max\{F_{i,k-1}, a_{i,k}\} + \frac{L_{i,k}}{r_i}, \qquad (4.14)$$

where $a_{i,k}$ is the arrival time of the kth packet of session i. Figure 4.12 shows the curves $V(t)$ and $V^{VC}(t)$, and Figure 4.13 shows the service curves and the departure order of packets for the previous example.

Since the real time is always less than or equal to the virtual time, the VC scheduler can always provide the newly backlogged session with a latency smaller than or equal to that provided by the WFQ server [6]. However, the VC service discipline is defined with reference to the static time system, and the calculation of the timestamp is independent of the behavior of other sessions. As a result, if a connection has sent more packets than specified, it may be punished by the VC, regardless of whether such misbehavior affects the performance of other connections. For example, suppose there are two sessions 1 and 2 as shown in Figure 4.14. All packets from both sessions are the same size. The link capacity is normalized as 1 packet per time slot. Let $r_1 = r_2 = 0.5$ packet/slot, so the timestamp for a session will be advanced by 2 slots each time its HOL packet is sent, based on (4.14). Initially, at time 0, $F_{1,0} = F_{2,0} = 0$. The session 1 source continuously generates packets from time 0, while the session 2 source starts to send packets continuously from time 900 (in units of slots), as illustrated in Figure 4.14. Up to time 900, 900 packets from source 1 have been transmitted by the VC scheduler, and based on (4.14), $F_{1,901} = 1802$ at time 900, while $F_{2,1} = 902$. Therefore, the 901st

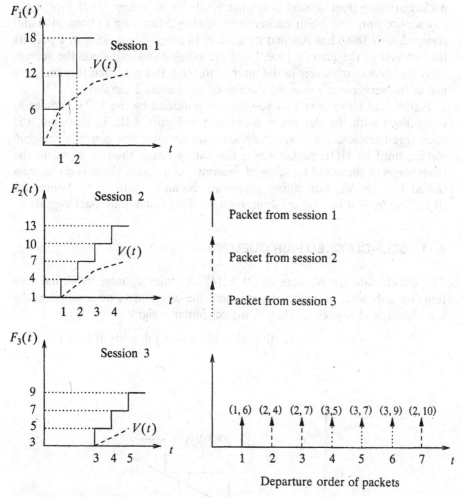

Fig. 4.13 Virtual finish timestamps and the departure order of packets in the virtual clock scheduler.

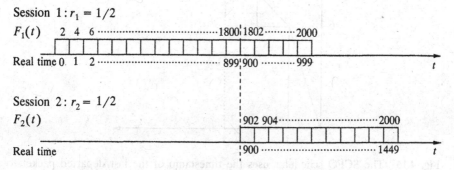

Fig. 4.14 An example showing unfairness of the virtual clock.

packet arriving from session 1 at time 900 (with its stamp 1802) cannot get any service until the 449th packet from session 2 (arriving at time 1449 and stamped with 1800) has finished its service. In other words, session 1 packets that arrived in the interval $[900, 1500)$ are being punished, since the session used the server exclusively in the interval $[0, 900)$. But note that this exclusive use of the server was *not at the expense of any session 2 packets*.

Figure 4.13 also shows that session 1 is punished by the VC scheduler in comparison with the departure order in the Figure 4.10. In this case, old backlogged sessions must wait for the server to serve the newly backlogged session until its HOL packet has a timestamp larger than or equal to the timestamps of those old backlogged sessions. As a result, there is no fairness bound for the VC scheduling algorithm, because there is no bound for $|W_i(\tau_1, \tau_2)/r_i - W_j(\tau_1, \tau_2)/r_j|$ when both sessions i and j are backlogged.

4.11 SELF-CLOCKED FAIR QUEUING

The self-clocked fair queuing (SCFQ) [14] scheduler updates its virtual time function only when a packet departs, and the assigned value is equal to the timestamp of that packet. That is, the scheduler assigns

$$V^{\text{SCFQ}}(t) = F_{j,l} \qquad \text{if the } l\text{th packet of session } j \text{ departs at time } t \geq 0.$$

$$(4.15)$$

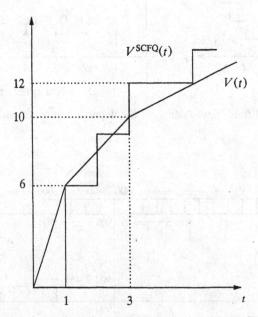

Fig. 4.15 The SCFQ scheduler uses the timestamp of the last departed packet to approximate the virtual time.

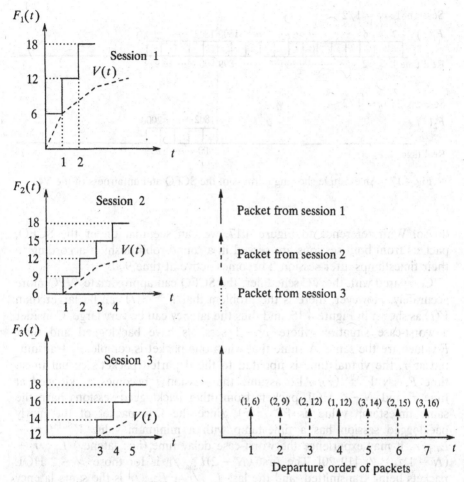

Fig. 4.16 Virtual finish timestamps and the departure order of packets in the SCFQ scheduler.

Similarly, the kth packet from session i will be assigned a timestamp $F_{i,k}$ from (4.9) and (4.15), namely,

$$F_{i,k} = \max\{F_{i,k-1}, V^{\text{SCFQ}}(a_{i,k})\} + \frac{L_{i,k}}{r_i}. \qquad (4.16)$$

Figure 4.15 shows the curves corresponding to $V(t)$ and $V^{\text{SCFQ}}(t)$, respectively, while Figure 4.16 shows the timestamps and the departure order of packets for the previous example.

Figure 4.17 demonstrates that how the SCFQ is able to provide a fairness guarantee for the same situation given in Figure 4.14 where the VC fails to

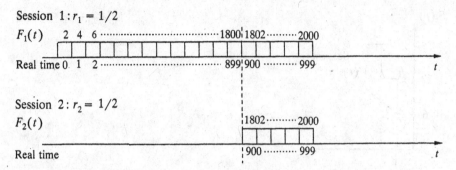

Fig. 4.17   An example showing fairness of the SCFQ and unfairness of the VC.

do so. With reference to Figure 4.17, we can see that under the SCFQ, packets from both sessions are served in a round-robin fashion according to their timestamps after session 2 becomes active at time 900.

Compared with the VC scheduler, the SCFQ can approximate WFQ more accurately. However, there is the problem that $V^{SCFQ}(t)$ can be larger than $V(t)$ as shown in Figure 4.15, and thus the latency can be very large. Consider a worst-case situation where $N - 1$ sessions have backlogged and their F-values are the same. Assume that when one packet is completely transmitted at τ, the virtual time is updated to the departed packet's virtual finish time F, say $V^{SCFQ}(\tau)$. Also assume that session i becomes backlogged at time τ, and $N - 2$ HOL packets from other backlogged sessions have the same timestamp value as $V^{SCFQ}(\tau)$. Since the first packet of the newly backlogged session has a timestamp with a minimum value $V^{SCFQ}(\tau) + L_{i,1}/r_i$, it may experience the worst-case delay time (i.e., latency) $L_{i,1}/r_i + (N - 1)L_{max}/r$ [19, 20]. The first $(N - 2)L_{max}/r$ is for those $N - 2$ HOL packets being transmitted, and the last $L_{i,1}/r_i + L_{max}/r$ is the same latency as in WFQ. As a result, the latency of the SCFQ scheduler is $L_{i,1}/r_i + (N - 1)L_{max}/r$.

4.12   WORST-CASE FAIR WEIGHTED FAIR QUEUING

The results given by (4.10) and (4.11) can be easily interpreted to show that WFQ and GPS provide almost identical service except for one packet. What Parekh has proven is that WFQ cannot fall behind GPS with respect to the service given to a session by one maximum-size packet. However, packets can leave much *earlier* in a WFQ system than in a GPS system, which means that WFQ can be far ahead of GPS in terms of the number of bits served for a session.

Consider the example [26] illustrated in Figure 4.18(a), where there are 11 sessions sharing the same link [26]. The horizontal axis shows the time line

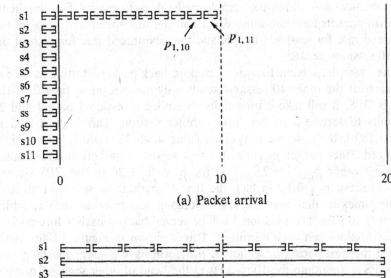

(a) Packet arrival

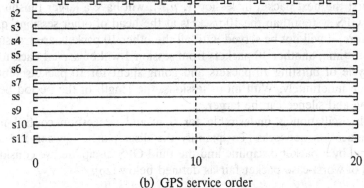

(b) GPS service order

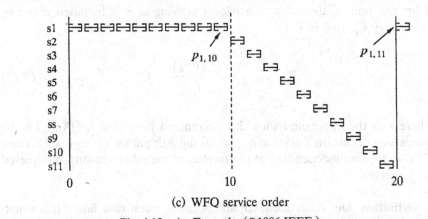

(c) WFQ service order

Fig. 4.18   An Example. (©1996 IEEE.)

and the vertical axis shows the sample path of each session. For simplicity, assume all packets have the same size 1 and the link speed is 1. Also, let the guaranteed rate for session 1 be 0.5, and the guaranteed rate for each of the other 10 sessions be 0.05.

In the example, session 1 sends 11 back-to-back packets starting at time 0, while each of the other 10 sessions sends only one packet at time 0. If the server is GPS, it will take 2 time units to service a session 1 packet and 20 time units to service a packet from another session. This is illustrated in Figure 4.18(b). If the server is WFQ, at time 0, all 11 sessions have packets backlogged. Since packet $p_{1,1}$ (i.e., the first session 1 packet) finishes at time 2 while all other $p_{i,1}$ ($i = 2, \dots, 11$) finish at time 20 in the GPS system, WFQ will serve $p_{1,1}$ first. In fact, the first 10 packets on session 1 all have finishing times smaller than packets belonging to any other session, which means that 10 packets on session 1 will be served back to back before packets on other sessions can be transmitted. This is shown in Figure 4.18(c). After the burst the next packet in session 1, $p_{1,11}$, will have a larger finishing time in the GPS system than the 10 packets at the head of other sessions' queues. Therefore, it will not be served until all the other 10 packets are transmitted, at which time, another 10 packets from session 1 will be served back to back. This cycle of bursting 10 packets and going silent for 10 packets times can continue indefinitely. With more sessions, the length of the period between bursting and silence can be larger.

Such oscillation is undesirable for flow and congestion control in data communication networks. To quantify the discrepancy between the service provided by a packet discipline and the fluid GPS discipline, we consider the notion of worst-case packet fair as defined below [26].

Definition 4.5   A service discipline s is called **worst-case fair for session** i if for any time τ, the delay of a packet arriving at τ is bounded above by $Q_i^s(\tau)/r_i + C_i^s$, that is,

$$ d_{i,k}^s < a_{i,k} + \frac{Q_i^s(a_{i,k})}{r_i} + C_i^s, \tag{4.17} $$

where r_i is the minimum bandwidth guaranteed to session i, $Q_i^s(a_{i,k})$ is the queue size of session i at time $a_{i,k}$ when the kth packet of session i arrives, C_i^s is a constant independent of the queues of the other sessions multiplexed at the server.

Definition 4.6   A service discipline is called **worst-case fair** if it is worst-case fair for all sessions.

Definition 4.7 C_i^s is called the **worst-case fair index** (WFI) for session i at server s.

Since C_i^s is measured in absolute time, it is not suitable for comparing C_i^s's of sessions with different r_i's. To perform such a comparison, the normalized WFI for session i at server s can be defined as

$$c_i^s = \frac{r_i C_i^s}{r}. \tag{4.18}$$

For a server that is worst-case fair, we define its normalized WFI to be

$$c^s = \max_i \{c_i^s\}. \tag{4.19}$$

Notice that GPS is worst-case fair with $c^{\text{GPS}} = 0$. Thus, we can use c^s as the metric to quantify the service discrepancy between a packet discipline s and GPS. It has been shown in [26] that c^{WFQ} may increase linearly as a function of the number of sessions N.

To minimize the difference between a packet system and the fluid GPS system, another class of scheduling algorithms called *shaper–schedulers* [22, 24, 26, 27, 33] has been proposed to achieve minimum WFI and have better worst-case fairness properties. With these algorithms, when the server is picking the next packet to transmit, it chooses, among all the *eligible* packets, the one with the smallest timestamp. A packet is eligible if its virtual start time is no greater than the current system virtual time. This is called the *eligibility test* or *smallest eligible virtual finish time first* (SEFF) policy [22, 26].

Worst-case fair weighted fair queuing or WF^{2}Q [26], is one such example. Recall that in a WFQ system, when the server chooses the next packet for transmission at time τ, it selects, among all the packets that are backlogged at τ, the first packet that would complete service in the corresponding GPS system. In a WF^{2}Q system, when the next packet is chosen for service at time τ, rather than selecting it from among all the packets at the server as in WFQ, the server only considers the set of packets that have started (and possibly finished) receiving service in the corresponding GPS system at time τ, and selects the packet among them that would complete service first in the corresponding GPS system.

Now consider again the example discussed in Figure 4.18, but in light of WF^{2}Q policy. At time 0, all packets at the head of each session's queue, $p_{i,1}$, $i = 1, \ldots, 11$, have started service in the GPS system [Fig. 4.18(a)]. Among them, $p_{1,1}$ has the smallest finish time in GPS, so it will be served first in WF^{2}Q. At time 1, there are still 11 packets at the head of the queues: $p_{1,2}$ and $p_{i,1}$, $i = 2, \ldots, 11$. Although $p_{1,2}$ has the smallest finish time, it will not start service in the GPS system until time 2; therefore, it won't be eligible for transmission at time 1. The other 10 packets have all started service at time 0 at the GPS system; thus, they are eligible. Since they all finish at the same time in the GPS system [Fig. 4.18(b)], the tie-breaking rule of giving highest priority to the session with the smallest number yields $p_{1,2}$ as the next packet

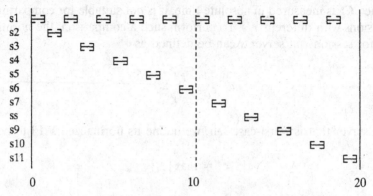

Fig. 4.19 WF^{2}Q service order: an example. (©1996 IEEE.)

for service. In contrast, if a WFQ server is used, then, rather than selecting the next packet from among the 10 packets that have started service in the GPS system, it will pick the packet among all 11 packets, which results in packet $p_{1,2}$. At time 3, $p_{1,2}$ becomes eligible and has the smallest finish time among all backlogged packets; thus it starts service next. The rest of the sample path for the WF^{2}Q system is shown in Figure 4.19 [26].

Therefore, even in the case when session 1 is sending back-to-back packets, its output from the WF^{2}Q system is rather smooth, as opposed to the bursty output under a WFQ system. The following theorem summarizes some of the most important properties of WF^{2}Q [26].

Theorem 4.1 *Given a WF^{2}Q system and a corresponding GPS system, the following properties hold for any i, k, τ:*

$$d_{i,k}^{\text{WF}^2\text{Q}} - d_{i,k}^{\text{GPS}} \leq \frac{L_{\max}}{r}, \tag{4.20}$$

$$W_i^{\text{GPS}}(0, \tau) - W_i^{\text{WF}^2\text{Q}}(0, \tau) \leq L_{\max}, \tag{4.21}$$

$$W_i^{\text{WF}^2\text{Q}}(0, \tau) - W_i^{\text{GPS}}(0, \tau) \leq \left(1 - \frac{r_i}{r}\right) L_i. \tag{4.22}$$

4.13 WF^{2}Q+

While WF^{2}Q provides the tightest delay bound and smallest WFI among all PFQ algorithms, it has the same worst-case time complexity, $O(N)$, as WFQ, because they both need to compute the virtual time or the system virtual time $V(t)$ by tracing the fluid GPS system.

WF^{2}Q + [27] and SPFQ [19] have been shown to have worst-case fairness properties similar to WF^{2}Q, but they are simpler to implement, by introducing the following system virtual time function:

$$V(t + \tau) = \max\left\{V(t) + \tau, \min_{i \in \hat{B}(t)} \{S_i(t)\}\right\}, \tag{4.23}$$

where $\hat{B}(t)$ is the set of sessions that are backlogged in the system at time t, and $S_i(t)$ is the virtual start time of backlogged session i's HOL packet. Let $W(t, t + \tau)$ be the total amount of service provided by the server, or the number of bits that have been transmitted during a time interval $(t, t + \tau]$. In the special case of a fixed-rate server, $\tau = W(t, t + \tau)/r$, where r is the link capacity. The time complexity is reduced to $O(\log N)$, attributed to the operations of searching for the minimum start time value among all N sessions.

To approximate the GPS, a PFQ algorithm, such as WF^2Q + and SPFQ, maintains a system virtual time function $V(t)$, a virtual start time $S_i(t)$, and a virtual finish time (or timestamp) $F_i(t)$ for each queue i. $S_i(t)$ and $F_i(t)$ are updated on the arrival of the HOL packet for each queue. A packet departure occurs when its last bit is sent out, while an HOL packet arrival occurs in either of two cases: (I) a previously empty queue has an incoming packet that immediately becomes the HOL; or (II) the packet next to the previous HOL packet in a nonempty queue immediately becomes the HOL after its predecessor departs. Obviously, a packet departure and a packet arrival in case II could happen at the same time. Therefore,

$$S_i(t) = \begin{cases} \max\{V(t), F_i(t^-)\} & \text{for packet arrival in case I,} \\ F_i(t^-) & \text{for packet arrival in case II,} \end{cases} \quad (4.24)$$

$$F_i(t) = S_i(t) + \frac{L_i^{\text{HOL}}}{r_i}, \quad (4.25)$$

where $F_i(t^-)$ is the finish time of queue i before the update, and L_i^{HOL} is the length of the HOL packet for queue i. The way of determining $V(t)$ is the major distinction among proposed PFQ algorithms [19, 29]. The next chapter describes how to implement a PFQ scheduler.

4.14 MULTIPLE-NODE CASE

It is important to note that a PFQ algorithm, such as WFQ with leaky bucket traffic access control, can provide worst-case end-to-end delay guarantees. To see this, suppose session i's traffic is regulated by a leaky bucket with parameters (σ_i, ρ_i), where σ_i is the maximum burstiness and ρ_i is the average source rate. That is, session i's arrivals at the input of the network during the interval $(\tau, t]$ satisfy the inequality [15]

$$A_i(\tau, t) \le \sigma_i + \rho_i(t - \tau).$$

(More details on the leaky bucket are in Chapter 3.) There are K PFQ schedulers along the path; each of the schedulers has the same link rate r and provides a minimum guaranteed bandwidth, $r_i \ge \rho_i$, for this session. Let

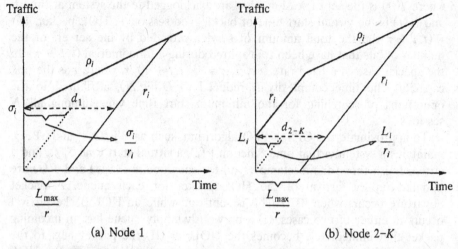

Fig. 4.20 Delay bound in multiple-node case (bold lines represent arrival curves, while light lines represent departure curves per node; link propagation delay excluded).

L_i and L_{max} be the largest packet allowed on session i and among all the sessions in the network, respectively. Then, independent of the behavior of the other sessions (even if they are not leaky-bucket-constrained), the worst-case end-to-end queuing and transmission delay D_i (excluding link propagation delay) are bounded by

$$D_i \leq \frac{\sigma_i}{r_i} + (K - 1)\frac{L_i}{r_i} + K\frac{L_{max}}{r}. \qquad (4.26)$$

Figure 4.20 illustrates how to compute the delay bound in (4.26). As shown in Figure 4.20(a), the maximum delay of the packet in consideration (*tagged packet*) at node 1, d_1, is $\sigma_i/r_i + L_{max}/r$. At node k, $k = 2, 3, \ldots, K$, we have $d_k = L_i/r_i + L_{max}/r$, as shown in Figure 4.20(b). Therefore, we have $D_i = d_1 + (K - 1)d_k$, which gives us the bound in (4.26).

Intuitively, (4.26) means that, although the session actually traverses a series of schedulers, it behaves as if it were served by a single scheduler with rate r_i, so that when the source sends a burst of length σ_i, it experiences a worst-case delay σ_i/r_i, which is the same as that in a GPS server. The second term models the situation at each scheduler where another packet from session i will receive its service before the tagged packet, so the tagged packet is delayed by at most L_i/r_i. The third term reflects the fact that if the tagged packet arrives at a busy scheduler, it may have to wait up to L_{max}/r before it is served. The inequality (4.26) can easily be extended to allow for more general situations, such as heterogeneous link rates [18].

Parekh and Gallager's theorem [15, 16] shows that, with a suitable choice of parameters, a network of WFQ schedulers can provide worst-case end-to-end delay guarantees. A session j requiring a particular worst-case end-to-end delay bound need only choose an appropriate value for r_j. This is the basic idea of the guaranteed service in the Internet integrated service architecture, which uses resource reservation protocol (RSVP) and allows the receiver to decide the level of bandwidth reservation to achieve a particular delay bound [18, 34].

4.15 COMPARISON

WFQ or packet GPS (PGPS) is probably the first PFQ algorithm [15] in which the state of the GPS is precisely tracked. Although, in terms of the number of bits served for a session, the WFQ has proven that it will not fall behind the GPS by one maximum size packet, it can be far ahead of the GPS. In other words, it is not *worst-case fair*, as indicated by a large worst-case fair index (WFI) [26]. Motivated by this, an eligibility test was introduced in WF^{2}Q [26] (and also in SPFQ [19]). In this test, when the next packet is chosen for service, it is selected from those eligible packets whose start times are not greater than the system virtual time. It has been proven that WF^{2}Q can provide almost identical service to that of GPS, differing by no more than one maximum-size packet. However, a serious limitation to WF^{2}Q (and WFQ) is its computational complexity arising from the simulation of the

TABLE 4.1 Latency, Fairness, and Time Complexity of Several Scheduling Algorithms[a]

Scheduler	Reference	Latency	WFI	Complexity
GPS	[15]	0	0	—
WFQ	[15]	$\dfrac{L_i}{r_i} + \dfrac{L_{max}}{r}$	$O(N)$	$O(N)$
SCFQ	[14]	$\dfrac{L_i}{r_i} + \dfrac{L_{max}}{r}(N-1)$		$O(\log N)$
Virtual clock	[6]	$\dfrac{L_i}{r_i} + \dfrac{L_{max}}{r}$	∞	$O(\log N)$
Deficit round-robin	[32]	$\dfrac{3F - 2\phi_i}{r}$	$O(N)$	$O(1)$
WF^{2}Q	[26]	$\dfrac{L_i}{r_i} + \dfrac{L_{max}}{r}$	$O\left(\dfrac{L_i}{r_i}\right)$	$O(N)$
WF^{2}Q +	[27]	$\dfrac{L_i}{r_i} + \dfrac{L_{max}}{r}$	$O\left(\dfrac{L_i}{r_i}\right)$	$O(\log N)$

[a]In deficit round-robin, F is the frame size and ϕ_i is session i's weighting factor in bandwidth allocation.

GPS. A maximum of N events may be triggered in the simulation during the transmission of a packet. Thus, the time for completing a scheduling decision is $O(N)$.

Table 4.1 summarizes the latency, fairness measures, and time complexity of several scheduling algorithms [19]. Note that the $O(\log N)$ complexity of most of the sorted-priority algorithms arises from the complexity of priority queue operations, as explained in detail in the following chapter.

4.16   A CORE-STATELESS SCHEDULING ALGORITHM

A shaped virtual clock [22] is a simple implementation of a combined shaper-scheduler to achieve the minimum WFI as in WF^{2}Q [26]. In this section, we study a simple core-stateless scheduling scheme, called the core-stateless shaped virtual clock (CSSVC) algorithm, to approximate the behavior of a shaped virtual clock network without keeping per-flow state information at core nodes.

In a CSSVC network, when packets arrive at an ingress node, where per-flow state information is maintained, their associated state variables are initialized by the ingress node and carried in the packet headers. Interior nodes in the core network (i.e., core nodes) do not keep per-flow state information, but rather use the state variables carried in the packet header to schedule the incoming packets. In addition, the interior nodes update the state variables in the packet headers before they are sent to the next node.

It is shown that the CSSVC scheme can provide the same end-to-end delay bound and provides the minimum WFI for the network as the shaped virtual clock scheme does. Also, we assume that a WFI packet shaper, introduced in Section 3.3, is implemented in the edge routers of CSSVC networks as shown in Figure 4.21.

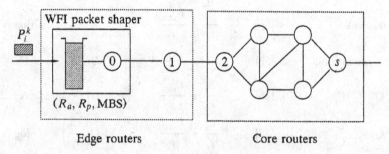

Fig. 4.21   Edge routers and core routers in the CSSVC network.

4.16.1 Shaped Virtual Clock Algorithm

By using the virtual clock [6] algorithm as the underlying scheduler, a shaped virtual clock server uses a real-time clock as the virtual time function and serves the eligible packet that has the smallest virtual finishing time F. The kth packet of session i is eligible to get served at time t if and only if

$$S_{i,s}^k \le V_s(t), \tag{4.27}$$

where $S_{i,s}^k$ is the virtual starting time of the kth packet of session i at node s, and $V_s(t)$ the system virtual time of node s at time t. When a packet arrives at time $a_{i,s}^k$, $S_{i,s}^k$ is defined as

$$S_{i,s}^k = \max\left[V_s(a_{i,s}^k), F_{i,s}^{k-1}\right] = \max\left[a_{i,s}^k, F_{i,s}^{k-1}\right]. \tag{4.28}$$

Meanwhile, the timestamp or virtual finishing time of the kth packet of session i is defined as

$$F_{i,s}^k = S_{i,s}^k + \frac{l_i^k}{r_i}, \tag{4.29}$$

where l_i^k is the packet length, and r_i the allocated rate of session i.

By employing the shaped virtual clock algorithm to schedule packets, we can guarantee the minimum WFI of session i as

$$\text{WFI}_{i,s} = \frac{L_{i,\max}}{r_i} + \frac{(L_{\max} - L_{i,\max})}{r_s} \tag{4.30}$$

where r_s is the service rate of server s, $L_{i,\max}$ the maximum packet length of session i, and $L_{\max}$ the maximum packet length that can be seen by server s.

Theorem 4.2 *In a network of two shaped virtual clock servers, if servers 1 and 2 can guarantee the WFI for session i as $\text{WFI}_{i,1}$ and $\text{WFI}_{i,2}$, respectively, the end-to-end WFI of this network is equal to $\text{WFI}_{i,1} + \text{WFI}_{i,2}$.*

Proof   Note that the unit of WFI can be either the bit or the second, where $\text{WFI(bits)} = r_i \times \text{WFI(s)}$. Here, we use WFI^b to denote WFI(bits) and WFI to denote WFI(s). From [43] and [27], if shaped virtual clock server s guarantees a delay bound of $D_{i,s}$ to a session i constrained by a leaky bucket (σ_i, r_i), it also guarantees a WFI^b of $r_i \times D_{i,s} - \sigma_i$. That is

$$\text{WFI}_{i,s}^b = r_i \times D_{i,s} - \sigma_i, \tag{4.31}$$

where $\text{WFI}_{i,s}^b = r_i \times \text{WFI}_{i,s}$. Therefore, (4.31) becomes

$$\text{WFI}_{i,s} = D_{i,s} - \frac{\sigma_i}{r_i}. \tag{4.32}$$

By applying s = 1 and 2 in the above equation, we get

$$D_{i,1} = \frac{\sigma_i}{r_i} + \text{WFI}_{i,1} \tag{4.33}$$

and

$$D_{i,2} = \frac{\sigma_i}{r_i} + \text{WFI}_{i,2}, \tag{4.34}$$

where $D_{i,1}$ and $D_{i,2}$ are the delay bounds of session i provided by server 1 and 2, respectively. Therefore, the end-to-end delay bound $\overline{D}_i$ at the end of the second server is equal to

$$\overline{D}_i = D_{i,1} + D_{i,2} = 2 \times \frac{\sigma_i}{r_i} + \text{WFI}_{i,1} + \text{WFI}_{i,2}. \tag{4.35}$$

The first term in the above equations, σ_i/r_i, comes from the delay of leaky bucket shaping and should only count once in a network. Therefore, an improved delay bound $\overline{D}_{i,s}$ at the end of the second server is equal to

$$\overline{D}_{i,s} = \overline{D}_i - \frac{\sigma_i}{r_i} = D_{i,1} + D_{i,2} - \frac{\sigma_i}{r_i}. \tag{4.36}$$

Meanwhile, the end-to-end WFI, denoted $\overline{\text{WFI}}_i$, can be expressed as

$$\overline{\text{WFI}}_i = \overline{D}_{i,s} - \frac{\sigma_i}{r_i}, \tag{4.37}$$

By (4.36), we obtain from

$$\overline{\text{WFI}}_i = D_{i,1} + D_{i,2} - \frac{\sigma_i}{r_i} - \frac{\sigma_i}{r_i}. \tag{4.38}$$

From (4.33) and (4.34), we get

$$\overline{WFI}_i = \text{WFI}_{i,1} + \text{WFI}_{i,2}. \tag{4.39}$$

4.16.2 Core-Stateless Shaped Virtual Clock Algorithm

As shown by (4.28) and (4.29), a shaped virtual clock algorithm needs two state variables for each flow i: the reserved rate r_i, and the previous packet's virtual finishing time $F_{i,s}^{k-1}$. Since all nodes along the path use the same r_i-value for flow i, it is easy to eliminate r_i by putting it in the packet header.

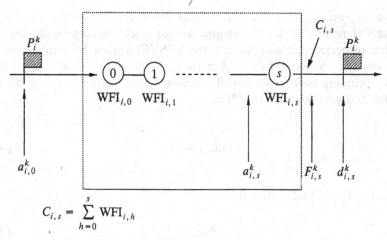

$$C_{i,s} = \sum_{h=0}^{s} \mathrm{WFI}_{i,h}$$

Fig. 4.22   End-to-end WFI bound, $C_{i,s}$, in the CSSVC network.

However, $F_{i,s}^{k-1}$ is a dynamic value that is computed iteratively at each node and cannot be simply eliminated. Therefore, we need to design an algorithm that can calculate virtual finishing time in core nodes without keeping the information of $F_{i,s}^{k-1}$. From (4.28), since $F_{i,s}^{k-1}$ is only used in a max operation, we can eliminate it by adding a state variable to a packet's arrival time, and ensure that the sum is never less than $F_{i,s}^{k-1}$. Here, we denote this state variable as $X_{i,s}{}^{k}$, so that for every core node s along the path, the following holds

$$S_{i,s}^{k} = a_{i,s}^{k} + X_{i,s}^{k} \ge F_{i,s}^{k-1}. \qquad (4.40)$$

Our goal, as shown in Figure 4.22, is to use the CSSVC network to approximate the behavior of a shaped virtual clock network without keeping per-flow state information at core nodes. When the kth packet of session i arrives at an edge node in a CSSVC network at time $a_{i,1}^{k}$, and departs from node s at time $d_{i,s}^{k}$, it should experience the same end-to-end WFI bound, $C_{i,s}$, as it experiences in a shaped virtual clock network. Therefore, from Theorem 4.2, it's easy to get the result

$$C_{i,s} = \sum_{h=0}^{s} \mathrm{WFI}_{i,h}, \qquad (4.41)$$

where $\mathrm{WFI}_{i,h}$ is the WFI for session i guaranteed by server h.
Note that from [15] and [22],

$$d_{i,s}^{k} - d_{i,s}^{k}(\mathrm{fluid}) \le \frac{L_{\max}}{r_s}, \qquad (4.42)$$

where $d_{i,s}^k$ and $d_{i,s}^k$(fluid) is the time at which the kth packet of session i departs at node s under the shaped virtual clock's packet model and fluid model, respectively. Since we use the CSSVC scheme to approximate the behavior of a shaped virtual clock network, (4.42) should hold at any CSSVC node, including node s. Meanwhile, session i gets service rate r_i in the fluid model, and we obtain $d_{i,s}^k$(fluid) as

$$d_{i,s}^k(\text{fluid}) = S_{i,s}^k + \frac{l_i^k}{r_i}. \tag{4.43}$$

From (4.29) and (4.43), we get

$$d_{i,s}^k(\text{fluid}) \leq F_{i,s}^k. \tag{4.44}$$

Combining (4.42) and (4.44), we get

$$d_{i,s}^k - F_{i,s}^k \leq \frac{L_{\max}}{r_s}, \tag{4.45}$$

By subtracting $a_{i,1}^k$ on both sides of (4.45), we obtain

$$d_{i,s}^k - F_{i,s}^k - a_{i,1}^k \leq \frac{L_{\max}}{r_s} - a_{i,1}^k, \tag{4.46}$$

or

$$d_{i,s}^k - a_{i,1}^k \leq F_{i,s}^k + \frac{L_{\max}}{r_s} - a_{i,1}^k \tag{4.47}$$

The right-hand side of the above inequality is a tight delay bound on session i. From this, we obtain

$$\overline{D}_{i,s}^k = F_{i,s}^k + \frac{L_{\max}}{r_s} - a_{i,1}^k. \tag{4.48}$$

Meanwhile, from the result of (4.37), when the kth packet of session i, which is leaky-bucket-smoothed (σ_i, r_i), departs from server s in the CSSVC network, it should get an end-to-end delay bound $\overline{D}_{i,s}$ equal to

$$\overline{D}_{i,s}^k = \frac{\sigma_i}{r_i} + C_{i,s}. \tag{4.49}$$

Considering propagation delay, (4.49) becomes

$$\overline{D}_{i,s}^k = \frac{\sigma_i}{r_i} + C_{i,s} + \sum_{h=1}^{s-1} \pi_h. \tag{4.50}$$

where π_h is the propagation delay between node $h-1$ and h.

The first term σ_i/r_i in the above equation comes from a leaky-bucket-smoothed incoming traffic stream, since we implement a WFI packet shaper in the edge router. This term is adjusted to the delay caused by traffic shaper, i.e., $D_{i,\text{shaper}}^k$. That is

$$\overline{D}_{i,s}^k = D_{i,\text{shaper}}^k + C_{i,s} + \sum_{h=1}^{s-1} \pi_h. \tag{4.51}$$

As shown in previous chapter, for a given set of traffic descriptors (R_a, R_p, and MBS), the delay caused by a traffic shaper in the edge router is equal to $L_{i,\max}/Rp$, where $L_{i,\max}$ is the maximum packet length of session i. From the following derivation, we will find that the traffic shaper only increases the delay of the edge router and is irrelevant to that of core routers in the CSSVC scheme.

Note that the result (4.51) comes from the end-to-end WFI bound of the shaped virtual clock algorithm (i.e., $C_{i,s}$), and is also a tight delay bound on session i. From (4.48) and (4.51), we get

$$\overline{D}_{i,s}^k = F_{i,s}^k + \frac{L_{\max}}{r_s} - a_{i,1}^k = D_{i,\text{shaper}}^k + C_{i,s} + \sum_{h=1}^{s-1} \pi_h. \tag{4.52}$$

By replacing $F_{i,s}^k$ with $S_{i,s}^k + l_i^k/r_i$, we get

$$S_{i,s}^k = a_{i,1}^k - \frac{L_{\max}}{r_s} + D_{i,\text{shaper}}^k + C_{i,s} + \sum_{h=1}^{s-1} \pi_h - \frac{l_i^k}{r_i}. \tag{4.53}$$

From (4.53), we represent $S_{i,s-1}^k$ as

$$S_{i,s-1}^k = a_{i,1}^k - \frac{L_{\max}}{r_{s-1}} + D_{i,\text{shaper}}^k + C_{i,s-1} + \sum_{h=1}^{s-2} \pi_h - \frac{l_i^k}{r_i}. \tag{4.54}$$

Combining (4.53) and (4.54), we obtain the relationship between $S_{i,s}^k$ and $S_{i,s-1}^k$ to guarantee $C_{i,s}$,

$$S_{i,s}^k = S_{i,s-1}^k + \text{WFI}_{i,s} + \frac{L_{\max}}{r_{s-1}} - \frac{L_{\max}}{r_s} + \pi_{s-1}. \tag{4.55}$$

By iterating the above equation, we get

$$S_{i,s}^k = S_{i,1}^k + C_{i,s} - \text{WFI}_{i,1} + \frac{L_{\max}}{r_1} - \frac{L_{\max}}{r_s} + \sum_{h=1}^{s-1} \pi_h. \tag{4.56}$$

Also from (4.40) and (4.29), we have $S_{i,s}^k \geq F_{i,s}^{k-1} = S_{i,s}^{k-1} + l_i^{k-1}/r_i$. By using (4.56), we get the inequality between the kth and $k-1$th packets at edge node 1,

$$S_{i,1}^k \geq S_{i,1}^{k-1} + \frac{l_i^{k-1}}{r_i}. \tag{4.57}$$

The right-hand side in the above inequality is the bound for $S_{i,1}^k$ to guarantee $S_{i,s}^k \geq F_{i,s}^{k-1}$. We can get several pairs of solutions ($S_{i,1}^k$ and $S_{i,1}^{k-1}$) as long as (4.57) is satisfied. As a result, we can set

$$S_{i,1}^k = S_{i,1}^{k-1} + \frac{l_i^{k-1}}{r_i}. \tag{4.58}$$

From (4.40),

$$S_{i,s}^k = a_{i,s}^k + X_{i,s}^k = d_{i,s-1}^k + \pi_{s-1} + X_{i,s}^k, \tag{4.59}$$

where $d_{i,s-1}^k$ is the packet's departure time at node $s-1$. Also, from (4.45), $d_{i,s-1}^k \leq F_{i,s-1}^k + L_{\max}/r_{s-1}$, so (4.59) becomes

$$S_{i,s}^k \leq F_{i,s-1}^k + \frac{L_{\max}}{r_{s-1}} + \pi_{s-1} + X_{i,s}^k, \tag{4.60}$$

or

$$S_{i,s}^k \leq S_{i,s-1}^k + \frac{l_i^k}{r_{s-1}} + \frac{L_{\max}}{r_{s-1}} + \pi_{s-1} + X_{i,s}^k. \tag{4.61}$$

The right-hand side in the above inequality is the bound for $S_{i,s}^k$ to guarantee end-to-end WFI. We can get several pairs of solutions ($S_{i,s}^k$ and $S_{i,s-1}^k$) as long as (4.61) is satisfied. As a result, we can set $S_{i,s}^k$ as

$$S_{i,s}^k = S_{i,s-1}^k + \frac{l_i^k}{r_{s-1}} + \frac{L_{\max}}{r_{s-1}} + \pi_{s-1} + X_{i,s}^k. \tag{4.62}$$

Combining (4.55) and (4.62), we get

$$S_{i,s-1}^k + \frac{l_i^k}{r_{s-1}} + \frac{L_{\max}}{r_{s-1}} + \pi_{s-1} + X_{i,s}^k$$

$$= S_{i,s-1}^k + \text{WFI}_{i,s} + \frac{L_{\max}}{r_{s-1}} - \frac{L_{\max}}{r_s} + \pi_{s-1}. \tag{4.63}$$

Rearranging terms, we get the state variable

$$X_{i,s}^k = \text{WFI}_{i,s} - \frac{l_i^k}{r_{s-1}} - \frac{L_{max}}{r_s}.$$ (4.64)

By combining the WFI value from (4.30) with the above equation, we get

$$X_{i,s}^k = \frac{L_{i,max}}{r_i} + \frac{L_{max} - L_{i,max}}{r_s} - \frac{l_i^k}{r_{s-1}} - \frac{L_{max}}{r_s}$$

$$= \frac{L_{i,max}}{r_i} - \frac{l_i^k}{r_{s-1}} - \frac{L_{i,max}}{r_s}.$$ (4.65)

4.16.3 Encoding Process

4.16.3.1 *State Variable Reduction* When all flows in the network have the same maximum packet length L, we can further reduce the state variables carried in the packet's header. From (4.30), we get

$$\text{WFI}_{i,s} = L_{i,max} \times \left(\frac{1}{r_i} - \frac{1}{r_s}\right) + \frac{L_{max}}{r_s}$$

$$= L \times \left(\frac{1}{r_i} - \frac{1}{r_s}\right) + \frac{L}{r_s} = \frac{L}{r_i}.$$ (4.66)

From (4.64),

$$X_{i,s}^k = \text{WFI}_{i,s} - \frac{l_i^k}{r_{s-1}} - \frac{L_{max}}{r_s} = \frac{L}{r_i} - \frac{l_i^k}{r_{s-1}} - \frac{L}{r_s}.$$ (4.67)

4.16.3.2 Encoding State Variables In order to save the bits to represent the state variable, we encode state variables $X_{i,s}^k$ as follows:

$$X_{i,s}^k = Y_1 - Y_1 Y_2 - \frac{L}{r_s},$$ (4.68)

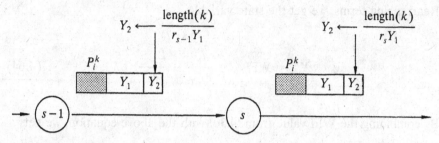

Fig. 4.23 Implementing the CSSVS scheme based on Y_1 and Y_2 in the packet's header by assuming all flows have the same maximum packet length.

where

$$Y_1 = \frac{L}{r_i}, \tag{4.69}$$

$$Y_2 = \frac{l_i^k/r_{s-1}}{Y_1}. \tag{4.70}$$

As shown in Figure 4.23, packet p_i^k arrives at node s at time "now" with packet length l_i^k. Based on the Y_1 and Y_2 in its header, we can calculate its virtual starting time and virtual finishing time as follows: From (4.68)

$$S_s = \text{now} + X_s = \text{now} + Y_1 \times (1 - Y_2) - \frac{L}{r_s}. \tag{4.71}$$

and

$$F_s = S_s + \frac{l_i^k}{r_i} = S_s + \frac{l_i^k \times Y_1}{L}. \tag{4.72}$$

Along the path traversed by flow i, Y_1 is a constant. However, Y_2 is a function of the rate r_{s-1} and is updated at each node. Meanwhile, Y_1 and Y_2 are not functions of node number s. Therefore, implementing the CSSVC scheme does not require knowing the path traversed by a flow.

4.16.3.3 Pseudo-code for the CSSVC Algorithm with Encoded Variables (Y_1, Y_2) The implementation of the CSSVC scheme based on encoded (Y_1, Y_2) values is described in the following pseudo-code:

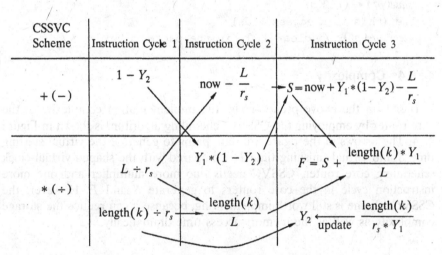

Shaped Virtual Clock Scheme	Instruction Cycle 1	Instruction Cycle 2
$+ \, (-)$	$S(k) = \max[\text{now}, F(k-1)]$	$F(k) = S(k) + \dfrac{\text{length}(k)}{r_i}$
$* \, (\div)$	$\text{length}(k) \div r_i$	

CSSVC Scheme	Instruction Cycle 1	Instruction Cycle 2	Instruction Cycle 3
$+ \, (-)$	$1 - Y_2$	$\text{now} - \dfrac{L}{r_s}$	$S = \text{now} + Y_1 * (1 - Y_2) - \dfrac{L}{r_s}$
$* \, (\div)$	$L \div r_s$	$Y_1 * (1 - Y_2)$ $\text{length}(k) \div r_s \longrightarrow \dfrac{\text{length}(k)}{L}$	$F = S + \dfrac{\text{length}(k) * Y_1}{L}$ $Y_2 \xleftarrow[\text{update}]{} \dfrac{\text{length}(k)}{r_s * Y_1}$

Fig. 4.24 Implementation complexities of the core routers by employing shaped virtual clock and CSSVC scheduling schemes.

```
All flows have the same maximum packet length L,
ingress node, rate = r₁
On packet pᵢᵏ arrival
  i = flowID(pᵢᵏ); /* The function flowID() gets the packet's flowID */
  length(k) = getlength(pᵢᵏ) /* The function getlength() gets the packet's
  length */
  if (HOL(pᵢᵏ,i))
    Sᵢ = now; /* now is the current real time */
  else
    Sᵢ = Sᵢ + lᵢ/rᵢ; /* From (4.58) */
```

$l_i = length(k)$
$F_i = S_i + l_i/r_i;$ /* From (4.29) */
$Y_1 = L/r_i;$ /* From (4.69) */
$Y_2 = (l_i/r_1)/Y_1;$ /* From (4.70) */
On packet p_i^k transmission
$state(p_i^k) \leftarrow (Y_1, Y_2);$ /* insert (Y_1, Y_2) into the packet's header */

core node/egress node, rate $= r_s$
On packet p_i^k arrival
$(Y_1, Y_2) \leftarrow state(p_i^k);$ /* get (Y_1, Y_2) from the packet's header */
$S = now - Y_1 \times (1 - Y_2) - L/r_s;$ /* From (4.71) */
$F = S + length(k) \times Y_1/L;$ /* From (4.72) */
$Y_2 = length(k)/r_s/Y_1;$ /* From (4.70) */
On packet p_i^k transmission
  if (*core node*)
    $state(p_i^k) \leftarrow (Y_1, Y_2);$
  else (this is a egress node)
    clearstate$(p_i^k);$ /* clear (Y_1, Y_2) in the packet's header */

4.16.4 Complexity

Based on the above pseudo-code, the implementation complexity of the core router by employing the CSSVC scheduling algorithm is shown in Figure 4.24. The arrows in the figure are data paths to generate the virtual starting time S and virtual finishing time F. Compared with the shaped virtual clock scheduling core router, CSSVC needs one more multiplier and one more instruction cycle in the core routers to generate S and F. However, the CSSVC scheme is still worth implementing, because it can reduce the storage complexity as well as the memory access time dramatically.

REFERENCES

1. S. Keshav, *An Engineering Approach to Computer Networking*: *ATM Networks, the Internet, and the Telephone Network*, Addison-Wesley, 1997.
2. S. Keshav and R. Sharma, "Issues and trends in router design," *IEEE Commun. Mag.*, pp. 144–151, May 1998.
3. V. P. Kumar, T. V. Lakshman, and D. Stiliadis, "Beyond best effort: router architectures for the differentiated services of tomorrow's Internet," *IEEE Commun. Mag.*, pp. 152–164, May 1998.
4. R. L. Cruz, "A calculus for network delay, part I: network elements in isolation," *IEEE Trans. Inf. Theory*, vol. 37, no. 1, pp. 114–131, Jan. 1991.
5. R. L. Cruz, "A calculus for network delay, part II: network analysis," *IEEE Trans. Inf. Theory*, vol. 37, no. 1, pp. 132–141, Jan. 1991.
6. L. Zhang, Virtual clock: a new traffic control algorithm for packet switching networks," *Proc. ACM SIGCOMM*, pp. 19–29, Sep. 1990.

7. C. Kalmanek, H. Kanakia, and S. Keshav, "Rate controlled servers for very high-speed networks," *Proc. IEEE GLOBECOM*, Dec. 1990.

8. C. L. Liu and J. W. Wayland, "Scheduling algorithms for multi-programming in a hard real-time environment," *J. ACM*, pp. 46–61, Jan. 1973.

9. D. Ferrari and D. Verma, "A scheme for real-time channel establishment in wide-area networks," *IEEE J. Select. Areas Commun.*, pp. 368–379, Apr. 1990.

10. D. Verma, H. Zhang, and D. Ferrari, "Guaranteeing delay jitter bounds in packet switching networks," *Proc. Tricomm*, pp. 35–46, Apr. 1991.

11. S. J. Golestani, "A framing strategy for congestion management," *IEEE J. Select. Areas Commun.*, vol. 9, no. 7, pp. 1064–1077, Sep. 1991.

12. H. Zhang and D. Ferrari, "Rate-controlled static priority queuing," *Proc. IEEE INFOCOM*, pp. 227–236, Apr. 1993.

13. H. Zhang and D. Ferrari, "Rate-controlled service disciplines," *J. High Speed Netw.*, vol. 3, no. 4, pp. 389–412, 1994.

14. S. J. Golestani, "A self-clocked fair queuing scheme for broadband applications," *Proc. IEEE INFOCOM*, pp. 636–646, Apr. 1994.

15. A. K. Parekh and R. G. Gallager, "A generalized processor sharing approach to flow control in integrated services networks: the single node case," *IEEE/ACM Trans. Netw.*, vol. 1, no. 3, pp. 344–357, Jun. 1993.

16. A. K. Parekh and R. G. Gallager, "A generalized processor sharing approach to flow control in integrated services networks: the multiple node case," *IEEE/ACM Trans. Netw.*, vol. 2, no. 2, pp. 137–150, Apr. 1994.

17. A. Demers, S. Keshav, and S. Shenker, "Analysis and simulation of a fair queueing algorithm," *Proc. ACM SIGCOMM*, pp. 1–12, Sep. 1989.

18. L. Georgiadis, R. Guerin, V. Peris, and R. Rajan, "Efficient support of delay and rate guarantees in an internet," *ACM SIGCOMM*, pp. 106–116, Aug. 1996.

19. D. Stiliadis, "Traffic scheduling in packet-switched networks: analysis, design, and implementation", Ph.D. dissertation, Computer Science Department, University of California at Santa Cruz, June 1996.

20. D. Stiliadis and A. Varma, "Latency-rate servers: a general model for analysis of traffic scheduling algorithms," *Proc. IEEE INFOCOM*, pp. 111–119, Mar. 1996.

21. D. Stiliadis and A. Varma, "Design and analysis of frame-based fair queueing: a new traffic scheduling algorithm for packet-switched networks," *Proc. ACM SIGMETRICS*, pp. 104–115, May 1996.

22. D. Stiliadis and A. Varma, "A general methodology for design efficient traffic scheduling and shaping algorithms," *Proc. IEEE INFOCOM*, Kobe, Japan, Apr. 1997.

23. D. Stiliadis and A. Varma, "Rate-proportional server: a design methodology for fair queueing algorithms," *IEEE/ACM Trans. Netw.*, vol. 6, no. 2, pp. 164–174, Apr. 1998.

24. D. Stiliadis and A. Varma, "Efficient fair queueing algorithms for packet-switched networks," *IEEE/ACM Trans. Netw.*, vol. 6, no. 2, pp. 175–185, Apr. 1998.

25. H. Zhang and S. Keshav, "Comparison of rate based service disciplines," *Proc. ACM SIGCOMM*, pp. 113–122, 1991.

26. J. C. R. Bennett and H. Zhang, "WF^{2}Q: worst-case fair weighted fair queueing," *Proc. IEEE INFOCOM*, Mar. 1996.

27. J. C. R. Bennett and H. Zhang, "Hierarchical packet fair queueing algorithms," *IEEE/ACM Trans. Netw.*, vol. 5, no. 5, pp. 675–689, Oct. 1997; *Proc. SIGCOMM*, Aug. 1996.

28. P. Goyal, H. M. Vin, and H. Cheng, "Start-time fair queueing: a scheduling algorithm for integrated services packet switching networks," *IEEE/ACM Trans. Netw.*, vol. 5, no. 5, pp. 690–704, Oct. 1997.

29. J. C. R. Bennett, D. C. Stephens, and Hui Zhang, "High speed, scalable, and accurate implementation of fair queueing algorithms in ATM networks," *Proc. ICNP*, 1997.

30. D. C. Stephens and H. Zhang, "Implementing distributed packet fair queueing in a scalable switch architecture," *Proc. INFOCOM*, 1998.

31. S. Y. Liew, "Real time scheduling in large scale ATM cross-path switch," Ph.D. dissertation, Department of Information Engineering, The Chinese University of Hong Kong, Jul. 1999.

32. M. Shreedhar and G. Varghese, "Efficient fair queueing using deficit round-robin," *IEEE/ACM Trans. Netw.*, vol. 4, no. 3, pp. 375–385, Jun. 1996.

33. S. Suri, G. Varghese, and G. Chandranmenon, "Leap forward virtual clock: a new fair queueing scheme with guaranteed delays and throughput fairness," *Proc. IEEE INFOCOM*, Apr. 1997.

34. S. Shenker, C. Partridge, and R. Guerin, "Specification of Guaranteed Quality of Service," RFC 2212, Internet Engineering Task Force (IETF), Sep. 1997.

35. D. D. Clark, S. Shenker, and L. Zhang, "Supporting real-time applications in an integrated services packet network: architecture and mechanism," *Proc. ACM SIGCOMM*, pp. 14–26, Aug. 1992.

36. R. Braden, D. Clark, and S. Shenker, "Integrated services in the Internet architecture: an overview," Informational, RFC 1633, Internet Engineering Task Force (IETF), Jun. 1994.

37. R. Braden, l. Zhang, S. Berson, S. Herzog, and S. Jamin, "Resource ReSerVation Protocol (RSVP)—version 1 functional specification," RFC 2205, Internet Engineering Task Force (IETF), Sep. 1997.

38. J. Wroclawski, "Specification of the controlled-load network element service," RFC 2211, Internet Engineering Task Force (IETF), Sep. 1997.

39. S. Floyd and V. Jacobson, "Link-sharing and resource management models for packet networks," *IEEE/ACM Trans. Netw.*, vol. 3, no. 4, pp. 365–386, Aug. 1995.

40. I. Stoica and H. Zhang, "Providing guaranteed services without per flow management," *Proc. ACM SIGCOMM'99*, Sep. 1999.

41. H. J. Chao, Y. R. Jenq, X. Gao, and C. H. Lam, "Design of packet fair queueing schedulers using a RAM-based searching engine," *IEEE J. Select. Areas Commun.*, pp. 1105–1126, Jun. 1999.

42. J. C. R. Bennett and H. Zhang, "Why WFQ is not good enough for integrated services networks?" *Proc. NOSSDAV'96*, Springer-Verlag, Heidelberg, Germany, Apr. 1996.

43. R. Cruz, "Service burstiness and dynamic burstiness measures: a framework," *J. High Speed Network*, vol. 1, no. 2, pp. 105–127, 1992.

CHAPTER 5

PACKET FAIR QUEUING IMPLEMENTATIONS

In Chapter 4, several packet fair queuing (PFQ) algorithms were described. Some of the schemes promise valuable results, such as guaranteeing the reserved bandwidth independent of other connections' behavior, meeting the connection's delay bound, and having smooth output packet streams. The challenge is to implement these algorithms, especially when the number of connections is large or the link rate is high, e.g., to 10 or 40 Gbit/s. Most of the algorithms require packets with timestamps and choose a packet with the smallest timestamp to depart. The problem is that when there are many connections, choosing the smallest value may take too much time to meet the timing requirement. This chapter describes several architectures to implement the algorithms for a large number of connections and high-speed links.

Section 5.1 gives a conceptual framework of the PFQ implementation and points out the design issues. Section 5.2 describes a very large scale integration (VLSI) chip called Sequencer that can be used to sort the timestamp values at very high speed. Section 5.3 describes another VLSI chip, the *priority content addressable memory* (PCAM) chip, which can also be used to find the smallest timestamp value. Section 5.4 describes a searching engine based on random access memory (RAM). Since it is based on the memory structure, it has better scalability and cost-effectiveness. Section 5.5 presents an implementation architecture for a general shaper−scheduler. The last section depicts mechanism to handle the timestamp aging problem due to the limited number of bits representing timestamp values.

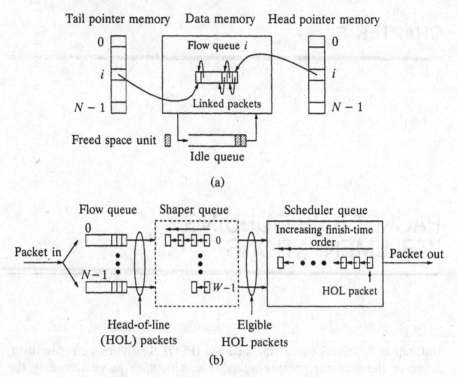

Tail pointer memory   Data memory   Head pointer memory

(a)

(b)

N : the total number of flows or sessions in the system

Fig. 5.1   A PFQ scheduler: (a) logical queue per session; (b) conceptual framework (shaper queue is only needed for implementing shaper–schedulers, where $W = \max\{\text{start time}\}$).

5.1   CONCEPTUAL FRAMEWORK AND DESIGN ISSUES

Generally, a PFQ scheduler maintains a logical queue for each flow or session in the data memory, as shown in Figure 5.1(a). Each queue can be implemented in a linked list with head and tail pointers pointing to its HOL and tail-of-line (TOL) packets. An idle queue may also be needed to maintain the idle space in the data memory. There is a head pointer memory, which stores each queue's head pointer, and a tail pointer memory, which stores each queue's tail pointer, as shown in Figure 5.1(a).

When a packet arrives at the system, it is first stored in the corresponding queue, as shown in Figure 5.1(b). The scheduler queue prioritizes all HOL packets (or all eligible HOL packets if a shaper–scheduler is implemented), based on their finish times, as shown in Figure 5.1(b), and chooses the packet with the smallest finish time to transmit first. Referred to as *design issue I*, an efficient hardware-based priority queuing architecture [15, 16–30], where packet transmissions are arranged based on their timestamp values, is required for high-speed networks. A binary tree of comparators [18, 19, 20] is the most straightforward way to implement the priority queue with $\log_2 N$ levels, where N is the number of sessions in the system. But its time complexity is high for large N, and it is expensive to implement. An application-specific integrated circuit (ASIC) called a *Sequencer* chip [27] was used to facilitate the priority queue with a time complexity of $O(1)$, independent of the number of sessions (N) in the system. However, each Sequencer chip can only handle up to 256 sessions. For a practical application where there are hundreds of thousands of sessions, the number of required Sequencer chips would be too large to be cost-effective. Section 5.2 describes the Sequencer chip.

A searching-based, rather than sorting-based, approach has been presented in [21], where a number of timing queues are maintained for distinct timestamp values, resulting in a calendar queue. The HOL packets from different sessions that have the same timestamp value are linked together to form a timing queue. The priority queue selects a packet with the smallest timestamp to send out. This can create a system bottleneck when the number of distinct timestamp values is large. A new ASIC, called a *priority content addressable memory* (PCAM) chip [29], can search for the minimum timestamp value at a very high speed. Section 5.3 describes the PCAM chip. However, due to a sizable on-chip memory requirement, the PCAM is still too expensive to implement. This motivates us to use off-chip memory and implement hierarchical searching to further reduce the hardware cost.

Jeng [30] generalizes the two-level searching in the PCAM chip to hierarchical searching with a tree data structure [17] (p. 189 in [6]) and proposes a novel RAM-based searching engine (RSE) for efficient implementation, as compared to the brute-force approach using a tree of priority encoders/decoders (Section V in [11]). The time to seek the smallest timestamp value in the RSE is independent of the number of sessions in the system and is only bounded by the memory accesses needed. It can be implemented with commercial memory and field-programmable gate array (FPGA) chips. Section 5.4 describes the RSE in detail.

Recall that a class of shaper–schedulers [9, 11, 12, 13, 39] has been proposed to achieve minimum WFI and to have better worst-case fairness properties. However, these shaper–schedulers are difficult to implement, mainly because of the eligibility test. In particular, whenever the server selects the next packet for service, it needs first to move all the eligible packets from the priority queue based on eligibility times (called the shaper queue) to the priority queue based on timestamps (called the scheduler queue). In the worst case, a maximum of N packets must be moved from the shaper queue to the scheduler queue before selecting the next session for service. In [9], by taking advantage of the fixed-length ATM cells, a simple shaper–scheduler implementation architecture was proposed for ATM switches. Since the system virtual time function can be increased by one after sending a cell, only two eligible cells must be moved in the worst case. However, this is generally not true in packet networks, due to the variable packet size.

As shown in Figure 5.1(b),if an incoming packet is an HOL packet, it (its session index actually) is placed in the scheduler queue—or the shaper queue first, in the shaper–scheduler case. In the latter case, if this packet is eligible, it is moved to the scheduler queue. In general, all the HOL packets are first stored in the shaper queue. Only those that are currently eligible can be moved to the scheduler queue. Some efficient mechanism is needed to compare the system virtual time with the start times of the packets in the shaper queue (i.e., performing the eligibility test) and then move eligible packets to the scheduler queue. In the worst case, there may be a maximum of N packets that become eligible. To answer this challenge (referred to as *design issue II*), Section 5.5 proposes a general shaper–scheduler for both ATM switches and IP routers. A slotted mechanism is introduced to update the system virtual time. With the extension of the RSE, [30] proposes a two-dimensional RSE (2-D RSE) architecture to facilitate the operations, and shows that only *two* eligible packets at most need to be transferred to the scheduler queue in each time slot. The inaccuracy caused by the slotted scheme can be reduced by choosing a smaller update interval. The problems of timestamp overflow and aging due to finite bits are also addressed. Throughout this section, we use packet schedulers as a general term to refer to both stand-alone schedulers and shaper–schedulers.

Refer to Figure 5.1(b). Suppose the scheduler queue selects the HOL packet of queue i. It uses i to fetch the head pointer associated with queue i and then reads out the packet using the head pointer, as illustrated in Figure 5.1(a). The pointers, if necessary, are updated after being used. There are more design issues, such as handling timestamp overflow and timestamp aging problems. The former is discussed in Sections 5.4.2 and 5.5.2.3, the latter in Section 5.6. The next section describes the Sequencer chip.

5.2 SEQUENCER

This section describes two possible architectures used to implement queue management for ATM switches or statistical multiplexers. The first architecture is an intermediate one that suffers some implementation constraints. It is described in an order that leads up to the procedure for generating the second, and final, architecture. The example used to illustrate the architecture in this section has P priority levels, N inputs, and 1 output, but it can be generalized to more outputs. The hardware complexity of the two architectures, in terms of memory requirements and implementation constraints, is compared and discussed.

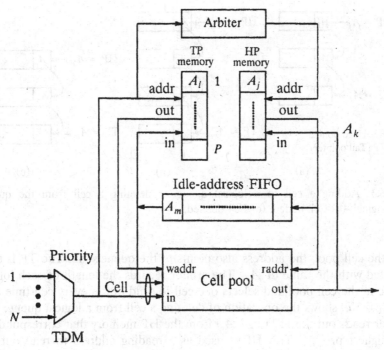

Fig. 5.2 Architecture 1: storing cells in a logical queue. Each logical queue associated with each priority level is confined by two pointers: head and tail pointers. (©1991 IEEE.)

5.2.1 Store Cells in Logical Queues

Instead of using P FIFOs to store cells' addresses according to their priority levels [27], the architecture shown in Figure 5.2 uses two small memories having P entries for each. Each entry has both head and tail pointers (HP and TP) indicating the starting and ending addresses of a logical queue. The logical queue is stored in the cell pool and is associated with its priority level. Here, instead of storing each cell's address in the FIFOs, two addresses are stored for every logical queue, which results in a big memory saving. Every cell in the pool is attached with a pointer to point to the next cell that is linked in the same queue, as shown in Figure 5.3. A similar approach for implementing a priority queue that can handle only a small number of priority levels is presented in [21].

When a cell arrives, it is added to a logical queue based on its priority level. As shown in Figure 5.3(b), the cell's priority field is extracted first and used as an address to read out the TP from the TP memory (e.g., A_l). The TP is then used as a writing address for the cell to be written into the pool. In the meantime, an idle address (e.g., A_m) attached to the cell is written

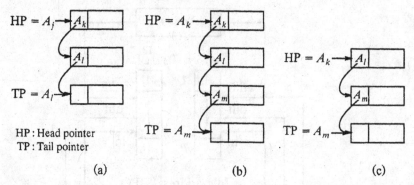

Fig. 5.3 Adding a cell to a logical queue or deleting a cell from the queue. (a) Original. (b) Cell added. (c) Cell deleted.

into the cell pool; the address also points to the queue's tail. The TP is then updated with the value of A_m. The arbiter records the length of each logical queue in the cell pool and selects one cell to send out in every cell time slot. Figure 5.3(c) shows the operation of deleting a cell from a logical queue. The arbiter reads out the HP (e.g., A_j) from the HP memory that corresponds to the highest priority. This HP is used as a reading address to read out the corresponding cell from the pool. Once the cell is read out, its pointer (e.g., A_k) is written into the HP memory to update the old HP.

This architecture obviously saves considerable memory. But it also adds complexity to the arbiter, because it has to record the occupancy status of all logical queues with counters and, in the worst case, has to examine all counter (up to 16,384 in this case) to choose a single cell in one cell time (2.83 μs). This is very difficult to achieve with state-of-the-art hardware. Since all of these functions are performed centrally by the arbiter, its processing speed limits the number of priority levels. Furthermore, if any pointer in the TP or HP memory or in the cell pool is somehow corrupted, the linkage between cells in the logical queues will be wrong, and cells will be accessed mistakenly. Although this can be checked by adding an extra parity bit to the pointers, it is still not easy to recover from faults once errors occur in the pointer, unless the entire cell pool memory is reset.

5.2.2 Sort Priorities Using a Sequencer

The architecture described above limits the number of priorities because of its centralized processing characteristics. A novel architecture proposed in [26], as shown in Figure 5.4, requires less memory and is not limited by the number of priority levels, because it uses the concepts of fully distributed and highly parallel processing to manage cells' sending and discarding sequences. The write/read controllers generate proper control signals for all other functional blocks. The data and address buses are indicated with bold lines.

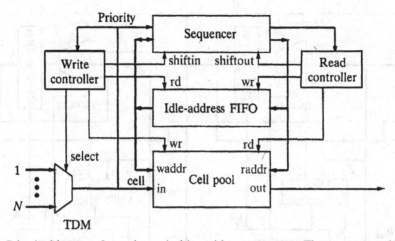

Fig. 5.4 Architecture 2: sorting priorities with a sequencer. The sequencer determines the departure sequence for the cells that are stored in the cell pool. (©1992 IEEE.)

As shown in Figure 5.4, the cells are time-division multiplexed first and then written into the cell pool with idle addresses stored in a FIFO. A pair composed of a cell's priority field and its corresponding address, denoted as PA, is stored in the sequencer in such a way that higher-priority pairs are always at the right of lower-priority ones, so they will be accessed sooner by the read controller. Once the pair has been accessed, the address is used to read out the corresponding cell in the cell pool.

The concept of implementing the Sequencer is very simple, as illustrated in Figure 5.5. Assume that the value of P_n is less than that of P_{n+1} and has a

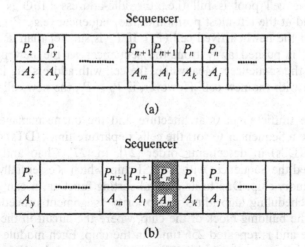

Fig. 5.5 Operations of the sequencer. As a pair of new priority and address is inserted, all pairs with lower priority are pushed to the left.

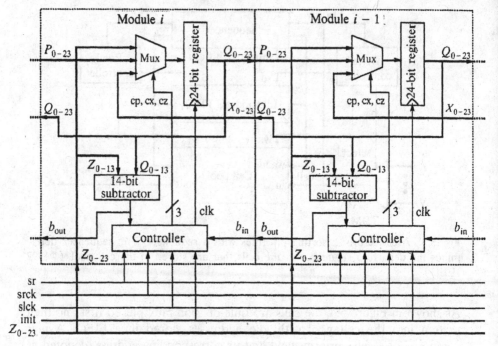

Fig. 5.6 Block diagram of the Sequencer chip. (©1991 IEEE.)

higher priority. When a new cell with priority P_n arrives, all pairs on the right of A_k, as well as A_k itself, remain at their positions while others are shifted to the left. The vacant position is replaced with the pair composed of the new cell's priority field (P_n) and address (A_n).

When the cell pool is full (i.e., the idle-address FIFO is empty), the priority field at the left-most position of the sequencer (e.g., P_z) is compared with that of the newly arrived cell (P_n). If P_n is smaller than P_z, the pair of P_z and A_z is pushed away from the sequencer as the new pair $P_n A_n$ is inserted in the sequencer. Meanwhile, the cell with address A_z in the pool is overwritten with the new cell. However, if $P_n \geq P_z$, the new cell is discarded instead.

Both the traffic shaper's architecture and the queue manager's architecture require a Sequencer to sort the cells' departure times (DTs) or departure sequences (DSs) in descending order [27]. In [27], Chao and Uzun have implemented the Sequencer with a VLSI chip, which is essentially a 256-word sorting-memory chip. Due to its general sorting function, it can also be used for other scheduling algorithms and priority assignment procedures. Figure 5.6 shows the building block of the chip, where the circuit in the dashed box is a module and is repeated 256 times in the chip. Each module has a 24-bit register, which stores the 14-bit DT or DS values and the 10-bit address. A single chip can accommodate a cell pool capacity of up to 256 cells, and DT

TABLE 5.1   Three Possible Actions Performed by the Controller

Case	b_{out}	b_{in}	Action Performed by the Controller
(a) $X_{0-13} \le Z_{0-13} < Q_{0-13}$	1	0	Module i shifts its contents to the left, and $Q_{0-23} = Z_{0-23}$
(b) $Z_{0-13} < X_{0-13} \le Q_{0-13}$	1	1	Both modules i and $i-1$ shift their contents left, and $Q_{0-23} = X_{0-23}$
(c) $X_{0-13} \le Q_{0-13} \le Z_{0-13}$	0	0	Retain the Q_{0-23}

or DS values (or numbers of priority levels in some applications) up to $2^{14} - 1$. This provides DT or DS values ranging from 0 to 4095, with the ability to handle services with bit rates equal to or higher than (48 bytes × 8 bits/byte)/(2.83 μs/cell × 4096 cells), or 33 Kbit/s, if each cell's 48-byte payload is filled with service information. Any services with bit rates lower than 33 Kbit/s will require a larger DT or DS range. By cascading multiple Sequencer chips in series or in parallel, a larger cell pool (e.g., a few thousand cells) or a larger DT or DS value can be supported.

Since every module is identical, let us examine the operations of an arbitrary module, say module i. When a new pair of the DT or DS and the address field, denoted by Z_{0-23}, is to be inserted into the Sequencer, it is first broadcast to every module. On comparing the DT or DS values (Q_{0-13}) of module $i-1$ and module i, with the new broadcast value (Z_{0-13}), the controller generates signals, cp, cx, cz, and clk, to shift the broadcast value (Z_{0-23}) into the 24-bit register in module i, shift module $i-1$'s Q_{0-23} to the register, or retain the register's original value. Table 5.1 lists these three possible actions performed by the controller, where X_{0-13} is module ($i-1$)'s Q_{0-13}. Here b_{out} is the borrow-out of $Z_{0-13} - Q_{0-13}$), and b_{in} is the borrow-out of $Z_{0-13} - X_{0-13}$). Since the smaller DT or DS is always on the right of the larger one, the case where $Q_{0-13} \le Z_{0-13} < X_{0-13}$, or $b_{out}b_{in} = 01$, will not happen.

When a cell with the smallest DT or DS value is to be transmitted, its corresponding address will be shifted out from the Sequencer chip, and the

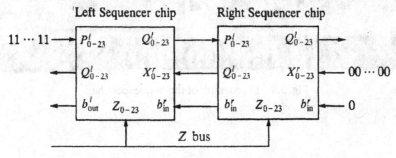

Fig. 5.7   Interconnection signals of two cascaded Sequencer chips. (©1991 IEEE.)

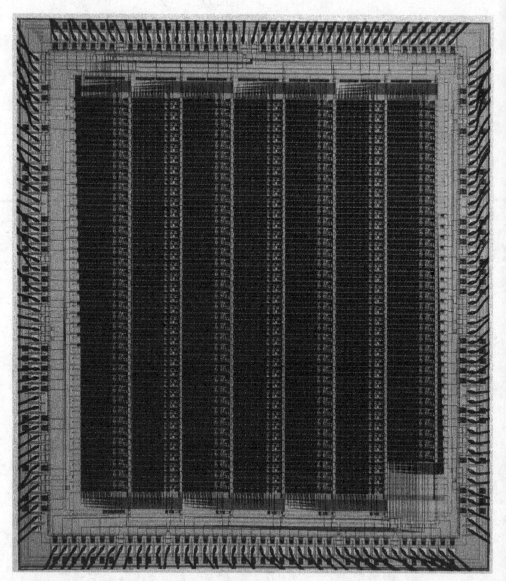

Fig. 5.8 Photograph of the Sequencer chip.

data of all registers will be shifted one position to the right. For instance, Q_{0-23} in module i will be shifted to the register in module $i - 1$. Figure 5.7 shows the connection of signals between two cascaded Sequencer chips. Note that P_{0-3}^l of the left sequencer chip are all connected to 1's; X_{0-23}^r and b_{in}^l of the right Sequencer chip are all connected to 0's. The superscripts l and r indicate, respectively, the leftmost and the rightmost modules of the Sequencer chip. At the initialization, all the registers inside the chip are loaded with the largest DT or DS values, i.e., all 1's, so that new arrival cells with DT or DS values between 0 and $2^{14} - 1$ can be inserted into the Sequencer. The initialization is done by asserting the init and srck signals and setting Z_{0-23} to 11...11. More details are in [26, 27]. Figure 5.8 shows a photograph of the sequencer chip.

5.3 PRIORITY CONTENT-ADDRESSABLE MEMORY

Priority content-addressable memory (PCAM) [29] mainly differs from the Sequencer in that it performs the searching function, rather than the sorting function as the Sequencer does. The PCAM chip arranges the timestamp values in fixed locations and starts searching for the first one that has a valid bit. As long as the searching speed is not a bottleneck, this approach is not limited by the VC number or by the buffer size.

5.3.1 Searching by the PCAM Chip

The PCAM chip consists of many entries that are addressed by the F (virtual finish time) values. Each entry contains a *zone bit* (Z) and a *validity bit* (V), denoted as a (Z, V) pair, as shown in Figure 5.9. The Z-bit is used to resolve the overflow problem of the F-values (to be explained later). The V-bit indicates whether there is any cell assigned to the F.

The (Z, V) pairs in the PCAM chip are arranged in such a way that pairs with smaller F-values are on top of those with the larger F-values. For those cells with identical F-values, a logical queue is formed to link all of them and is called the *timing queue*. Since cells are arranged by their F-values, the PCAM chip facilitates the required search function by identifying the first (Z, V) pair that has the V-bit set to 1. Cells that are associated with the identified (Z, V) pair will be transmitted to the network. For instance, in Figure 5.9, when the (Z, V) at location 2 is found, cells a, b, and c of the timing queue at $F = 2$ are transmitted in sequence. Once they are all transmitted, the pair's V-bit is reset to zero, indicating that no more backlogged cells are assigned with $F = 2$. Thus, during the next round of searching, cell d at location t will be chosen and transmitted.

On some occasions, the calculated F may overflow, i.e., exceed the maximum number the hardware can handle, say $2^{14} - 1$ for 14-bit F-values.

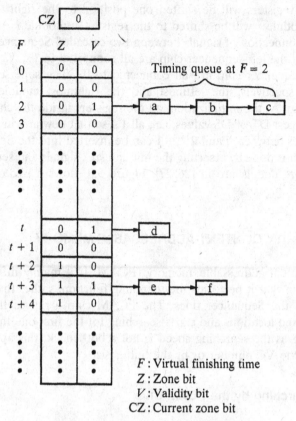

Fig. 5.9 (Z, V) pairs in the PCAM chip.

This is because as time passes, F increases monotonically and eventually exceeds its maximum. To overcome the overflow problem, we have previously proposed using two sorting devices to store nonoverflow and overflow time-stamps separately [26]. Here, to save hardware, we store them in the same device by using the Z-bit to indicate whether the calculated timestamp is overflow or not. The definition of overflow here is different from the traditional one. We use a CZ (*current zone*) bit to indicate the zone of the cells that are currently being served. Whenever the 15th bit of the calculated F (i.e., the overflow bit) has the same value as the CZ, the F is classified as nonoverflow. However, if its 15th bit has a different value from the CZ bit, the F is classified as overflow. Thus, when seeking the V-bits in the PCAM chip, the CZ enables the PCAM chip to choose the first V-bit from an appropriate zone. When all the V-bits in the current zone are zero and there is at least one nonzero V-bit in the other zone, the CZ will be toggled after sending a cell from the other zone, indicating that the service zone is flipped. For example, when all cells a, b, c, and d are transmitted and the V-bit at the other zone ($Z = 1$) is found at $t + 3$, the CZ bit is toggled from zero to one. From then on, cells in zone 1 will be scheduled before those in zone 0. In conclusion, the searching zone in the PCAM chip alternates between CZ = 0 and 1. As long as the 15th bit of the calculated F does not change more than once when serving in the current zone, no cell out-of-sequence problem will occur.

5.3.2 Block Diagram

Figure 5.10 shows the PCAM chip's block diagram. It consists of a CAM array, a few selectors, one input/output (I/O) interface, two inhibit circuits, two 7-to-128 decoders, two 128-to-7 encoders, and the chip controller.

The CAM array has 32K bits arranged into 16K modules, each module with 2 bits. The input data (2 bits) are written to the array through the input bus IN[0:1] with an address provided on the input address bus, X[0:13].

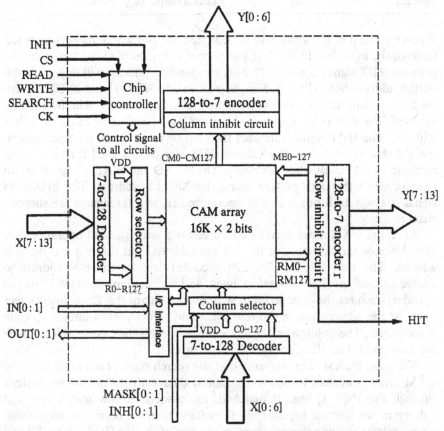

Fig. 5.10 Block diagram of the PCAM chip.

R0-R127 : Row select bit
C0-127 : Column select bit
RM0-RM127 : Row match line
CM0-CM127 : Column match line
ME0-ME127 : Match enable line

TABLE 5.2   I/O Pin Signals of the PCAM Chip

Pin	I/O	Description
X[0 : 13]	In	Address input (e.g., *F*-value)
IN[0 : 1]	In	Data input or searching pattern input
MASK[0 : 1]	In	Mask bits for searching pattern
INH[0 : 1]	In	Inhibit bits for writing data
INIT	In	Initialization signal
WRITE	In	Write enable
SEARCH	In	Search enable
READ	In	Read enable
CS	In	Chip select
CK	In	Clock signal
HIT	Out	Indicate pattern found
OUT[0 : 1]	Out	Data output for read operation
Y[0 : 13]	Out	Address output (e.g., *F*-value)

When a pattern in the array is to be searched for, the searching pattern is fed through the input bus IN[0 : 1]. If a pattern that is closest to the top of the list is found (HIT signal is asserted), its corresponding address will appear at the output address bus, Y[0 : 13]. Signals such as MASK[0 : 1] and INH[0 : 1] are used when part of the data pattern in the array is to be searched for or updated. The data pattern can also be directly accessed using the address bus X[0 : 13]. The INIT signal initializes the chip by resetting all data patterns in the CAM array to zero. The CS and HIT signals are used for connecting multiple PCAM chips when necessary. The READ and WRITE signals are to read or write the data pattern using the X[0 : 13] address. The SEARCH signal is asserted during a search operation. These I/O signals are summarized in Table 5.2.

When the CAM array is accessed for either a write or read operation, the first 7 bits of the address are the column address, and the rest are the row address. The row decoder (7-to-128 decoder) decodes the row address to enable one of the 128 row select lines, and the column decoder (7-to-128 decoder) decodes the column address to enable one of the 128 column select lines. At the intersection of the row select and column select lines, a module is identified. The column select line enables the bit–line pair, which routes the addressed data through the I/O data bus.

When the PCAM chip is operated at the search mode, each module in the CAM array compares its internally stored data with the broadcast pattern through the IN[0 : 1] bus. If matched, the associated row match line and column match line will be asserted. If multiple matches occur, the row inhibit circuit selects the one that is closest to the top of the list (RM0–RM127) and forwards the match enable lines (ME0–ME127), among them only one being enabled, to the CAM array so that only those modules on the enabled line will be able to participate in the column search. Again, all 128 column match lines (CM0–CM127) are sent to the column inhibit circuit, which then chooses only the one that is leftmost in the list (CM0–CM127). Two 128-to-7 encoders simultaneously encode the results from both row and column inhibit circuits to the address outputs Y[7 : 13] and Y[0 : 6]. The PCAM chip uses parallel hardware in the inhibit circuit to identify the topmost or the leftmost bit, resulting in only a few gates' delay in the inhibit circuit.

Note that the PCAM chip is different from the traditional CAM chip. First, the PCAM chip is deep and narrow (16K × 2), while the traditional CAM chip is shallow and wide (e.g., 1K × 64). Second, the PCAM chip gives out the address of the pattern that is found, while the traditional CAM gives out the data field associated with the pattern. Third, the PCAM chip can write the pattern to a specific location associated with the timestamp, while the traditional CAM chip cannot. Fourth,. the PCAM reduces wiring complexity by seeking and encoding the first matched pattern through vertical and horizontal dimensions, thus only processing $2\sqrt{M}$ match lines instead of M match lines (where M is the size of the PCAM chip).

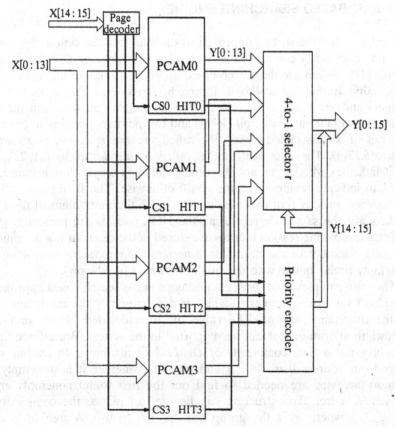

Fig. 5.11 Parallel connection of four PCAM chips.

5.3.3 Connecting Multiple PCAM Chips

Multiple PCAM chips can be connected to accommodate a larger F-value. The connection is achieved by using the CS and HIT signals. Figure 5.11 shows a system with four PCAM chips connected in parallel, which can accommodate a maximum F-value up to 64K (2^{16}, or 4×2^{14}), with each chip handling 16K(2^{14}) F-entries. The most-significant bits, X[14 : 15], are decoded to determine the page to which the timing queue belongs. X[0 : 13] is connected to all chips. During the read-out and write-in operations, only the PCAM chip that has the CS signal asserted through the page decoder will be enabled. However, during the search operation, four PCAM chips are enabled to search for a pattern simultaneously. The output HIT signal from each chip indicates if there is any pattern matched in the chip. These four HIT signals are encoded by a priority encoder. The output of the priority encoder, Y[14 : 15], selects a proper F-value, Y[0 : 13], through the 4-to-1 selector. They are then combined to form a final result Y[0 : 15].

5.4 RAM-BASED SEARCHING ENGINE

Instead of using the sorting approach to find the smallest timestamp, where the time complexity can be $O(\log N)$ for binary sorting or $O(1)$ for parallel sorting [27], we can use the search-based approach to reduce implementation complexity. In the search-based approach, timestamps are quantized into integers and are used as the address for the priority queue. Each memory entity may contain a validity bit (V-bit) and two pointers pointing to the head and tail of an associated linked list called the *timing queue*, as shown in Figure 5.12(a). The data structure is called a *calendar queue* [21, 22]. The V-bit indicates whether or not the timing queue is empty. For instance, we use 1 to indicate nonempty status and 0 otherwise. The timing queue links the indexes, such as i, of each session, for which the timestamps of the HOL packets are the same. Therefore, all the HOL packets are presorted when their corresponding session indexes are stored in the calendar queue. Finding the next packet with the minimum timestamp is equivalent to finding the nonempty timing queue with the smallest address (see below).

The time complexity of sorting timestamps in the search-based approach is a tradeoff for space complexity, which is determined by the maximum value of the timestamp, say M. The value of M is decided by the minimum bandwidth allocation that can be supported in the system. Brute-force linear searching has a time complexity of $O(M)$. It is attributed to reading each entry from address 0 to $M - 1$ and checking whether it is nonempty. M memory accesses are needed to find out the first (only) nonempty entry. Obviously, a tree data structure can be used to reduce the complexity to $O(\log_g M)$, where g is the group size (p. 189 in [6]). A tree of priority encoders and decoders might be used to implement this data structure

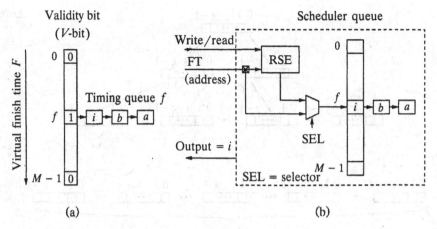

M : maximum value of timestamp F
i, b, a : session indices
RSE : RAM-based searching engine
FT : virtual finish time of an incoming head-of-line (HOL) packet

Fig. 5.12 Calendar queue: (a) data structure, (b) implementation in a scheduler queue using the RSE. (©1999 IEEE.)

(Section V in [11]). However, the cost would be prohibitively large. The PCAM chip [29] can search for the minimum timestamp value at very high speed. However, due to a sizable on-chip memory requirement, the PCAM is still too expensive to implement. Motivated by the need for efficient hardware implementation, [30] proposes a RAM-based search engine (RSE), as explained below.

The RSE reorganizes and stores all the V-bits in the calendar queue. As shown in Figure 5.12, its main function is to find a nonempty timing queue that has the smallest finish time (address) and output its address. Suppose f is the output of RSE (read operations). It is used to fetch the queue index, say i, as shown in Figure 5.12(b), which, in turn, locates the pointer pointing to queue i's HOL packet, as shown in Figure 5.1(a). Suppose FT is the virtual finish time of a new HOL packet. This packet's session index will be stored at the tail of the timing queue addressed by FT. In the meantime, the corresponding V-bit may be set to 1 (write operations) if the timing queue was empty previously. The selector is used to choose the appropriate memory address during write and read operations. A similar calendar queue based on start times of the HOL packets can be used to find the minimum value of start times (see Section 5.5.2.1), as required for updating $V(t)$ in (4.23). The RSE is detailed below.

5.4.1 Hierarchical Searching

In a calendar queue, a *validity bit* (V) is associated with each timing queue, indicating whether this queue is empty $(V = 0)$ or not $(V = 1)$. Since packets

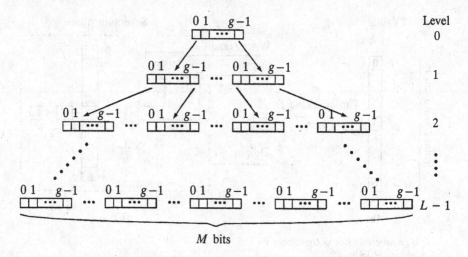

Fig. 5.13 Logical structure of hierarchical searching.

are automatically sorted based on their corresponding locations in the calendar queue, finding the next packet to be transmitted is equivalent to searching the first $V = 1$ bit in the calendar queue.

The key concept of hierarchical searching [30] is extended and generalized from the one for the PCAM chip by dividing the M validity bits in the basic searching into multiple groups, which form a tree data structure as shown in Figure 5.13, where M is the maximum value of timestamp F. Each group consists of a number of V-bits, so another bit string can be constructed at the upper level with its length equal to the number of groups at the bottom level. Each bit at the upper level represents a group at the bottom level with its value equal to the logical OR of all the g bits in the group. Further grouping can be performed recursively until the new string can be placed in a register.

Suppose L levels are formed from the original M-bit string. There are M_l bits at level l; each of its groups has g_l bits, where $l = 0, 1, \ldots, L - 1$. Another M_l/g_l-bit string can be constructed at upper level $l - 1$. So for $0 < l \leq L - 1$,

$$M_l = g_l M_{l-1}, \qquad M_{L-1} = M. \qquad (5.1)$$

Let us denote the M_l-bit string at level l as $\langle b_0^l b_1^l \ldots b_{M_l-1}^l \rangle$, where $b_i^l = \{0, 1\}$, $i = 0, 1, \ldots, M_l - 1$, and denote the g_l-bit string of the kth group as $\langle b_{kg_l}^l b_{kg_l+1}^l \ldots b_{(k+1)g_l-1}^l \rangle$. We have $b_k^{l-1} = b_{kg_l}^l + b_{kg_l+1}^l + \cdots + b_{(k+1)g_l-1}^l$, where $k = 0, 1, \ldots, M_{l-1} - 1$, and $+$ represents the logical OR operator. Figure 5.13 illustrates such a data structure with $g_l = g$ and $L = \log_g M$.

The string at level l ($l \neq 0$) can be stored in a RAM of size $M_{l-1} \times g_l$, while the string at the top level is stored in an M_0-bit register. Denote the m-bit address F as $\langle a_0 a_1 \ldots a_{m-1} \rangle$, where we assume $M = 2^m$. The address used to locate any one of the M_l bits at level l is $\langle a_0 a_1 \ldots a_{m_l-1} \rangle$, where

$m_l = \log_2 M_l$. Hence, it follows from (5.1) that

$$m_l = \log_2 M_l = m_{l-1} + \log_2 g_l = \sum_{i=0}^{l} \log_2 g_i, \qquad g_0 = M_0. \qquad (5.2)$$

This equation illustrates the principle of addressing in the hierarchical searching. That is, m_0 most-significant bits (MSBs) of F should be used at level 0. Then at level l, the complete address used at upper level $l - 1$ (which is m_{l-1} bits wide) will be used to locate the proper g_l-bit word in its $M_{l-1} \times g_l$ memory. Another $\log_2 g_l$ MSBs following the previous m_{l-1} MSBs are extracted from F and is used to locate the proper bit in the g_l-bit word that has just been identified.

The priority encoder and decoder are two basic modules in the RSE. Each level l requires a priority decoder with $\log_2 g_l$-bit input and g_l-bit output, which can be used to write or reset any bit of a g_l-bit word stored in its RAM. For that, we can simply OR the g_l-bit outputs from both the decoder and the $M_{l-1} \times g_l$ RAM (*write* operation) or AND the inverted output of the decoder with the RAM output (*reset* operation), and then write the result back to the memory. Each level l also requires a priority encoder with g_l-bit input and $\log_2 g_l$-bit output, which can be used to search the first MSB (equal to one) of any g_l-bit word stored in its RAM and provide $\log_2 g_l$ bits of the m-bit time stamp for the first $V = 1$ bit in the original string (*search* operation). Since the search works top down, according to (5.2) the m-bit timestamp should be the concatenated result of outputs from all L encoders, as illustrated in Figure 5.14.

The searches at all levels need to be carried out sequentially based on (5.2). The time to search for the first $V = 1$ bit in the original string is decided by finding the first $V = 1$ bit at each level. It needs one register reading at level 0 and $L - 1$ memory accesses at other levels, a requirement that is independent of the number of flows in the system (PCAM is a special case of the RSE with $g = \sqrt{M}$), as is the time to update (i.e., write or reset) each of the V-bits. The total memory requirement is $\sum_{l=1}^{L-1} M_l$ or $\sum_{l=1}^{L-1} \prod_{i=0}^{l} g_i$

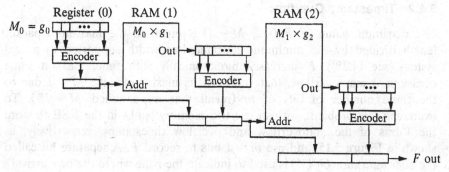

Fig. 5.14 Output of the m-bit time stamp F in hierarchical searching ($L = 3$). (©1999 IEEE.)

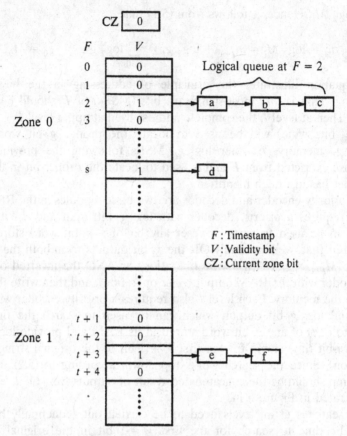

Fig. 5.15 Timestamp overflow control in the RSE.

bits according to (5.1). As an example, if $M = 32K$, $g_l = g = 32$, and $L = \log_g M = 3$, the required memory is 32×32 bits at level 1 and $32^2 \times 32$ or $1K \times 32$ bits at level 2 (total of 33,792 bits).

5.4.2 Timestamp Overflow

The maximum value of F (i.e., $M - 1$) is equal to the maximum packet length divided by the minimum allocated bandwidth supported in a real system [see (4.25)]. F increases monotonically with time, and on some occasions it may overflow, that is exceed its maximum value $M - 1$ due to the finite number of bits of m (recall that we assumed $M = 2^m$). To overcome this problem, we can use two memory banks in the RSE to store the V-bits of the nonoverflow and overflow timestamps, respectively, as shown in Figure 5.15, and use $m + 1$ bits to record F. A separate bit called the *zone indication bit* (Z) is used to indicate the zone where the new arrival's timestamp is to be stored; it is actually the MSB of the timestamp value, and is used to indicate overflow. The overflow here is the same as that defined in the previous section, which is different from the traditional one. That is, we use a *current zone* bit (CZ) to indicate the zone of the packets that are currently being served. Whenever the MSB of a calculated F (i.e., its overflow bit) has the same value as the CZ, the F is classified as nonoverflow; otherwise, it is classified as overflow. Thus, when searching for the V-bits in the RSE, the CZ enables the RSE to choose the first V-bit from an appropriate zone. When all the V-bits in the current zone are zero and there is at least one nonzero V-bit in the other zone, the CZ will be toggled after sending a packet from the other zone, indicating the service zone is flipped. The timestamp is nondecreasing within each zone; the system virtual time with recalibration is at least equal to the minimum start time of the HOL packet among all currently backlogged sessions and thus is also nondecreasing. New HOL packets with their timestamps derived from (4.25) will be placed either in the current zone or in the other zone (which is now regarded as the overflow zone) due to timestamp overflow, as indicated by CZ, ensuring the correct sequence of packet transmission. As shown in Figure 5.15, when all packets a, b, c, and d are transmitted and the V-bit at the other zone ($Z = 1$) is found at $t + 3$, the CZ bit is toggled from zero to one. From then on, packets in zone 1 will be scheduled before those in zone 0. The searching in the RSE chip alternates between the two zones with CZ = 0 and 1. As long as the MSB of the calculated F does not change more than once when serving in the current zone, no packet out-of-sequence problem will occur.

5.4.3 Design of the RSE

Figure 5.16 shows a block diagram of the RSE that can handle 32K (2^{15}) timestamp values. The RSE consists of a controller and a RAM that is divided into two banks. The input data (15 bits) are written to the RSE with the WRITE signal through the bus IN[14 : 0] to set the validity bit. The SEARCH signal is asserted when searching for a $V = 1$ bit in the RSE. If the $V = 1$ bit that is closest to the top of the list is found, the HIT signal is asserted and its corresponding timestamp value appears at the output bus, OUT[14 : 0]. The MODE signal is used to determine whether the RSE is configured to one 32K-bit zone or two 16K-bit zones (to deal with the timestamp overflow problem). The CZ signal is used to indicate the zone of the V-bits that is currently being searched. At the initialization, the INIT signal is asserted, and all of the data (i.e., V-bits) in the REG and RAM are set to zero. These I/O signals of the RSE are summarized in Table 5.3. To support a larger timestamp value, we can increase the group size, which in turn increases the width of the register and the size of the memory.

5.4.4 RSE Operations

Consider that the RSE in Figure 5.16 is configured to accommodate up to $M = 32K$ different timestamp values. The group size g is set to 32 bits. Thus, the total number of searching levels, L, is $\log_g M$ or 3. The V-bits of level 0

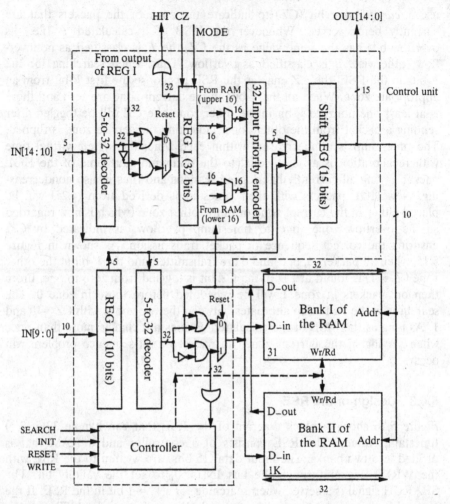

Fig. 5.16 Block diagram of the RAM-based searching engine (RSE). (©1999 IEEE.)

TABLE 5.3 I/O Signals of the RSE

Pin	I/O	Description
IN[14:0]	IN	Data input (e.g., timestamp value)
INIT	IN	Initialization
WRITE	IN	Write enable
SEARCH	IN	Search enable
RESET	IN	Reset enable
MODE	IN	Mode select
CZ	IN	Current zone bit
HIT	OUT	Pattern found indication
OUT[14:0]	OUT	Data output (e.g., timestamp value)

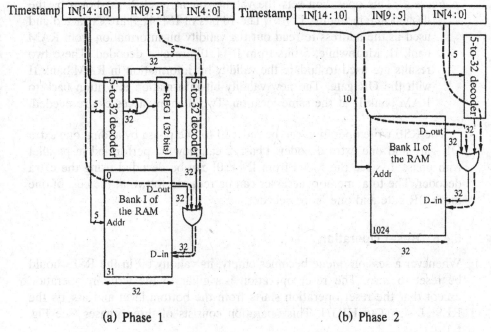

(a) Phase 1   (b) Phase 2

Fig. 5.17   Data paths in the RSE write operation. (©1999 IEEE.)

are stored in REG I, while those of levels 1 and 2 are stored in two banks of the RAM, bank I (32 × 32) and bank II (1K × 32),respectively.

5.4.5 Write-in Operation

When a new packet arrives, its 15-bit timestamp is divided into three parts, 5 bits each. The write-in operation sets the validity bits properly in the register and memory banks. This operation consists of two phases (see Fig. 5.17):

Phase 1: This operation sets the validity bits at levels 0 and 1, which can be done at the same time because those bits are stored at different places. In Figure 5.17(a), the solid line indicates the data path for setting the validity bit at level 0. The first 5 bits of the timestamp (IN[14 : 10]) are extracted, decoded, and ORed with the old value in REGI. Meanwhile, the validity bit at level 1 is also set to one, as shown by the dashed lines. During this operation, the 5 bits of IN[14 : 10] are used as the address to read the old validity bit information (actually the 32-bit word containing the V-bit) from RAM bank I. At the same time, the 5 bits of IN [9 : 5] are also decoded. These two results are then used to update the validity bit information in RAM bank I with an OR gate. The new validity bit information is written back to RAM bank I at the same location. Two memory accesses (write and read) are needed.

Phase 2: This operation sets the validity bit at level 2, as shown by the dashed line in Figure 5.17(b). The 10 bits of IN[14 : 5] are extracted and used as the address to read out the validity bit information from RAM bank II. Meanwhile, 5 bits from IN[4 : 0] are also decoded. These two results are used to update the validity bit information in RAM bank II with the OR gate. The new validity bit information is written back to RAM bank II at the same location. Two memory accesses are needed.

The RSE write operation can be reduced to one phase by adding one extra OR gate and one extra decoder. Phase 2 can now be performed in parallel with phase 1, since the 5 bits from IN[4 : 0] can be decoded using the extra decoder. The total memory accesses can be reduced to two at the cost of one more OR gate and one more decoder.

5.4.6 Reset Operation

Whenever a session queue becomes empty, its validity bit in the RSE should be reset to zero. The reset operation is similar to the write-in operation except that the reset operation starts from the bottom level and asserts the RESET signal to HIGH. This operation consists of three phases (see Fig. 5.18).

Phase 1: This operation resets the validity bit at level 2, as shown by the dashed line in Figure 5.18(a). The 10 bits of IN[14 : 5] are extracted and used as the address to read out the validity bit information (32 bits) from RAM bank II. Meanwhile, 5 bits from IN[4 : 0] are decoded to a 32-bit word, which is then inverted and ANDed with the 32-bit output from RAM bank II. The newly updated validity bit information is written back to RAM bank II at the same location. In addition, the new 32 validity bits are ORed. If the result is zero, meaning that all validity bits in this group are all zero, we proceed to phase 2. Two memory accesses are needed in this phase.

Phase 2: This operation resets the validity bit at level 1 of the corresponding group at level 2, as shown by the dashed line in Figure 5.18(b). The 5 bits of IN[14 : 10] are used as the address to read the old validity bit information from RAM bank I. At the same time, the 5 bits of IN[9 : 5] are decoded and then inverted. These two results are ANDed to obtain the new validity bit information, which is written back to RAM bank I at the same location. In addition, the new validity bits are ORed. If the result is zero, meaning that all validity bits in this group are zero, we proceed to phase 3. Two memory accesses are needed in this phase.

Phase 3: This operation resets the validity bit at level 0 of the corresponding group at level 1. In Figure 5.18(c), the heavier line indicates the data path for resetting the validity bit at level 0. The first 5 bits of the time stamp (IN[14 : 10]) are extracted, decoded, inverted, and ANDed with the old value in REG I. The result is written back to REG I.

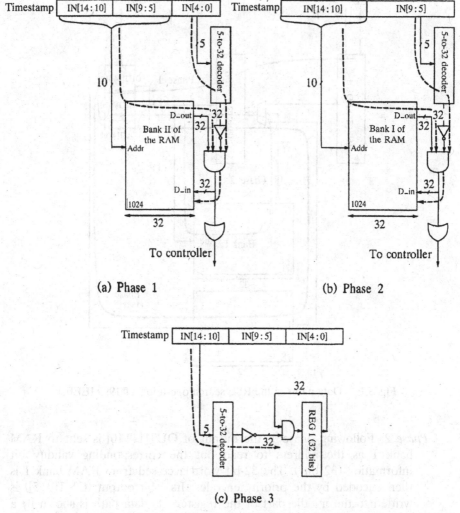

Fig. 5.18 Data paths in the RSE reset operation.

5.4.7 Search Operation

The search operation consists of three phases (see Fig. 5.19).

> *Phase 1:* If there is at least one bit in REG I that is set to one, the HIT
> signal is asserted, indicating a match is found and the OUT[14 : 0] signal
> is valid. The 32-bit data from REG I are then encoded by the priority
> encoder into a 5-bit output (OUT[14 : 10]), which is then written to the
> upper part of a register. The data path is shown by a solid line in Figure
> 5.19.

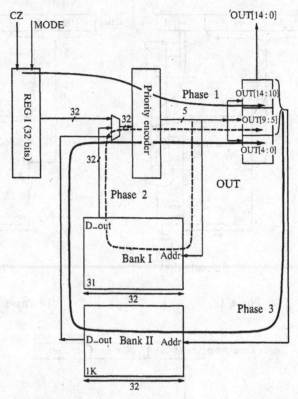

Fig. 5.19 Data paths in the RSE search operation. (©1999 IEEE.)

Phase 2: Following phase 1, the output of OUT[14 : 10] is sent to RAM bank I as the address to read out the corresponding validity bit information (32 bits). The 32-bit word accessed from RAM bank I is then encoded by the priority encoder. Its 5-bit output (OUT[9 : 5]) is written to the middle part of the register. Its data path is shown by a dashed line in Figure 5.19. One memory access is needed in this phase.

Phase 3: The outputs from phases 1 and 2 are combined to form OUT[14 : 5], which is used as the address to read out the validity bit information from RAM bank II. The 32-bit word accessed from RAM bank II is then encoded by the priority encoder. Its 5-bit output (OUT[4 : 0]) is written to the lower part of the register. The data path is shown by a bold solid line in Figure 5.19. One memory access is needed in this phase.

The register content, OUT[14 : 0], is the final result, indicating that the location of the first validity bit has been found.

5.5 GENERAL SHAPER–SCHEDULER

A shaper–scheduler, such as the $WF^2Q +$ and SPFQ, also needs to maintain another priority queue, called the *shaper queue*, for performing the eligibility test, as shown in Figure 5.1(b). A shaper–scheduler was proposed in [9] for ATM switches, in which the shaper queue is implemented as a multitude of priority lists. Refer to Figure 5.1(b). Each list is associated with a distinct start time common to all queued packets in it. Using the search-based approach, we can construct a 2-D calendar queue based on the start times of the queued packets, as shown in Figure 5.20, where the start time and timestamp are used as the column and row addresses, respectively, and W is the maximum value of the start times. All packets with the same start time, S, are placed in the same column addressed by S, and also are sorted according to their timestamps F. Hence, each column represents a priority list. Each V-bit in a column can be located by its unique address (S, F).

Performing the eligibility test is equivalent to using the system virtual time as the column address to find the nonempty column(s) with their addresses ranging from the previous value up to the current value of the system virtual time, and then moving packets in these column(s) to the scheduler queue.

According to (4.23), $WF^2Q +$ and SPFQ advance the system virtual time by the amount of work the server performs. It has been shown in [9] that for ATM switches, the system virtual time is advanced by one after the transmission of each cell, because a cell is a constant unit of work. Only two cells at most need to be transferred to the scheduler queue. In packet networks, the system virtual time may be advanced by more than one, due to the variable size of packets. Refer to Figure 5.1(b). In the worst case, a maximum of N

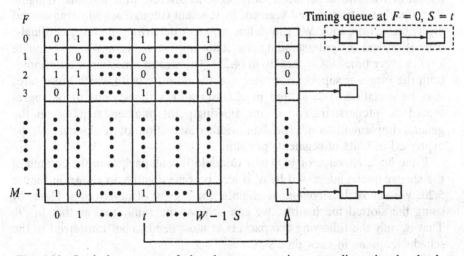

Fig. 5.20 Logical structure of the shaper queue in a two-dimensional calendar queue. (©1999 IEEE.)

eligible packets need to be moved from the shaper queue to the scheduler queue. No solutions in hardware implementation have appeared so far. Furthermore, it remains unclear whether the implementation of a large number of priority lists as W increases would be feasible. Simply extending the concept of the RSE allows us to construct a 2-D RSE [30] to implement the 2-D calendar queue. Below we present a *general shaper–scheduler* for both ATM and packet networks, together with the 2-D RSE, and we show that only two packets at most need to be transferred to the scheduler queue.

5.5.1 Slotted Updates of System Virtual Time

According to (4.23), the system virtual time $V(t)$ in shaper–schedulers, such as WF^{2}Q + and SPFQ, is advanced by the amount of work the server performs during the time interval $(t, t + \tau]$. Conceptually, the length of each packet serves as a measure of work in a packet system, so $V(t)$ could be advanced quite differently from time to time due to variable packet lengths. However, for efficient hardware handling, packets are usually divided into fixed-length segments in IP routers and packet switches before they are stored in the buffer and forwarded to the next node(s). As a result, a packet scheduler can be viewed as a slotted (synchronous) system. A time slot T corresponds to the time (or number of system clock cycles) that is needed to transmit a segment at the link capacity r. Assume a fixed-rate server and $\tau = T$ in (4.23), the work that the packet server performs in a time slot is a constant $W(t, t + \tau) = rT$, which we can normalize to one, as is done for ATM switches [9].

Without introducing much inaccuracy (explained below), we assume that packet arrival and departure events occur at discrete time instants. Imagine that a packet is transmitted segment by segment (though actually transmitted bit by bit) on the link. We can define segment departure events accordingly. This is critical to understanding the following mechanism. We may update $V(t)$ in every time slot according to (4.23). The size of timestamps determines both the range of supportable rates and the accuracy with which those rates may be specified. The slotted mechanism allows timestamps to be represented as integers, instead of the floating-point numbers required in the general implementation. The bandwidth reservation of a session can be expressed in units of segments per slot.

If the updated value of $V(t)$ is α (modulo W), all packets in the column of the shaper queue addressed by α, if any, become eligible, as shown in Figure 5.20. We say such a *column* is eligible, for brevity of description. Here, by using the slotted mechanism, we can achieve the same goal as that in [9]. That is, only the following two packets at most need to be transferred to the scheduler queue in each time slot:

- the packet with the smallest timestamp in column α;
- if the packet being transmitted or just sent out has a start time value of β (modulo W), the packet with the smallest time stamp in column β, if any, is also moved to the scheduler queue.

Each packet is referred to as the *first* packet in the corresponding column. Based on the above operations, the first packet of each nonempty and eligible column is moved to the scheduler queue, and is prioritized by its finish time. Since each eligible column has at least one packet in the scheduler queue, after the above operations, the remaining packet(s) in the same column, if any, will all be moved to the scheduler queue eventually.

According to (4.23), $V(t)$ may be advanced by more than one in case several columns in the shaper queue are empty. These empty columns fall between the column pointed to by the previous system virtual time and that pointed to by the minimum value of start times among all currently nonempty sessions, which is the current system virtual time based on (4.23). This can help immediately find the next nonempty and eligible column so that the work-conserving property of the packet scheduler is preserved [6]. Still, only two packets at most need to be transferred to the scheduler queue. Since this mechanism can be used in both ATM and packet switches, we call it the *general shaper–scheduler*.

Figure 5.21 illustrates the basic operations of our slotted mechanism when packets enter and leave the scheduler queue. The length of a time slot is regarded as one. From time t to $t+3$, there are four arriving HOL packets

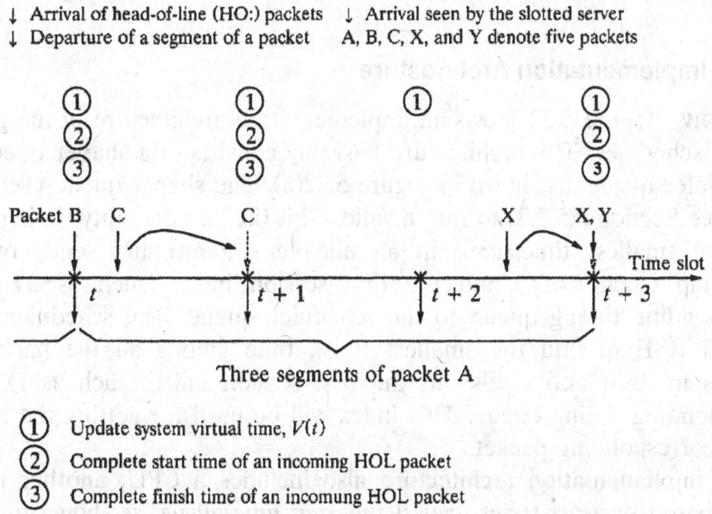
↓ Arrival of head-of-line (HO:) packets
↓ Arrival seen by the slotted server
↓ Departure of a segment of a packet
A, B, C, X, and Y denote five packets
Packet B
C
C
X
X, Y
Time slot
t
t + 1
t + 2
t + 3
Three segments of packet A
① Update system virtual time, V(t)
② Complete start time of an incoming HOL packet
③ Complete finish time of an incomung HOL packet

Fig. 5.21 An example of illustrating the slotted mechanism.

(B, C, X, Y), and one departing packet (A). Since the server is busy all the time, the system virtual time $V(t)$ is updated at each of these time instants (step 1), as indicated by the circled 1 in Figure 5.21. Since packet B arrives at time t, its virtual start time and virtual finish time are also computed (steps 2 and 3), as indicated by the circled 2 and 3 in Figure 5.21, respectively. Although packet C arrives in the time interval $(t, t + 1)$, the server assumes it receives this packet at time $t + 1$, as illustrated in Figure 5.21. This also applies for packet X. At time $t + 2$, no HOL packet arrives, but a segment departs, and only $V(t)$ is updated. Packet A leaves the system at time $t + 3$; packet Y could be the successor to packet A in the same column or a packet from another eligible column with the smallest finish time, so it is loaded (as an arrival) to the scheduler queue.

The slotted mechanism basically is an approximation. It can simplify hardware implementation. However, it also introduces some inaccuracy by assuming packet departures and arrivals at discrete times. Since not every packet can be divided into an exact number of segments on account of the variable packet length, the last segment of a packet can have fewer bits of data than a complete segment. We call it an *incomplete* segment. The server is not work-conserving, since some bit times within a time slot for transmitting an incomplete segment are wasted, regardless of other HOL packets in the system. As a result, a physical link could be underutilized. This also happens when an HOL packet arrives in a time slot but has to wait for the end of this slot before it is processed. However, the maximum time that the server is idle while the system is not empty is bounded by one slot time. For instance, suppose the system is empty at time t, there is no packet B, and packet C arrives in $(t, t + 1)$, as shown in Figure 5.21. The server won't handle packet C until time $t + 1$. Its idle time is at most one slot time.

5.5.2   Implementation Architecture

Overview   Figure 5.22 shows an implementation architecture of the general shaper–scheduler. The architecture basically consists of a shaper queue and a scheduler queue, as shown in Figure 5.22(a). The shaper queue uses a 2-D RSE (see Section 5.5.2.2) to find a valid V-bit (i.e., a nonempty timing queue with the smallest timestamp in an eligible column) and sends out this timestamp (such as f) and the first session index (such as a) in the corresponding timing queue to the scheduler queue. The scheduler queue uses an RSE to find the smallest finish time among all the backlogged sessions in itself and sends out the first session index (such as i) in the corresponding timing queue. This index will be used for getting the address of the corresponding packet.

The implementation architecture also includes a CPU; another priority queue based on start times, called the *start time queue*, as shown in Figure 5.22(b); and a regular queue called the *finish time queue* that stores finish times of HOL packets of every session. The start time queue uses another

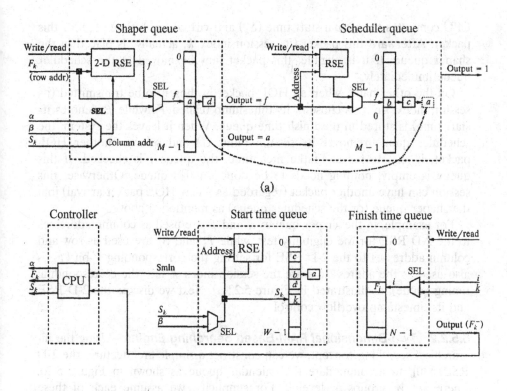

(a)

RSE = RAM-based searching engine W = max(start time) β = start time of head-of-line (HOL) packet of system i
2-D RSE = two-dimensional RSE N = number of sessions in the system k = session index of an incoming HOL packet p_k
SEL = selector, f is a finish time $i, k, a, b, c,$ and d are session indices S_k, F_k = start time and finish time of P_k respectively
M = max(finish time) α = updated system virtual time F_k^- = finish time of previous HOL packet of session k

Fig. 5.22 A general shaper-scheduler: (a) main priority queues, (b) auxiliary queues. (©1999 IEEE.)

RSE to find the smallest start time among all backlogged sessions in the system, i.e., $\min_{i \in \hat{B}(t)}\{S_i(t)\}$, for updating $V(t)$ according to (4.23). This value, denoted by $S_{\min}$ in Figure 5.22(b), is provided as the output of the RSE. The finish time queue uses the session index as the address for direct access to the finish time of each HOL packet. Since there are a total of N sessions in the system, the queue has N entries. The stored timestamp (such as F_k^-) is used by the CPU to compute the virtual start time of a new HOL packet (such as P_k for session k) according to (4.24).

The basic operations can be briefly described as follows. At every time slot, the start time queue provides the minimum start time $S_{\min}$ to the CPU, and the CPU keeps updating the system virtual time, denoted by α. The shaper queue performs the eligibility test and sends those eligible packet(s), if any, and their time stamps (such as a and f) to the scheduler queue. When a new HOL packet (such as P_k) comes, its predecessor's timestamp (such as F_k^-) is first fetched from the finish time queue and sent to the CPU; then the

CPU computes the virtual start time (S_k) and virtual finish time (F_k) for this packet. Afterward this packet (its session index k, actually) is placed in the shaper queue. If it is eligible, this packet may be moved to the scheduler queue immediately.

On the other hand, when an HOL packet is chosen to be transmitted (its session index, say i, is chosen), its timestamp (called F_i, while β denotes its start time) is stored in the finish time queue. When it leaves the system, the scheduler queue removes its session index (i) and selects another HOL packet, if any, to serve. In the meantime, the queue i is checked. If this queue is empty, nothing needs to be done on this queue. Otherwise, this session can have another packet (regarded as a new HOL packet arrival) join the shaper queue (or the scheduler queue) as mentioned above.

For example, in the shaper queue, α and β are used as column addresses to the 2-D RSE for the eligibility test, while F_k and S_k are used as row and column addresses to the 2-D RSE for setting the corresponding V-bit (F_k is also used as the address to store the session index k into the corresponding timing queue), as illustrated in Figure 5.22(a). Next we discuss the 2-D RSE and its timestamp overflow control.

5.5.2.2 Two-Dimensional RAM-Based Searching Engine

Using the hierarchical searching concept, we can construct a simple architecture, the 2-D RSE) [30], to accommodate the calendar queue as shown in Figure 5.20. There are W groups at level 0. For simplicity, we assume each of these groups has g bits. Each group at level 0, equivalent to the register at level 0 of the RSE, is associated with a column in the calendar queue shown in Figure 5.20 and represents all V-bits in the column. The RSE grouping operation is then applied recursively to the M-bit string of each column and the total level $L = \log_g M$. All the Wg bits at level 0 can be placed in a

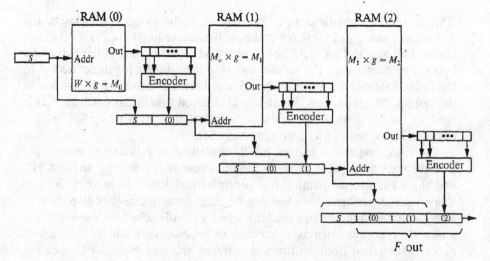

Fig. 5.23 Output F for the first $V = 1$ bit in the column pointed to by S ($L = 3$).

$W \times g$ memory, as shown in Figure 5.23. Except for this extra memory, the 2-D RSE architecture is similar to that of the RSE.

Figure 5.23 also shows how to find out the first $V = 1$ bit in a column of the calendar queue given its S. Since the system virtual time maintained as an integer in the integrated shaper–scheduler is advanced by one after the transmission of each packet, only two packets at most need to be transferred to the scheduler afterward [9]: the packet at the head of column I with its associated start time equal to the updated system virtual time, and the packet at the head of column II from which the transmitted packet departed. The 2-D RSE will be searched twice at most in each time slot. In each searching cycle, only V-bits in columns I and II will be selected to participate in the search. Since packets in the same column are arranged in monotonic order of their time stamps, the 2-D RSE can easily find any of the two eligible packets by identifying the first $V = 1$ bit in the proper column.

Therefore, the time to find the first $V = 1$ bit in a column, given its S, is decided by finding the first $V = 1$ bit at each level, and the search needs L memory accesses, which is independent of the number of sessions in the system, as is the time to update each of the V-bits. As an example, if $M = 32K$ and $g = 32$, then $L = \log_g M = 3$.

5.5.2.3 *Timestamp Overflow*

Both the timestamp F and the start time S in the 2-D RSE can have an overflow problem. To solve it, we can follow the method described in Section 5.4.2. Since $F = S + D$ according to (4.25), where we can omit the super- and subscripts without confusion and D corresponds to L_i^k/r_i, the range of F is bounded by the maximum of D, which equals the maximum packet length divided by the minimum allocated bandwidth supported in a real system.[1] We can choose M such that the maximum of D, rounded to an integer in implementation, is equal to $M - 1$. So F ranges from 0 to $M - 1$. D should also be at least one, because zero packet length or infinite reserved bandwidth, is impossible, conceptually and practically.

Below we discuss two methods to control the timestamp overflow. In theory, F should always be greater than S, because $F = S + D$ with $D > 0$. In implementation, each of them is represented with finite bits, so even if S does not overflow, F could overflow after adding D to S. Whether F overflows or not should be determined with respect to S as a result. Henceforth, unless otherwise stated, we always discuss F, S, and D within a real system.

Consider $M = 8$. Two banks of memory are needed to solve the overflow of S, as shown in Figure 5.24, within each of which there are two zones for F: overflow and nonoverflow zones. With respect to S, F is defined as nonoverflowing if $S < F$, or overflowing if $S > F$. Since $1 \leq D < M$, F will never be equal to S. The bits with their row–column addresses $F = S$ are dummies in

---

[1]Note that a burst of arrived packets for the same session will join the flow queue first, and there is no need to compute their timestamps. Only the HOL packet of each session can join the scheduler and is timestamped.

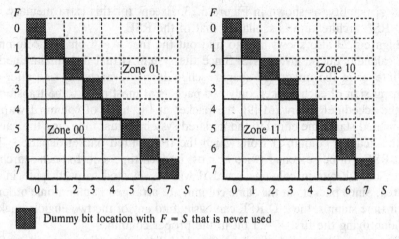

Fig. 5.24   Timestamp overflow control in the 2-D RSE ($M = 8$): method A. (©1999 IEEE.)

the calendar queue (see Fig. 5.20) and are never used, as signified by shaded blocks in Figure 5.24. They simply serve as a zone boundary dividing each bank of memory into two zones for F. Thus, there are a total of four zones (00_2–11_2), each of triangular form. It follows that $W = M$ to provide equal-sized zones in each bank of memory, as shown in Figure 5.24, with each zone containing $M(M - 1)/2$ V-bits

A 2-bit zone indication (Z) and a 2-bit current zone indication (CZ) are needed. Recall that we assume $M = 2^m$. We use an $(m + 1)$-bit word to represent F as well as S with the MSB defined as its overflow bit. The leftmost bit of Z or CZ corresponds to S's overflow bit, while the rightmost bit corresponds to F's overflow bit. Suppose CZ $= 00_2$ initially, indicating the current zone to be served, as shown in Table 5.4. Whenever the S's MSB is equal to one, we define it to be overflowing, indicating zones 10_2 and 11_2;

TABLE 5.4   Zone Indication CZ versus F and S

MSB(S) \ MSB(F)	0	1
0	00 ⟶ 01	
1	10 ⟵ 11	

Initially
CZ = 00

⟶ Zone service sequence

otherwise, we define it to be nonoverflowing, indicating zones 00_2 and 01_2. Whenever the F's MSB is different from the Ss MSB, we define it to be overflowing (with respect to S), indicating zones 01_2 and 10_2; otherwise, we define it to be nonoverflowing, indicating zones 00_2 and 11_2. The four zones are served in the order $00_2 \rightarrow 01_2 \rightarrow 11_2 \rightarrow 10_2 \rightarrow 00_2$. The current zone indication CZ will be flipped according to the arrowed sequence in Table 5.4 only after sending a packet from another zone, indicating the service zone is changed. As long as the overflow bit of F does not change more than once while the current zone is being served, as described in Section 5.4.2, no packet out-of-sequence problem will occur. This is also true for S, since $F = S + D$.

All V-bits in zones 00_2 and 01_2 satisfy the conditions that $S < F$ and $S > F$, respectively. This is also the case with zones 11_2 and 10_2. Figure 5.24 illustrates how in each bank of memory, we need to mask off the V-bits of each column that don't belong to the current zone in order to perform correctly the memory read, write, and reset operations. Consider $CZ = 00_2$. When we search the column of $S = 1$ top down for a $V = 1$ bit with the smallest F, only those bits with $F = 2\text{–}7$ are possible candidates. The others, with $F = 0, 1$, should be masked off. If $CZ = 01_2$, then those bits with $F = 1\text{–}7$ should be masked off. This is also the case for zones 11_2 and 10_2. We can XOR the two bits of CZ and use the result to decide the proper masking operation when we search a column of S generally. A result of zero (zones 00_2 and 11_2) indicates the V-bits in the column with $F \leq S$ will be masked off; otherwise (zones 01_2 and 10_2) the V-bits with $F \geq S$ will be masked off. To perform the masking operations, we need to find the boundary between the region that is to be masked and the region that is not to be masked, using dividers or a table where the precomputed results are stored. Extra priority decoders and gates are also needed.

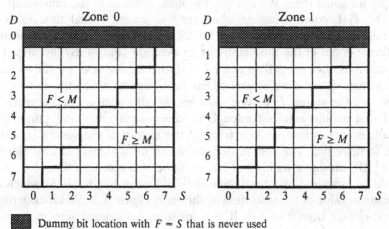

Dummy bit location with $F = S$ that is never used

Fig. 5.25 Timestamp overflow control in the 2-D RSE ($M = 8$): method B. (©1999 IEEE.)

A more desirable alternative to the above approach is to use D instead of F, as shown in Figure 5.25. There are still two zones, 0 and 1, for S. The row of $D = 0$ is a dummy as explained before. The attractive point of this method is that we only consider the overflow of S, since D has no overflow problem. Besides, no masking operations are needed, because each V-bit in a column is ranked unambiguously by D. Regardless of that, each bank of memory can also be divided into two areas for F: overflow with $F = S + D \geq M$, and nonoverflow with $F < M$. One bit is needed for both Z and CZ. The tradeoff is that one extra adder or adding operation is needed to recover the value of F from (4.25). Besides, the scheduler has to take care of those eligible packet(s) that are read out and have their F's overflow. In other words, the handling of F-overflow is actually moved from the 2-D RSE in the shaper queue to the RSE in the scheduler queue, as shown in Figure 5.22(a).

5.6 TIMESTAMP AGING PROBLEM

Recall that, as shown in Figure 5.22(b), when a packet of session i departs, its finish time F_i is stored in a lookup table (i.e., finish time queue) for later use according to (4.24). Information other than F_i for session i, such as S_i and r_i, can also be stored in the same location (addressed by i). Later, when a new packet of this session arrives at the head of its queue and immediately becomes the HOL packet, the F_i needs to be read out and compared with the current system virtual time according to (4.24) to decide the new S_i. However, since the system virtual time is also represented with finite bits in implementation, it can overflow in the same way as the timestamp, as explained before. It is impossible to decide which is greater without any previous history or certain constraints, especially when the queue has been empty for some time. We call this the *aging problem* of the timestamp.

The F_i becomes *obsolete* whenever the system virtual time exceeds it. Recall that in Sections 5.4.2 and 5.5.2.3, an extra bit is used to indicate two different zones of the timestamp for its overflow control. By the same token, we can introduce more than one bit to record a number of overflow events of the system virtual time as well as the time zone to which the system virtual time and the stored F_i belong. Besides, we also need a purging mechanism [31] that should run fast enough to check each entry and purge all the obsolete ones before we lose track of the history of the system virtual time due to finite bits. The system virtual time $v(t)$ is updated per time slot based on (4.23), causing a number of entries to be checked. In a time slot, the $v(t)$ could be increased by a maximum of $W - 1$ according to (4.23), which is the maximum value of the start time as shown in Figure 5.20. As an example, $v(t)$ could change from 0 to $S_j = W - 1$ for some backlogged session j, while all the other sessions are empty, generating a maximum of $N - 1$ entries that would need to be checked.

Each purging operation has at most two memory accesses: one is to read the F_i, which is always needed, and if the F_i is obsolete, the other marks it with a write operation. Due to the limited memory speed, it may not be possible to perform many purging operations in a time slot when N is large. However, since the $v(t)$ can overflow at most once in every time slot, we use a multibit counter to keep track of its overflow in a period of multiple time slots, so it is possible to purge all obsolete entries by carefully designing a purging scheme.

Below we introduce a periodic purging mechanism that is required to check A entries in the lookup table in T consecutive time slots with

$$A \geq N - 1. \tag{5.3}$$

The lower limit is $N - 1$ rather than N, because when all session queues are empty (the system is empty), all N entries in the lookup table become obsolete. As a result, the system virtual time is simply reset to zero. Since in the worst case there could be A purging operations, plus $2T$ regular memory accesses (write F_i after current HOL packet of session i departs, and read F_j when a new packet from session j becomes HOL) in the T time slots, the value of T must thus satisfy

$$T \times \text{slot time} \geq (2T + 2A) \times \text{memory cycle}, \tag{5.4}$$

which guarantees that all A entries can be purged, if obsolete, within T time slots. $2T + 2A$ is the maximum number of memory accesses during this time. As an example, let a slot time be equal to that needed to transmit a 64-byte packet segment at a speed of 10 Gb/s, that is, 51.2 ns; memory cycle is 10 ns, and $N = 32K$. According to (5.3), we choose $A = N = 32,768$. It follows that $T = 21,006$, based on (5.4). There will be $A/T \approx 1.56$ purging operations (3.12 memory accesses) per time slot.

To ensure unambiguous comparisons between the system virtual time and each stored timestamp in any of the T time slots, the multibit counter must be able to record at least $T + 1$ times of overflow for the system virtual time, as explained later. We therefore introduce a $\lceil \log_2(T + 1) \rceil$-bit counter variable $C_v(t)$ for the system virtual time $v(t)$, which is increased by one each time $v(t)$ overflows. To facilitate the purging operations, we also define for each entry F_i an *obsolete* bit (O_i) and a similar $\lceil \log_2(T + 1) \rceil$-bit counter variable (C_i). The obsolete bit is similar to the time-out bit in [14]. Figure 5.26 shows the format of an entry in the table. Similarly to the overflow bit introduced for the timestamp in Sections 5.4.2 and 5.5.2.3, $C_v(t)$ and C_i can be regarded as the time zone indicators for $v(t)$ and F_i, respectively. So the $v(t)$ and F_i can be compared directly if both are in the same time zone. Otherwise, simply comparing their time zone values indicates which is larger. This is the basic idea of the purging scheme.

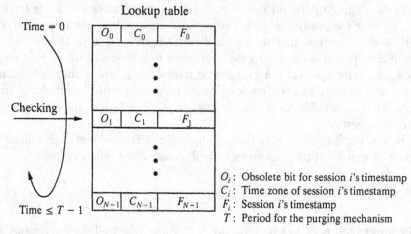

Fig. 5.26 Organization of the table for F_i in the purging mechanism. (©1999 IEEE.)

Whenever we store F_i in the table at time slot s, as shown in Figure 5.26, we set its O_i-bit to zero, since it is not obsolete at that time, and store its time zone in the C_i-field of the entry. C_i could be $C_v(s) + 1$, because even if S_i and $v(s)$ are in the same time zone, F_i may overflow and is one zone ahead of $v(s)$ according to (4.25), as indicated by $F_i < v(s)$ in contrast to $F_i \geq v(s)$. Thus

$$C_i = \begin{cases} C_v(s) & \text{if } F_i \geq v(s), \\ C_v(s) + 1 & \text{otherwise.} \end{cases} \tag{5.5}$$

As mentioned before, since F_i cannot overflow more than once while the system is serving the current zone, there will be no packet out-of-sequence problem.

Suppose all sessions are visited in the order of their indexes as shown in Figure 5.26, and F_i is examined at time slot $t > s$. Figure 5.27 is a flowchart of the periodic purging algorithm, which is explained as follows. If the O_i-bit is one, indicating that F_i is obsolete at this time, we are done for this entry and proceed to F_{i+1} for the next; otherwise, we compare the C_i with the current value of $C_v(t)$. Note that $C_v(t) = 0, 1, \ldots, T$. There are three cases:

Case 1: If $C_v(t) = C_i$, indicating both $v(t)$ and F_i are in the same time zone, then they are compared directly. If $v(t) \geq F_i$, then F_i is obsolete from this time on and its O_i-bit is set to one; otherwise, without further operations we proceed to F_{i+1}.

Case 2: If $C_v(t) = C_i - 1$ or $C_v(t) = T$ with $C_i = 0$, F_i is one zone ahead of $v(t)$ from (5.5) and is not obsolete. We proceed to F_{i+1} for the next step. The latter subcase occurs when $C_v(s) = T$, $C_i = [C_v(s) + 1] \mod (T + 1) = 0$ from (5.5).

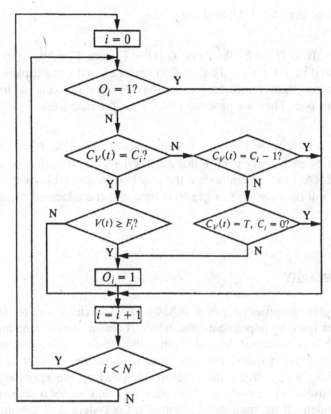

Fig. 5.27 Flowchart of the periodic purging mechanism.

Because $A \geq N - 1$, sessions can be checked within T time slots according to (5.3), this guarantees that F_i can be checked (and purged, if eligible) at
least once since time s, when it is stored until time t inclusive, with
$t - s \leq T$. Remember that F_i is not obsolete at time s when it was stored.
The $v(t)$ can overflow at most T times during the period $(s, t]$, while the
$C_v(t)$ can overflow (be wrapped around) at most once. The critical question is
whether, under the above assumptions, it can be guaranteed that a $C_v(t)$
exceeding the C_i, if wrapped around, will never be above $C_v(s) - 1$. If that is
true, this case can never be misinterpreted as case 1 or 2. Otherwise,
unsuccessful purging operations will result. Suppose $C_i = C_v(s)$, which is
worse than $C_i = C_v(s) + 1$ because it would be closer to the wrapped $C_v(t)$.
Since

$$C_v(t) = [C_v(s) + (t - s)] \bmod (T + 1), \tag{5.6}$$

if wrapped around, $C_v(t)$ will be $C_v(s) - 1$ when $t - s = T$, which ensures
correct purging operations. This is why we require the counter variables $C_v(t)$

and C_i to be $\lceil \log_2(T+1) \rceil$ bits wide. Thus,

> *Case 3:* If $C_v(t) > C_i$ [but not $C_v(t) = T$ with $C_i = 0$], or if wrapped around $[C_v(t) < C_i - 1]$, then $C_v(t)$ must exceed C_i as explained above. As a result, $v(t)$ must be greater than F_i. F_i is obsolete, so its O_i-bit is set to one. Then we proceed to F_{i+1} for the next step.

Let $M = 32K$ in the previous example, where $N = 32K$, $T = 21,006$. Then the width of each entry is $(1 + \lceil \log_2(T+1) \rceil + \lceil \log_2 M \rceil)$, that is, 31 bits. A $32K \times 32$ RAM can accommodate the required lookup table (without including other information such as the start time and the allocated bandwidth for each session).

5.7 SUMMARY

This chapter proposes a novel RAM-based searching engine (RSE) for packet fair queuing implementation, which is aimed at designing an efficient and scalable architecture that can support hundreds of thousands of sessions in a cost-effective manner. The basic idea is to use the concept of hierarchical searching with a tree data structure to speed up the searching process. The RSE uses commercially available memory chips, its total time complexity is independent of the number of sessions in the system, and it is limited only by the memory accesses needed. This is achieved by trading off the time complexity (e.g., sorting) with the space complexity (e.g., memory size). With an extension of the RSE, we have proposed a 2-D RSE architecture to implement a general shaper–scheduler. By introducing a slotted mechanism for updating the system virtual time, we have shown that only two eligible packets at most are transferred to the scheduler queue in each time slot.

We have demonstrated the feasibility of our approach by presenting an implementation architecture for the RSE. The finite-number-bit overflow, such as timestamp overflow, is a generic problem in implementing GPS-related scheduling algorithms. We suggest dividing the memory into two zones in the RSE to handle the overflow of timestamp F. For the 2-D RSE, conceptually four zones are needed, since both F and S could overflow. As an alternative, we suggest using D instead of F to organize the memory. Since an extra adder or adding operation is needed to restore F from $F = S + D$, the handling of timestamp overflow is moved from the 2-D RSE to the scheduler queue in the corresponding shaper–scheduler. Another kind of finite-number-bit overflow problem is the aging of the timestamp, which makes its comparison with the system virtual time difficult in a real system. To solve this problem, we suggest using a counter variable to record the evolution of the system virtual time as well as the timestamp within a finite period, and introduce a periodic purging mechanism that is required to clear all the obsolete timestamps (below current system virtual time) within this period, so there will be no ambiguity in comparison.

REFERENCES

1. L. Zhang, "Virtual clock: a new traffic control algorithm for packet switching networks," *Proc. ACM SIGCOMM*, pp. 19–29, Sep. 1990.

2. D. Verma, H. Zhang, and D. Ferrari, "Guaranteeing delay jitter bounds in packet switching networks," *Proc. Tricomm*, pp. 35–46, Apr. 1991.

3. S. J. Golestani, "A self-clocked fair queuing scheme for broadband applications," *Proc. IEEE INFOCOM*, pp. 636–646, Apr. 1994.

4. A. K. Parekh and R. G. Gallager, "A generalized processor sharing approach to flow. control in integrated services networks: the single node case," *IEEE/ACM Trans. Netw*, vol. 1, no. 3, pp. 344–357, Jun. 1993.

5. A. K. Parekh and R. G. Gallager, "A generalized processor sharing approach to flow control in integrated services networks: the multiple node case," *IEEE/ACM Trans. Netw.*, vol. 2, no. 2, pp. 137–150, Apr. 1994.

6. D. Stiliadis, "Traffic scheduling in packet-switched networks: analysis, design, and implementation," Ph. D. dissertation, Computer Science Department, University of California at Santa Cruz, Jun. 1996.

7. D. Stiliadis and A. Varma, "Latency-rate servers: a general model for analysis of traffic scheduling algorithms," *Proc. IEEE INFOCOM*, pp. 111–119, Mar. 1996.

8. D. Stiliadis and A. Varma, "Design and analysis of frame-based fair queueing: a new traffic scheduling algorithm for packet-switched networks," *Proc. ACM SIGMETRICS*, pp. 104–115, May 1996.

9. D. Stiliadis and A. Varma, "A general methodology for design efficient traffic scheduling and shaping algorithms," *Proc. IEEE INFOCOM*, Kobe, Japan, Apr. 1997.

10. D. Stiliadis and A. Varma, "Rate-proportional server: a design methodology for fair queueing algorithms," *IEEE/ACM Trans. Netw.*, vol. 6, no. 2, pp. 164–174, Apr. 1998.

11. D. Stiliadis and A. Varma, "Efficient fair queueing algorithms for packet-switched networks," *IEEE/ACM Trans. Netw.*, vol. 6, no. 2, pp.175–185, Apr. 1998.

12. J. C. R. Bennett and H. Zhang, "WF^{2}Q: worst-case fair weighted fair queueing," *Proc. IEEE INFOCOM*, Mar. 1996.

13. J. C. R. Bennett and H. Zhang, "Hierarchical packet fair queueing algorithms," *IEEE/ACM Trans. Netw.*, vol. 5, no. 5, pp. 675–689, Oct 1997; *Proc. ACM SIGCOMM*, Aug. 1996.

14. J. C. R. Bennett, D. C. Stephens, and H. Zhang, "High speed, scalable, and accurate implementation of fair queueing algorithms in ATM networks," *Proc. ICNP*, 1997.

15. D. C. Stephens and H. Zhang, "Implementing distributed packet fair queueing in a scalable switch architecture," *Proc. IEEE INFOCOM*, 1998.

16. F. M. Chiussi and A. Francini, "Implementing fair queueing in ATM switches: the discrete-rate approach," *Proc. IEEE INFOCOM*, 1998.

17. P. van Emde Boas, R. Kaas, and E. Zijlstra, "Design and implementation of an efficient priority queue," *Math. Syst. Theory*, vol. 10, no. 1, pp. 99–127, 1977.

18. D. Picker and R. Fellman, "A VLSI priority packet queue with inheritance and overwrite," *IEEE Trans. Very Large Scale Integration Syst.*, vol. 3, no. 2, pp. 245–252, Jun. 1995.

19. J. Rexford, J. Hall, and K.G. Shin, "A router architecture for real-time point-to-point networks," *Proc. Int. Symp. on computer Architecture*, pp. 237–246, May 1996.

20. S. W. Moon, K. G. Shin, and J. Rexford, "Scalable hardware priority queue architectures for high-speed packet switches," *Proc. Int. Symp. on Real-Time Applications*, Jun. 1997.

21. A. Lyengar and M. E. Zarki, "Switched prioritized packets," *Proc. IEEE GLOBE-COM*, pp. 1181–1186, Nov. 1989.

22. J. Rexford, A. Greenberg, and F. Bonomi, "Hardware-efficient fair queuing architectures for high-speed networks," *Proc. IEEE INFOCOM*, pp. 638–646, Mar. 1996.

23. J. W. Roberts, P. E. Boyer, and M. J. Servel, "A real time sorter with application to ATM traffic control," *Proc. ISS*, pp. 258–262, Apr. 1995.

24. M. R. Hashemi and Alberto Leon-Garcia, "A general purpose cell sequencer/scheduler for ATM switches," *Proc. IEEE INFOCOM*, Apr. 1997.

25. Y. H. Choi, "A queue manager for real-time communication in ATM networks," *Proc. ASAP*, 1997.

26. H. J. Chao, "A novel architecture for queue management in the ATM network," *IEEE J. Select. Areas Commun.*, vol. 9, no. 7, pp. 1110–1118, Sep. 1991.

27. H. J. Chao and N. Uzun, "A VLSI Sequencer chip for ATM traffic shaper and queue manager," *IEEE J. Solid-State Circuits*, vol. 27, no. 11, pp. 1634–1643, Nov. 1992.

28. H. J. Chao and N. Uzun, "An ATM queue manager with multiple delay and loss priorities," *IEEE/ACM Trans. Netw.*, vol. 3, no. 6, pp. 652–659, Dec. 1995.

29. H. J. Chao, H. Cheng, Y. R. Jenq, and D. Jeong, "Design of a generalized priority queue manager for ATM switches," *IEEE J. Select. Areas Commun.*, vol. 15, no. 5, pp. 867–880, Jun. 1997.

30. Y.-r. Jenq. "Design of a fair queueing scheduler for packet switching networks," Ph.D. dissertation. Electrical Engineering Department, Polytechnic University, Brooklyn, NY, Mar. 1998.

31. J. S. Hong. "Design of an ATM shaping multiplexer algorithm and architecture," Ph.D. dissertation, Electrical Engineering Department, Polytechnic University, Brooklyn, NY, Jan. 1997.

32. S. Shenker, C. Partridge, and R. Guerin, "Specification of guaranteed quality of service," RFC 2212, Internet Engineering Task Force (IETF), Sep. 1997.

33. R. Braden, D. Clark, and S. Shenker, "Integrated services in the internet architecture: an overview," Informational, RFC 1633, Internet Engineering Task Force (IETF), Jun. 1994.

34. R. Braden, L. Zhang, S. Berson, S. Herzog, and S. Jamin, "Resource ReSerVation Protocol (RSVP)—version 1 functional specification," RFC 2205, Internet Engineering Task Force (IETF), Sep. 1997.

35. J. Wroclawski, "Specification of the controlled-load network element service," RFC 2211, Internet Engineering Task Force (IETF), Sep. 1997.

36. S. Floyd and V. Jacobson, "Link-sharing and resource management models for packet networks," *IEEE/ACM Trans Netw*, Vol. 3, No. 4, pp. 365–386, Aug. 1995.

37. I. Stoica and H. Zhang, "Providing guaranteed services without per flow management," *Proc. ACM SIGCOMM'99*, Sep. 1999.

38. H. J. Chao, Y. R. Jenq, X. Gao, and C. H. Lam, "Design of packet fair queuing schedulers using a RAM-based searching engine," *IEEE J. Select. Areas Commun.*, pp. 1105–1126, Jun. 1999.

39. S. Suri, G. Varghese, and G. Chandranmenon, "Leap forward virtual clock: a new fair queueing scheme with guaranteed delays and throughput fairness," *Proc. IEEE INFOCOM*, Apr. 1997.

CHAPTER 6

BUFFER MANAGEMENT

Buffer management is the strategy for deciding when and how to discard packets to avoid network congestion. Its performance can be measured in terms of its ability to control traffic fairly and efficiently during periods of congestion. Typically, packet discard decisions are made either upon the arrival of a new packet, or at the onset of congestion, when currently stored packets may be discarded to accommodate a new, higher-priority packet. The information used to make these decisions can be per-flow or per-class buffer accounting, mainly depending on different tradeoffs between performance and implementation complexity [41]. Section 6.1 presents some buffer management mechanisms for the ATM network. The buffer management mechanisms for the Internet are described in Section 6.2.

6.1 A LOOK AT ATM NETWORKS

The buffers in an ATM network are designed to resolve congestion at the cell level. To satisfy cell delay requirements, it is necessary to have short queues within the ATM network. On the other hand, to guarantee the cell loss rate of each virtual channel (VC), there should be sufficient buffers to accommodate bursts from different VCs. Selective cell discarding is needed for the buffer management to drop cells on the basis of their assigned priority when congestion occurs.

In a given ATM connection and for each user-data cell in the connection, the ATM equipment that first emits the cell can set the cell loss priority (CLP) bit in the cell header equal to zero or one [1, 2]. The CLP bit is used to distinguish between cells of an ATM connection: a CLP bit equal to zero indicates a higher-priority cell, and a CLP bit equal to one indicates a lower-priority cell. Upon entering the network, a cell with CLP = 1 may be subject to discard, depending on network traffic conditions [1].

6.1.1 Overview

Three cell discarding strategies for congestion management have been proposed. They are *complete buffer sharing* with *pure pushout* [3], *partial buffer sharing* with *nested thresholds* for different loss priorities [4], and an *expelling policy* [5]. In the pure pushout strategy, all cells are admitted until the buffer is full, as shown in Figure 6.1, where N is the number of input ports of an ATM switch (assuming an output-buffered switch). When the buffer is full, an arriving high-priority cell can push out a low-priority cell. New arrivals may also push out the cells that have the same priority. It is desirable to push out the cells close to the head of queue. This results in smaller average queuing delay (older cells are discarded first), earlier detection of packet corruption, and quicker response for flow control (e.g., TCP).

The pushout scheme is more complicated than the tail-drop scheme (i.e., FIFO). But it still can be easily implemented by creating multiple logical queues in a physical memory, where cells that have the same priority are linked together and confined by the head and tail pointers. When pushout action is to be taken, a head-of-line (HOL) cell from the lowest non-empty priority logic queue is discarded by updating its head pointer. However, this scheme is not able to provide guaranteed cell loss rate for low priority traffic.

In [5], a variant pushout scheme called the *squeeze-out policy* has been proven to be optimal for a two priority system [5], where cells are placed in the buffer with high priority first. When the buffer is full and there are cell arrivals, the low-priority cells are pushed out of the buffer, starting with those closest to the head of the queue. Notice that low-priority cells are dropped to make space for other low-priority cells that are appended to the end of the queue. The squeeze-out policy minimizes the blocking probability of the high-priority (loss sensitive) class among all pushout policies. This result is true for any general arrival process.

In the nested threshold strategy, a threshold TH_p is assigned to each loss priority p, as shown in Figure 6.2, where $p = 1, 2, \ldots, P$, Q is the current

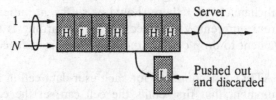

Fig. 6.1 Pure pushout for cell discarding.

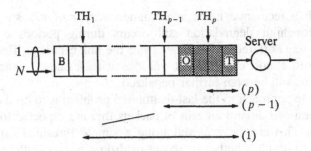

Fig. 6.2 Nested thresholds for cell discarding.

queue length, and B is the buffer size. Cells of a particular class are allowed to enter the buffer only if the current queue length is smaller than the associated threshold. If the queue length $Q \geq TH_p$, an arriving cell with priority p is discarded; otherwise, it is admitted. The thresholds are functions of the traffic characteristics and cell loss rate requirement of each class. This scheme can be easily implemented with a FIFO as the buffer and some registers to keep the threshold values. The drawback of it is that the buffer is not fully utilized and it is difficult to find proper threshold settings to meet each class's cell loss rate.

In the expelling policy [5], in addition to squeezing out low-priority cells, when the number of high-priority cells exceeds a predetermined threshold, all low-priority cells at the front of the queue are expelled until a high-priority cell is found and transmitted. This policy has been shown to perform better than the other two strategies in the case of two loss priority levels, but has higher implementation complexity.

The pushout schemes do not provide guaranteed cell loss rate to low-priority traffic, while the other two schemes have difficulty determining proper thresholds for different priorities under time-varying parameters, such as traffic burstiness and offered load. Since each service class has its own required cell loss rate, an acceptable discarding strategy must guarantee each class's cell loss rate requirement. Section 6.1.2 presents a cell discarding strategy—*self-calibrating pushout* (SCP) [6]—that, by incorporating some elements of the pushout policy, can balance the cell loss rates for every priority while still satisfying their respective loss requirements.

Generally, the means of detecting the onset of congestion and choosing the associated preventive measures vary with the link layer being used. For example, on ATM links an IP packet may be segmented into smaller units, e.g., 53-byte cells. In this case, congestion is measured with respect to the buffer's ability to store cells. A single cell loss will result in an entire packet being rendered useless, so it is advantageous to discard an entire packet's worth of cells at a time. One obvious approach is to drop all subsequent cells from a packet if there is no room in the buffer for an arriving cell. This is termed *packet tail discarding* in [24], and *partial packet discard* (PPD) in [12].

While such a technique helps avoid unnecessary buffer waste, substantial packet throughput degradation still occurs during periods of very high congestion (e.g., offered traffic above twice the link capacity). This is because in such cases, almost every packet will lose at least one cell, and flows with large packets will be even further penalized.

A possible approach to the last-mentioned problem is to predict the onset of congestion and discard all cells of packets that are expected to experience a cell loss. This can be achieved using a simple threshold on the buffer occupancy to decide whether to accept or drop a packet at the time its first cell is received. This is referred to as *early packet discard* (EPD) [13] and has the benefit that link capacity is not wasted in the transmission of partial packets. EPD performs better than PPD and results in an effective throughput close to the link capacity with fewer buffers [13, 23, 24].

However, EPD can be unfair to low-bandwidth flows. In particular, if the buffer occupancy is hovering around the threshold value, then high-rate flows are likely to have more opportunities to get their packets accepted (assuming identical packet sizes). Once a packet is accepted, it pushes the buffer occupancy above the threshold, and a subsequent packet from a low-bandwidth flow is then likely to be dropped. Some of these issues are discussed in [24], which presents a number of enhancements to the basic EPD scheme aimed at improving not only the *goodput*, i.e., the number of complete packets transmitted, but also fairness in selecting flows that can start buffering a new packet.

In general, it is desirable to devise buffer management schemes that preserve the goodput benefits of EPD while also ensuring fairness in how this goodput is distributed across flows. The fairness of a buffer management scheme is a function of how it penalizes packets from nonconforming flows, that is, flows sending at a higher rate than they are entitled to. Ideally, the scheme should ensure that no single flow can grab a disproportionate amount of the buffer space, thereby affecting the performance level of other flows. If per-flow buffer accounting is performed, it is relatively easy to identify misbehaving flows and take appropriate actions to ensure a fair allocation of the buffer space. However, there is a cost associated with per-flow buffer accounting, and in some environments where scalability is a concern, buffer management may need to be done at a coarser level. The penalty is that it is harder to ensure fairness when many flows share a common buffer. In some instances, such as adaptive control (e.g., TCP), it is possible to ensure some level of per-flow fairness without recording the per-flow state. This was one of the goals of the *early selective packet discard* (ESPD) [14] mechanism, which makes sessions (flows) take turns in accessing network capacity by discarding packets from selected sessions rather than randomly. In Section 6.1.3, we introduce the PPD, EPD, and ESPD.

As mentioned above, a static threshold places fixed limits on the amount of buffering available to any individual queue. It would be desirable to have a buffer management scheme with the simplicity of a static threshold (for ease of implementation) and the adaptivity of pushout (for robust performance). It would be better to achieve this adaptivity without explicitly monitoring the fluctuating arrival rate for each individual queue. Therefore, in [15] a scheme called *dynamic threshold* was proposed to fairly regulate the sharing of memory among different output queues for traffic of a single loss priority in a shared-memory packet switch. The dynamic threshold scheme deliberately wastes a small amount of buffer space, but attempts to share the remaining buffer space equally among the active output queues. In Section 6.1.4, we introduce this scheme.

6.1.2 Self-Calibrating Pushout

The main objective of the SCP scheme is to balance the cell loss rates by measuring, in real time, the numbers of discarded low- and high-loss-priority (LP) cells so that each priority will meet its cell loss rate requirement (e.g., 10^{-6} for low LP and 10^{-9} for high LP).

The primary advantage of the SCP scheme over the other proposed cell discarding schemes is its capability of automatically calibrating each service class's cell loss rate in real time so as to meet each class's QoS requirement; hence "self-calibrating." Many proposed cell discarding schemes rely on finding different discarding thresholds offline, based on traffic characteristics and QoS requirements. However, when cell discarding is considered jointly with cell scheduling, finding the thresholds may become very difficult (if not impossible). On the other hand, the SCP scheme can easily be combined with any cell scheduling scheme, because cell discarding is determined in real time by the measured data of the system, not by statistical prediction as in other schemes. In the following sub-subsections, we briefly describe the SCP cell discarding algorithms. SCP cell discarding combined with earliest-due-date (EDD) cell scheduling has been studied in [6], and has been designed using a hardware description language and simulated to function correctly [7].

6.1.2.1 *Mechanism* Let us denote the target cell loss rates of low- and high-LP classes by P_L and P_H, respectively, and their measured cell loss rates by $\hat{P}_L$ and $\hat{P}_H$, respectively. Our goal here is to selectively discard low- or high-LP cells so that the ratio $\hat{P}_L/\hat{P}_H$ approaches P_L/P_H. By balancing the cell loss rates between the low- and high-LP classes, each service class's cell loss requirement is met.

A control parameter, called {loss_weight}, is used in the SCP scheme. It is analogous to the bandwidth weight in weighted round-robin service, where the bandwidth received by each virtual channel (or number of cells transmitted) is proportional to the bandwidth weights. Similarly, in the SCP scheme, the number of discarded cells is proportional to the loss_weight of different loss priorities. Now let us consider two loss priorities, low and high. Let ρ_L and ρ_H be the offered load of low- and high-LP traffic, respectively. The total offered load ρ is equal to $\rho_L + \rho_H$. The loss–weight is defined as

$$\text{loss–weight} = \frac{P_L}{P_H} \times \frac{\rho_L}{\rho_H}. \tag{6.1}$$

For instance, if $P_L = 10^{-6}$, $P_H = 10^{-9}$, and $\rho_L = \rho_H$, then the loss–weight is equal to 1000. In other words, when 1000 low-LP cells have been discarded, a high-LP cell will be the next candidate to be discarded if the buffer is full. A variable called CNT is used to keep track of discarding. Initially, CNT is set to zero. It is incremented by 1 when a low-LP cell is discarded, and decremented by loss–weight when a high-LP cell is discarded.

Another control parameter is the threshold (TH). TH is compared with the number of high-priority cells currently in the buffer. If that number is less than TH, cells of both classes will be admitted into the buffer. If it is greater than TH and the CNT value is less than the loss–weight, we will discard low-LP cells, starting with those closest to the head of the queue. Our performance study shows that the target ratio $\hat{P}_L/\hat{P}_H$ can be kept constant for a wide range of TH-values. That means that the choice of a TH-value to ensure the SCP scheme works properly is not critical, which eliminates the necessity of adjusting TH for different traffic characteristics. This is because the control parameter TH only provides a coarse calibrating guideline to

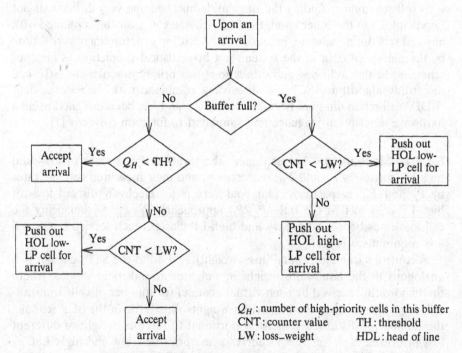

Q_H : number of high-priority cells in this buffer
CNT : counter value   TH : threshold
LW : loss–weight   HDL : head of line

Fig. 6.3 Self-calibrating pushout (SCP) algorithm.

bring $\hat{P}_L$ and $\hat{P}_H$ to the range of interest, while the CNT of discarded cells is the control parameter that fine-tunes the cell loss of low and high LPs to keep $\hat{P}_L/\hat{P}_H$ at a desired value. The algorithm is shown in Figure 6.3.

6.1.2.2 Loss Fairness

As-described above, the SCP scheme balances the cell loss rates of low- and high-loss-priority classes to meet their cell loss rate requirement. However, our ultimate goal is to ensure that each individual connection's cell loss rate requirement is met, and not just the one for each service class. Now that the SCP scheme determines a priority class from which a cell is to be discarded, the remaining problem is to select a connection in that priority class and discard its cell.

The idea is to achieve fair loss among the connections that have the same cell loss rate requirement by combining the SCP with a fair queuing scheduling policy such as generalized processor sharing (GPS) [10], which provides fair service among virtual connections on the same link. As a result, we can easily guarantee each individual connection's cell loss rate requirement under the SCP discarding policy.

In the following, we explain how fair loss can be automatically achieved by adopting the fair queuing scheduling policy. Suppose l ($\leq N$) cells arrive simultaneously when the buffer is full, that is, l is the number of data to be discarded due to buffer overflow. Then, for fairness, we expect the portion of loss from connection i (l_i) to be

$$l_i = \frac{\rho_i}{\sum_{j \in B} \rho_j} l, \tag{6.2}$$

where B is the set of connections that have at least one backlogged cell at the moment of buffer overflow, and ρ_i is the normalized offered load of connection i. What motivates us is the fact that fair queuing fundamentally conforms to (6.2) from the service point of view. The idea based on this observation is to select the cell that is in the position of next transmission as the one to be pushed out. If we regard the execution of discarding as a fictitious momentary transmission (with infinite link rate), then the total number of data served from each connection, which is the sum of the actually served number and the fictitiously transmitted (i.e.,discarded) number, conforms to the proportionality property. The detailed proof of the loss fairness property of the HOL pushout scheme can be found in [8].

6.1.2.3 Performance Study

6.1.2.3.1 Simulation Setup and Traffic Model

Four different service classes, I, II, III, and IV, are considered here and arranged in a priority matrix as shown in Figure 6.4(a). Each service class is associated with a logical queue. Cells in the same logical queue are first-come, first-served.

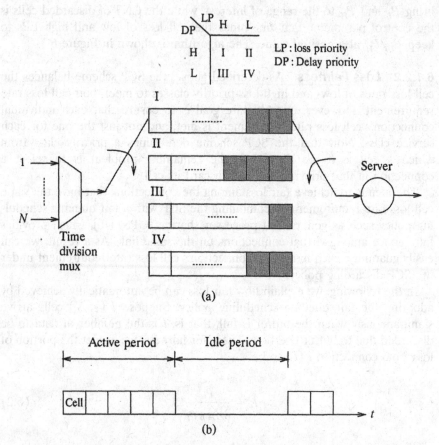

(a)

(b)

Fig. 6.4 (a) Simulated model, (b) cell arrival model.

According to the EDD scheduling policy, the kth cell from connection i, arriving at time a_i^k with local tolerable delay d_i, is assigned a timestamp value $a_i^k + d_i$, called the *due time* (T). The server serves the HOL cell from one of the four logical queues that has the smallest due time. A tie of the due time can be broken by random choice. The packet fair queuing (PFQ) algorithm is not considered here because we just want to investigate each service class's performance, whereas each virtual connection's performance will be met if the PFQ is applied. The total occupancy of these four logical queues is always less than or equal to the buffer size. The time division multiplexer (TDM) distributes cells from an output-buffered ATM switch to a particular logical queue according to their service classes. Cells are stored or discarded in the corresponding logical queue according to the rules described in the SCP algorithm. Namely, high-LP cells are always stored in the buffer when the buffer is not full. However, low-LP cells may be pushed out from the buffer (even if the buffer is not full) when the number of high-LP backlogged cells exceeds TH and CNT is less than the loss–weight.

The performance study assumes the source traffic is bursty and alternates between active and idle periods, as shown in Figure 6.4(b). More specifically, cells arrive in consecutive slots in an active period, and no cells arrive in an idle period. Traffic sources are assumed to be independent. Both the active and idle periods are assumed to be geometrically distributed with mean burst length L_{active} and mean idle time L_{idle}. Thus, the offered load ρ is given by

$$\rho = \frac{L_{active}}{L_{active} + L_{idle}} \tag{6.3}$$

6.1.2.3.2 Cell Loss Performance

Assume that the buffer size $B = 512$, the total offered load $\rho = \rho_H + \rho_L = 0.8$, where $\rho_H = \rho_I + \rho_{III}$ and $\rho_L = \rho_{II} + \rho_{IV}$, and the average burst length $L_{active} = 15$ cells, $P_H = 10^{-6}$, and $P_L = 10^{-3}$. The traffic mix ratio, ρ_H/ρ_L, varies from 0.1 to 10.

Figure 6.5 shows the loss probabilities, $\hat{P}_L$ and $\hat{P}_H$, vs. the traffic mix ratio ranging from 1 to 10 with different TH-values. With TH set to 64, 128, and 192, the ratio P_L/P_H can be maintained around 1000, shown by the equal spacing of the two curves in each case. This means when a cell of low or high LP is to be discarded, most of the time its associated logical queue is not empty. As a result, the measured loss probability ratio, $\hat{P}_H/\hat{P}_L$, can be kept to the target value. $\hat{P}_H$ is in the order of 10^{-6}, and $\hat{P}_L$ is in the order of 10^{-3}. Note that in Figure 6.5 both $\hat{P}_H$ and $\hat{P}_L$ increase when ρ_H/ρ_L is larger than 3. This is because when ρ_H/ρ_L becomes larger, high-LP cells will dominate. It is likely that cells in the buffer are all high-LP cells when the buffer is full. Thus, when a high-LP cell arrives and finds the buffer full, it will be discarded because there are no low-LP cells in the buffer; thus $\hat{P}_H$ will increase. This will make the CNT more negative, because the CNT is decremented by loss-weight when a high LP cell is discarded. On the other hand, since most of the time the number of high-LP backlogged cells exceeds TH, low-LP cells are blocked more often. Thus, the $\hat{P}_L$ is also increased. When ρ_H/ρ_L is larger than 4, $\hat{P}_H$ and $\hat{P}_L$ become more stable. Thus, we can conclude that because of TH, the ratio of loss probabilities, $\hat{P}_H/\hat{P}_L$, can be maintained as constant.

Figure 6.6 shows $\hat{P}_L$ and $\hat{P}_H$ vs. the traffic mix ratio varying from 0.1 to 1 with TH set to 64 and 128. As expected, the ratio still remains around 1000. The reason that the results for TH = 192 are not shown in Figure 6.6 is that $\hat{P}_H$ is too small to be obtained from computer simulations in a reasonable time. Also note that the smaller the traffic mix ratio ρ_H/ρ_L is, the less likely it is that the condition $Q_H \geq$ TH will occur. For ρ_H/ρ_L from 0.1 to 0.3, $\hat{P}_H$ and $\hat{P}_L$ fluctuate. The reason is the same as before, but now the dominating cells in the buffer are all of low LP. Note that both $\hat{P}_H$ and $\hat{P}_L$ in this case are smaller than in the case where the traffic mix ratio varies from 1 to 10. This is because low-LP cells prevail in the buffer and are always available when discarding is needed. Therefore, high-LP cells will not be mistakenly discarded due to the lack of low-LP cells in the buffer. This decreases the

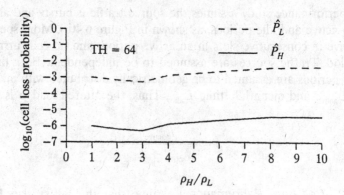

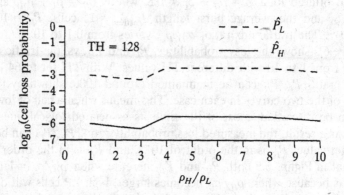

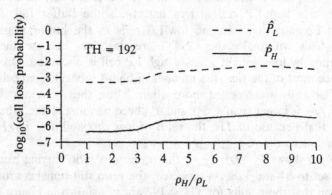

Fig. 6.5 Cell loss probabilities vs. traffic mix ratio ρ_H/ρ_L from 1 to 10, with $\rho_H + \rho_L = 0.8$, buffer size = 512, and $L_{\text{active}} = 15$. (©1994 IEEE.)

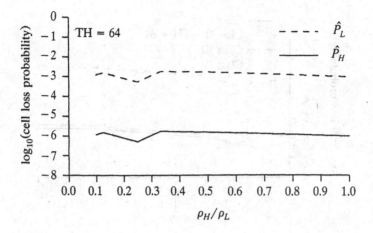

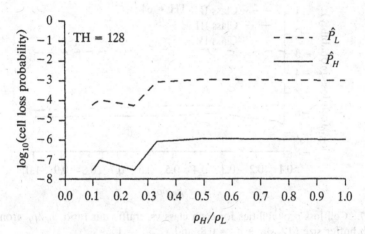

Fig. 6.6 Cell loss probabilities vs. trafic mix ratio ρ_H/ρ_L from 0.1 to 1, with buffer size 512, $\rho_H + \rho_L = 0.8$, and $L_{\text{active}} = 15$. (©1994 IEEE.)

number of discarded cells of high-LP and makes $\hat{P}_H$ smaller. Meanwhile, since low-LP cells dominate in the buffer, the number of high-LP backlogged cells will usually be below TH, and low-LP cells will not be blocked from entering the buffer. As a result, $\hat{P}_L$ is smaller than in the case of the traffic mix ratio ranging from 1 to 10.

From Figure 6.5 and Figure 6.6, we notice that for a traffic mix ratio ρ_H/ρ_L from 0.1 to 10, the target ratio $\hat{P}_L/\hat{P}_H$ can be kept constant for a large range of TH-values. That means that the choice of a TH-value to ensure the SCP scheme works properly is not critical, eliminating the necessity of adjusting the TH for different traffic distributions. This is because TH as a control parameter provides only a coarse calibration to bring $\hat{P}_L$ and $\hat{P}_H$ to the range of interest.

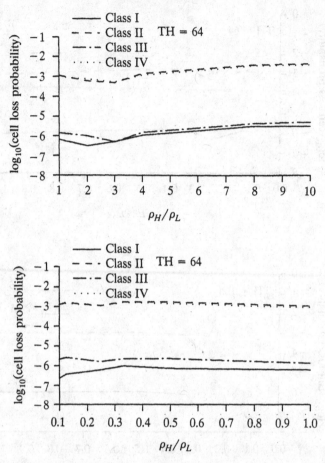

Fig. 6.7 Cell loss probabilities for each class vs. traffic mix ratio ρ_H/ρ_L from 0.1 to 10 with buffer size 512, $\rho_H + \rho_L = 0.8$, and $L_{active} = 15$.

Figure 6.7 shows the cell loss probabilities for each service class vs. the traffic mix ratio ranging from 1 to 10 and from 0.1 to 1. TH is set to 64. In both cases, we can see that class II and class IV (low-LP classes) have almost the same loss rates, while class I and class III (high-LP classes) are very close to each other. The discrepancy between class I and class III is due to the difference of their delay priorities. Since class I cells have higher delay priority, they have a larger chance of being served than class III cells have. As a result, class III cells will stay in the buffer longer than class I cells and have a higher probability of being pushed out upon a turn to discard a high-LP cell. This is why class III has a larger cell loss rate than class I. One way to eliminate the discrepancy between the service classes that have the same loss priority is to provide a separate counter for each class. By keeping track of each individual's counter value, we will be able to maintain the same cell loss rate for the service classes with the same LP. However, this will increase the implementation complexity. Since the discrepancy is small, it is appropriate to combine the cell losses from classes I and III, and those from classes II and IV, to simplify the implementation.

6.1.2.3.3 Cell Delay Performance Let us assume that the 99th-percentile delay requirement for high-DP cells is 150 μs and for low-DP cells it is 5 ms [9]. For an OC-3 (155.52 Mbit/s) transmission line, each cell time is about 2.83 μs. Then, the delay requirements for high and low DP are 53 and 1767 cell slots, respectively. Note that cells from the same class are served in a first-come, first-served manner because their due times are their arrival times plus the same tolerable delay.

We use the *delay probability mass function* (pmf) for each class to show its delay distributions and see the percentage of cells violating their delay requirements. The pmf usually gives us more useful information in practice than the average delay. Figure 6.8 shows the delay distributions of classes I, II, III, IV, where the traffic mix ratio ρ_H/ρ_L is equal to 1. For classes I and II with a higher DP, 2% of cells violate the delay requirement (150 μs, or 53 cell slots). For classes III and IV with a lower DP, less than 1% of the cells violate the delay requirement (5 ms, or 1767 cell slots). The delay distributions for the traffic mix ratio ρ_H/ρ_L varying from 0.1 to 10 are similar to Figure 6.8.

6.1.3 TCP/IP over ATM–UBR

For non-real-time data traffic (e.g., TCP/IP), the ATM Forum has defined two different services: available bit rate (ABR) and unspecified bit rate (UBR) [11]. Unlike ABR, which uses explicit feedback-based flow control, the only way UBR can influence the QoS is to discard the cells intelligently at the ATM buffer. Several schemes have been proposed, including partial packet discard (PPD) [12], early packet discard (EPD) [13], and early selective packet discard (ESPD) [14], as described below.

6.1.3.1 Partial Packet Discard In PPD, an ATM switch drops all subsequent cells from a packet as soon as one cell of the packet has been dropped. In ATM adaptation layer (AAL) 5, the last cell of a packet can be identified by the payload type identifier (3 bits) in the cell header and is regarded as an end of message (EOM) cell. Once the last cell of a packet is received, or at the beginning of the connection, the very first cell of the virtual channel (VC) is regarded as the first cell of the following packet as well as a beginning of message (BOM) cell. Any intermediate cell of a packet is regarded as a continuation of message (COM) cell.

The switch maintains a *drop list* containing a set of VCs for which any cell of a packet has been dropped already. When a cell arrives at an ATM buffer, if the cell's virtual path identifier (VPI) or virtual channel identifier (VCI) belongs to the drop list, that cell is discarded. If it is an EOM cell, its VPI or

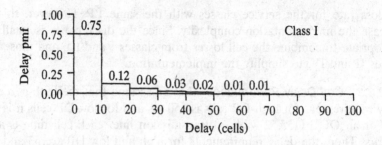

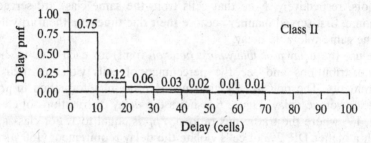

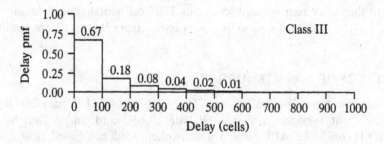

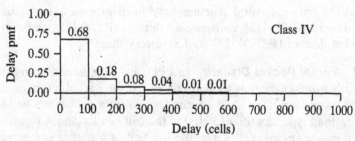

Fig. 6.8 Delay distributions for classes I, II, III, and IV, with buffer size 512, $\rho_H + \rho_L = 0.8$, $\rho_H / \rho_L = 1$, $L_{active} = 15$, and TH = 192. (©1994 IEEE.)

VCI can be removed from the drop list, indicating that all subsequent cells of a packet have been dropped. If the cell's VPI or VCI is not in the drop list, the cell is admitted into the buffer if the buffer is not full; otherwise, the cell is discarded and its VPI or VCI is put into the drop list, indicating that one cell of a packet for this VC has been dropped.

As we can see, the PPD discards an incoming cell when a buffer is full and also discards all the subsequent cells of a corrupted packet (i.e., a packet that has lost cells). In doing so, we can discard either the tail end of a packet or an entire packet. Thus, we can reduce the number of useless cells to a certain degree. However, since cells dropped by a switch might belong to a packet that contains cells already transmitted, a congested link can still transmit a significant portion of cells that belong to corrupted packets. A better policy is to drop cell(s) *earlier* than when the buffer is full, as described next.

6.1.3.2 Early Packet Discard

EPD was proposed [13] with the aim of discarding an entire packet prior to buffer overflow, so that corrupted packets will not be transmitted by the switches. The EPD mechanism sets a buffer threshold. As soon as the buffer occupancy exceeds the threshold, the ATM switch is *ready to discard* incoming cells. That is, instead of immediately discarding any incoming cell, the switch waits for an incoming BOM cell, drops it first, and continues to discard all the following cells of the same packet up to the EOM cell.

The switch also maintains a drop list, as in PPD. When a cell arrives at an ATM buffer and the cell's VPI or VCI belongs to the drop list, this cell (unless it is the EOM cell) is discarded. The EOM cell is discarded only if the buffer is full. If the incoming cell is an EOM cell, its VPI or VCI can be removed from the drop list, indicating all subsequent cells of a packet have been received or dropped properly.

Suppose the incoming cell's VPI or VCI is not in the drop list. Cells are always admitted to the buffer if the buffer occupancy is below the threshold. If the buffer occupancy exceeds the threshold but the buffer is not full, arriving BOM cells are discarded. Once a cell of a packet for the VC has been dropped, its VPI or VCI is put into the drop list. By doing so and by setting the proper threshold, we can emulate packet discarding almost perfectly.

It has been shown that the performance of TCP over ATM using EPD is better than that with PPD [23] (note that TCP congestion control relies on the information of packet loss to adjust the source's window size). However, when congestion occurs, EPD starts to drop the incoming packet irrespective of which session it belongs to, even though some highly active session is consuming more resources than others. ESPD [14], described below, is intended to solve this problem.

6.1.3.3 Early Selective Packet Discard

The ESPD scheme has a *drop timer*, a high buffer threshold (HB), a low buffer threshold (LB), and a drop list threshold (DL). HB is designed to capture VCs into the drop list, and LB is used to release VCs from the drop list. DL limits the number of VCs in the drop list during every congestion period. The drop timer is used to refresh the drop list during a congestion period. For example, HB can be several packets less than the buffer size, LB can be half the buffer size, DL can be set to 20% of the total number of sessions at a switch, and the drop timer can be set to half the minimum retransmission time-out value of TCP [14].

If the buffer occupancy exceeds HB, the VC is added to a drop list in the order of the corresponding BOM cell arrivals. All the subsequent cells except EOM cells are discarded, even though there is still buffer space available. When the drop list is full (i.e., the number of VCs reaches DL), no new VC is captured unless the buffer is full. When this happens, incoming cells are discarded, their VPIs or VCIs are captured into the drop list on a FIFO basis, and all subsequent cells belonging to a listed VC are discarded, as EPD does. Once the buffer occupancy falls below LB and an EOM cell arrives, the cell's VPI or VCI (if listed) is released from the drop list.

The drop timer is activated as soon as the first VC is captured in the drop list, and deactivated when any VC is released from the drop list. ESPD releases all the captured VCs simultaneously when the drop timer expires. The timer is designed to ensure fairness between sessions. Because ESPD is trying to discard packets from a few sessions only, it is possible that the buffer occupancy will not go below LB for an excessively long period, so that the targeted sessions in the drop list may lose fair access to network resources. For this reason, the target sessions are removed from the drop list if the drop timer expires. In [14], it has been shown that ESPD can enhance fairness and significantly improve throughput over EPD with only a modest increase of implementation complexity.

6.1.4   Dynamic Threshold with Single Loss Priority

The dynamic threshold with single loss priority in a shared memory [15] works as follows. Define the port-i occupancy $Q^i(t)$ as the length of the queue at port i at time t, and define the total occupancy $Q(t)$ as the sum of all the port queue lengths in the shared memory. The algorithm maintains a control threshold $T(t)$, whose value is some multiple α of the unused buffer space. In other words,

$$T(t) = \alpha[B - Q(t)] = \alpha\left[B - \sum_i Q^i(t)\right],$$

where B is the size of the shared buffer memory and $\alpha > 0$. Each output queue i attempts to limit its length to the control threshold by rejecting arriving cells whenever $Q^i(t) \geq T(t)$.

The dynamic threshold scheme adapts to changes in traffic conditions. Whenever the load changes, the system goes through a transient. For example, when a lightly loaded output port suddenly becomes active, its queue grows, the total buffer occupancy goes up, the control threshold goes down, and queues exceeding the threshold block their arrivals temporarily while they drain, freeing up more buffer space for the newly active queue. If there are S very active queues, then in *steady state*, their port occupancies Q^i and the control threshold T equal $\alpha(B - \Omega)/(1 + \alpha S)$, where Ω is the space occupied by queues below the control threshold. The amount of memory held in reserve by the algorithm will be $(B - \Omega)(1 + \alpha S)$. If $\alpha = 2$ and $S = 10$, for instance, then each of the ten queues takes $\frac{2}{21}$ of the buffer, and $\frac{1}{21}$ is left unallocated.

The dynamic threshold algorithm deliberately holds a small amount of buffer space in reserve in steady state. This "wasted" buffer space actually serves two useful functions. The first is that it provides a cushion during transient periods when an output queue first becomes active. This reduces cell loss for the newly active queue during such transients. Secondly, when an output queue has such a load increase and begins taking over some of the spare buffer space, this action signals the buffer allocation mechanism that the load conditions have changed and that a threshold adjustment is now required. If there were no built-in spare buffering, then the arrival rates and/or loss rates of the individual output queues would have to be monitored to determine when load conditions had changed. Such a monitoring scheme would be a complex undertaking.

While the dynamic threshold concept was originally applied to port queues, the concept can also be applied to finer-grained queues, such as those corresponding to priority classes or even individual connections.

6.2 A LOOK AT THE INTERNET

The Internet Protocol (IP) architecture is based on a connectionless end-to-end packet service using the IP. The advantages of its connectionless design, flexibility, and robustness, have been amply demonstrated. However, these advantages are not without cost: careful design is required to provide good service under heavy load. In fact, lack of attention to the dynamics of packet forwarding can result in severe service degradation, or *Internet meltdown*. This phenomenon was first observed during the early growth phase of the Internet of the mid 1980s [25], and is technically called *congestion collapse*.

The original fix for Internet meltdown was provided by Jacobson. Beginning in 1986, he developed the congestion avoidance mechanisms that are now required in TCP implementations [26, 27]. These mechanisms operate in the hosts to cause TC connections to back off during congestion. We say that TCP flows are *responsive* to congestion signals (i.e., dropped packets) from the network. It is primarily these TCP congestion avoidance algorithms that prevent the congestion collapse of today's Internet.

However, that is not the end of the story. Considerable research has been done on Internet dynamics since 1988, and the Internet has grown. It has become clear that the TCP congestion avoidance mechanisms [28], while necessary and powerful, are not sufficient to provide good service in all circumstances. Basically, there is a limit to how much control can be accomplished from the edges of the network. Beside packet scheduling algorithms, buffer management mechanisms are needed in the routers to complement the end point congestion avoidance mechanisms. Below we present some examples of buffer management in the Internet.

6.2.1 Tail Drop

Tail drop, the traditional technique for managing router queue lengths, sets a maximum length (in packets) for each queue, accepts packets for the queue until the maximum length is reached (i.e., we say the queue is full), then drops subsequent incoming packets until the queue decreases because a packet from the queue has been transmitted.

Tail drop is very simple, but it has two important drawbacks. First, in some situations tail drop allows a single connection or a few flows to monopolize queue space, preventing other connections from getting room in the queue. This *lockout* phenomenon is often the result of synchronization or other timing effects.

Second, tail drop allows queues to maintain a full (or almost full) status for long periods of time, since tail drop signals congestion (via a packet drop) only when the queue has become full. It is important to reduce the steady-state queue size for queue management, because even though TCP·constrains a flow's window size, packets often arrive at routers in bursts [30]. If the queue is full or almost full, an arriving burst will cause multiple packets to be dropped. This can result in a global synchronization of flows throttling back, followed by a sustained period of lowered link utilization, reducing overall throughput.

The point of buffering in the network is to absorb data bursts and to transmit them during the (hopefully) ensuing bursts of silence [29]. This is essential to permit the transmission of bursty data. According to [29], queue limits should not reflect the steady-state queues we want maintained in the network; instead, they should reflect the size of bursts we need to absorb.

6.2.2 Drop on Full

Besides tail drop, there are two alternative drop-on-full disciplines: *random drop on full* and *drop front on full*. Under the random drop-on-full discipline, a router drops a randomly selected packet from the queue when the queue is full and a new packet arrives. Under the drop-front-on-full discipline [51], the router drops the packet at the front of the queue when the queue is full and a new packet arrives. Both of these solve the lockout problem, but neither solves the full-queues problem described above.

In the current Internet, dropped packets serve as a critical mechanism of congestion notification to end nodes. The solution to the full-queues problem is for routers to drop packets before a queue becomes full, so that end nodes can respond to congestion before buffers overflow. This proactive approach is called *active queue management* according to [29]. The next section introduces an example—random early detection.

6.2.3 Random Early Detection

Random early detection (RED) [18] drops arriving packets probabilistically. The probability of drop increases as the estimated average queue size grows. Thus, if the queue has been mostly empty in the recent past, RED won't tend to drop packets unless the queue overflows. On the other hand, if the queue has recently been relatively full, indicating persistent congestion, newly arriving packets are more likely to be dropped [18, 29].

6.2.3.1 The RED Algorithm The RED algorithm itself consists of two main parts: estimation of the average queue size and the decision of whether or not to drop an incoming packet. The RED calculates the average queue size avg, using a low-pass filter with an exponentially weighted moving average. The average queue size is compared with two thresholds, a minimum threshold min_{th} and a maximum threshold max_{th}. When the average queue size is less than the minimum threshold, no packets are marked. When the average queue size is greater than the maximum threshold, every arriving packet is marked. If marked packets are in fact dropped, or if all source nodes are cooperative, this ensures that the average queue size does not significantly exceed the maximum threshold.

When the average queue size is between the minimum and the maximum threshold, each arriving packet is marked with probability p_a, which is a function of the average queue size avg. Each time a packet is marked, the probability that a packet is marked from a particular connection is roughly proportional to that connection's share of the bandwidth at the router. The general RED algorithm is given in Figure 6.9.

```
for each packet arrival
    calculate the average queue size avg
    if min_th ≤ avg < max_th
        calculate probability p_a
        with probability p_a:
            mark the arriving packet
    else if max_th ≤ avg
        mark the arriving packet
    else
        admit the arriving packet in the buffer.
```

Fig. 6.9 General algorithm for RED routers.

Thus the RED router has two separate algorithms. The algorithm for computing the average queue size determines the degree of burstiness that will be allowed in the router queue. The algorithm for calculating the packet-marking probability determines how frequently the router marks packets, given the current level of congestion. The goal is for the router to mark packets at fairly evenly spaced intervals to avoid biases and global synchronization, and to mark packets sufficiently frequently to control the average queue size.

The detailed algorithm for the RED is given in Figure 6.10. The router calculates avg at each packet arrival using

$$avg \leftarrow avg + w \cdot (q - avg)$$

When the queue is empty (the idle period), it considers this period by estimating the number m of small packets that *could* have been transmitted during this idle period,

$$avg \leftarrow (1 - w)^m \, avg,$$

where m is equal to the queue idle time ($time - q\_time$) divided by the small packet transmission time (s), as shown in Figure 6.10. That is, after the idle period, the router computes the average queue size as if m packets had arrived at an empty queue during that period.

As avg varies from min_{th} to max_{th}, the packet-marking probability (or drop probability) p_b varies linearly from 0 to P_{max}:

$$p_b \leftarrow P_{max}(avg - min_{th})/(max_{th} - min_{th}),$$

as illustrated in Figure 6.11. The final packet-marking probability p_a increases slowly as the count (the number of packets) increases since the last marked packet:

$$p_a \leftarrow p_b/(1 - count \cdot p_b),$$

which ensures that the router does not wait too long before marking a packet. The larger the count, the higher the marking probability. The router marks each packet that arrives at the router when the average queue size avg exceeds max_{th}.

One option for the RED router is to measure the queue in bytes rather than in packets. With this option, the average queue size accurately reflects the average delay at the router. When this option is used, the algorithm should be modified to ensure that the probability that a packet is marked is proportional to the packet size in bytes:

$$p_b \leftarrow P_{max}(avg - min_{th})/(max_{th} - min_{th})$$
$$p_b \leftarrow p_b \cdot PacketSize/MaximumPacketSize$$
$$p_a \leftarrow p_b/(1 - count \cdot p_b)$$

Saved Variables:

avg :	average queue size
q_time :	start of the queue idle time
$count$:	number of packets since last dropped packet

Fixed Parameters:

w :	queue weight
min_{th} :	minimum queue length threshold
max_{th} :	maximum queue length threshold
P_{max} :	maximum value for p_b
s :	typical transmission time (of a small packet)

Others:

q :	current queue size
p_a :	current packet-marking probability
$time$:	current time

Initialization:
    $count \leftarrow -1$
    $avg \leftarrow 0$
for each packet arrival
    calculate the new average queue size avg :
        if the queue is nonempty
            $avg \leftarrow avg + w \cdot (q - avg)$
        else
            $m = (time - q\_time)/s$
            $avg \leftarrow (1 - w)^m avg$
    if $min_{th} \leq avg < max_{th}$
        increment $count$
        calculate probability p_a :
            $p_b \leftarrow P_{max}(avg - min_{th})/(max_{th} - min_{th})$
            $p_a \leftarrow p_b/(1 - count \cdot p_b)$
        with probability p_a :
            mark the arriving the packet
            $count \leftarrow 0$
    else if $max_{th} \leq avg$
        mark the arriving packet
        $count \leftarrow 0$
    else $count \leftarrow -1$
when queue becomes empty
    $q\_time \leftarrow time$

Fig. 6.10 Detailed algorithm for RED routers.

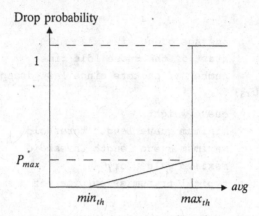

Fig. 6.11 Random early detection (RED) algorithm.

In this case, a large FTP packet is more likely to be marked than is a small TELNET packet.

The queue weight w is determined by the size and duration of bursts in queue size that are allowed at the router. The minimum and maximum thresholds min_{th} and max_{th} are determined by the desired average queue size. The average queue size that makes the desired tradeoffs (such as the tradeoff between maximizing throughput and minimizing delay) depends on network characteristics [18].

6.2.3.2 Performance This sub-subsection presents a simple simulation with RED. The model consists of a RED router fed by four FTP sources and connected to a common sink, where the link propagation delays between the FTP sources and the router are 1, 4, 5, and 8 ms, respectively, and that between the router and the sink is 2 ms. The link speed between each source and the router is 100 M/bits, while the link speed between the router and the sink is 45 M/bits. FTP sources always have a packet to send and always send a maximal-sized (1000-byte) packet as soon as the congestion control window allows them to do so. The sink immediately sends an ACK packet when it receives a data packet. The router uses FIFO queuing.

Source and sink nodes implement a TCP congestion control algorithm in which there are two phases to the window-adjustment algorithm. A threshold is set initially to half the receiver's advertised window. In the *slow-start* phase, the current window is doubled each round trip time until the window reaches the threshold. Then the *congestion-avoidance* phase is entered, and the current window is increased by roughly one packet each round-trip time. The window is never allowed to increase to more than the receiver's advertised window, which we refer to as the *maximum window size*. Packet loss (a dropped packet) is treated as a *congestion-experienced* signal. The source reacts to a packet loss by setting the threshold to half the current window,

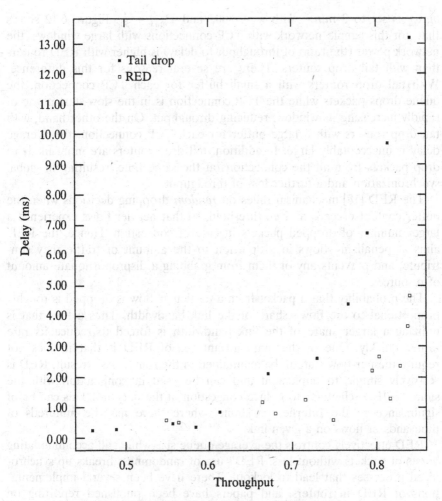

Fig. 6.12 Comparing tail drop and RED routers.

decreasing the current window to one packet, and entering the slow-start phase. More details about TCP congestion control are in Chapter 8.

In Figure 6.12 the x-axis shows the total throughput as a fraction of the maximum possible throughput on the congested link. The y-axis shows the average queuing delay. The simulations with Tail drop router were run with the buffer size ranging from 15 to 140 packets. As the buffer size is increased, the throughput and the average delay increase correspondingly. To avoid phase effects [17, 18] in the simulations with tail drop routers, the source node takes a random time drawn from the uniform distribution on $[0, t]$ to prepare an FTP packet for transmission, where t is the bottleneck service time of 0.17 ms [17].

The simulations with RED routers were all run with a buffer size of 100 packets, with min_{th} ranging from 3 to 50 packets. For the RED routers,

max_{th} is set to 3 min_{th}, with $w = 0.002$ and $P_{max} = \frac{1}{50}$. Figure 6.12 shows that, for this simple network with TCP connections with large windows, the network power (the ratio of throughput to delay) is higher with RED routers than with tail drop routers. There are several reasons for this difference. With tail drop routers with a small buffer for each TCP connection, the queue drops packets while the TCP connection is in the slow-start phase of rapidly increasing its window, reducing throughput. On the other hand, with tail drop routers with a large buffer for each TCP connection, the average delay is unacceptably large. In addition, tail drop routers are more likely to drop packets from all the connections at the same time, resulting in global synchronization and a further loss of throughput.

The RED [18] mechanism relies on *random* dropping decisions when the buffer content exceeds a given threshold, so that heavier flows experience a larger number of dropped packets in case of congestion. Hence, the RED aims at penalizing flows in proportion to the amount of traffic they contribute, and prevents any of them from grabbing a disproportionate amount of resources.

The probability that a packet from a particular flow is dropped is roughly proportional to the flow's share of the link bandwidth. Thus a flow that is utilizing a larger share of the link bandwidth is forced to reduce its rate rather quickly. One of the main advantages of RED is that it does not require the per-flow state to be maintained in the router. As a result, RED is relatively simple to implement and can be used in conjunction with the simple (FIFO scheduler to reduce congestion in the network. This can be of significance in the Internet backbone, where there may be hundreds of thousands of flows on a given link.

RED effectively controls the average queue size while still accommodating bursts of packets without loss. RED's use of randomness breaks up synchronized processes that lead to lockout. There have been several implementations of RED in routers, and papers have been published reporting on experience with these implementations [21, 33, 34]. For example, weighted RED (WRED) [21] combines the capabilities of the RED algorithm with IP precedence (a 3-bit field in the IP packet header). This combination provides for preferential traffic handling for high-priority packets. It can selectively discard lower-priority traffic when the router interface starts to get congested, and can provide differentiated performance characteristics for different classes of service. All available empirical evidence shows that the deployment of active queue management mechanisms in the Internet would have substantial performance benefits [29].

6.2.4 Differential Dropping: RIO

6.2.4.1 *Concept* RIO stands for RED routers with In-Out bit [35, 36]. It is designed to support assured service in the Diffserv Internet [37]. The general approach of this service mechanism is to define a *service allocation* *profile* for each user, and to design a mechanism (e.g., RIO) in the router that favors traffic that is within those service allocation profiles. The basic idea is to monitor the traffic of each user as it enters the network, tag packets as either In or Out of their service allocation profiles, and then, at each congested router, preferentially drop packets that are tagged as being Out. The idea of using In–Out is the same as used with the cell loss priority (CLP) bit in ATM networks.

Inside the network, at the routers, there is no separation of traffic from different users into different flows or queues. The packets of all users are aggregated into one queue, just as they are today. Different users can have very different profiles, which will result in different users having different numbers of In packets in the service queue. A router can treat these packets as a single common pool. This attribute of the scheme makes it very easy to implement.

6.2.4.2 *Mechanism*

RIO uses the same mechanism as RED, but is configured with two sets of parameters, one for In packets and the other for Out packets. By choosing the parameters for respective algorithms differently, RIO is able to discriminate against Out packets in times of congestion and preferentially drop Out packets.

In particular, upon each packet arrival at the router, the router checks whether the packet is tagged as In or Out. If it is an In packet, the router calculates avg_{in}, the average queue size for the In packets; if it is an Out packet, the router calculates avg_{total}, the average queue size for all (both In and Out) arriving packets. The probability of dropping an In packet depends on avg_{in}, and the probability of dropping an Out packet depends on avg_{total}.

As shown in Figure 6.13, there are three parameters for each of the twin algorithms. The three parameters, min_{in}, max_{in}, and $P_{max_{in}}$, define the

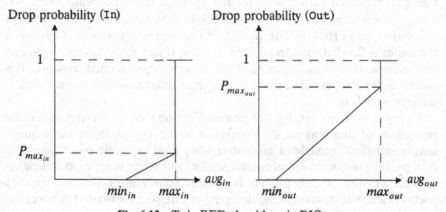

Fig. 6.13   Twin RED algorithms in RIO.

```
For each packet arrival
    if it is an In packet
        calculate the average In queue size avg_in;
    else calculate the average queue size avg_total;

If it is an In packet
    if min_in ≤ avg_in < max_in
        calculate probability P_in;
        with probability P_in drop this packet;
    else if max_in ≤ avg_in
        drop this packet.

If it is an Out packet
    if min_out ≤ avg_total < max_out
        calculate probability P_out;
        with probability P_out drop this packet;
    else if max_out ≤ avg_total
        drop this packet.
```

Fig. 6.14 RIO algorithm.

normal operation $[0, min_{in})$, congestion avoidance $[min_{in}, max_{in})$, and congestion control $[max_{in}, \infty)$ phases for In packets. Similarly, min_{out}, max_{out}, and $P_{max_{out}}$ define the corresponding phases for Out packets.

The discrimination against Out packets in RIO is created by carefully choosing the parameters. First, by choosing $min_{out} < min_{in}$ the RIO router drops Out packets much earlier than it drops In packets. Second, in the congestion avoidance phase, it drops Out packets with a larger probability, by setting $P_{max_{out}} > P_{max_{in}}$. Third, it goes into congestion control phase for the Out packets much earlier than for the In packets, by choosing $max_{out} \ll max_{in}$. In essence, RIO drops Out packets first when it detects incipient congestion, and drops all Out packets if the congestion persists. Only when the router is flooded with In packets, as a last resort, does it drop In packets in the hope of controlling congestion. In a well-provisioned network, this should never happen. If it does, it is a clear indication that the network is underprovisioned.

Figure 6.14 shows the RIO algorithm. By using avg_{total} to determine the probability of dropping an Out packet, a router can maintain short queue length and high throughput no matter what kind of traffic mix it has. The Out packets represent opportunistic traffic, and there is no valid indication of what number of Out packets is proper. Simply using the average Out packet queue size to control the dropping of Out packets would not cover the case where the whole queue is growing due to arriving In packets.

6.2.5 Fair Random Early Detection (FRED)

6.2.5.1 Concept   As pointed out in [20], RED does not always ensure all flows a fair share of bandwidth. In fact, RED is unfair to low-speed TCP flows. This is because RED randomly drops packets when the maximum threshold is crossed, and it is possible that one of these packets belongs to a flow that is currently using less than its fair share of bandwidth. Since TCP reacts rather strongly to packet loss, the lost packet will force further reduction in the congestion window resulting in an even lower rate. The fair random early detection (FRED) mechanism, presented in [20] as a modification to RED, intends to reduce some of its unfairness. Basically, FRED generates selective feedback to a filtered set of connections that have a large number of packets queued.

6.2.5.2 Mechanism   In brief, FRED acts just like RED, but with the following additions. FRED introduces the parameters min_q and max_q, which are the minimum and maximum number of packets each flow should be allowed to buffer. FRED introduces the global variable $avgcq$, an estimate of the average per-flow buffer count; flows with fewer than $avgcq$ packets queued are favored over flows with more. FRED maintains a count $qlen$ of buffered packets for each flow that currently has any packets buffered. FRED maintains a variable $strike$ for each flow, which counts the times the flow has failed to respond to congestion notification; FRED penalizes flows with high $strike$ values.

FRED allows each connection to buffer min_q packets without loss. All additional packets are subject to RED's random drop. An incoming packet is always accepted if the connection has fewer than min_q packets buffered and the average buffer size is less than max_{th}. Normally, a TCP connection sends no more than three packets back to back: two because of delayed ACK, and one more due to a window increase. Therefore, min_q is set to two to four packets.

When the number of active connections is small ($N \ll min_{th}/min_q$), FRED allows each connection to buffer min_q packets without dropping. It also dynamically raises min_q to the average per-connection queue length ($avgcq$). For simplicity, it calculates this value by dividing the average queue length (avg) by the current number of active connections. A connection is active when it has packets buffered, and is inactive otherwise.

FRED never lets a flow buffer more than max_q packets, and counts the times each flow tries to exceed max_q in the per-flow $strike$ variable. Flows with high $strike$ values are not allowed to queue more than $avgcq$ packets; that is, they are not allowed to use more packets than the average flow. This allows adaptive flows to send bursts of packets, but prevents nonadaptive flaws from consistently monopolizing the buffer space.

The original RED estimates the average queue length at each packet arrival. In FRED, the averaging is done at arrival and departure. Therefore, the sampling frequency is the maximum of the input and output rate, which helps reflect the queue variation accurately. In addition, FRED does not modify the average if the incoming packet is dropped unless the instantaneous queue length is zero. Without this change, the same queue length could be sampled multiple times when the input rate is substantially higher than the output link rate. This change also prevents an abusive user from defeating the purpose of the low-pass filter, even if all his/her packets are dropped. Interested readers are referred to [20] for further details of the FRED algorithm.

6.2.6   Stabilized Random Early Detection (SRED)

6.2.6.1   Concept   Similar to RED, stabilized RED (SRED) [22] preemptively discards packets with a load-dependent probability when a buffer in a router seems congested. SRED has an additional feature that, over a wide range of load levels, helps it stabilize its buffer occupation at a level independent of the number of active connections. SRED does this by estimating the number of active connections or flows. This estimate is obtained without collecting or analyzing state information on individual flows, as FRED does [20].

The main idea is to compare, whenever a packet arrives at some buffer, the arriving packet with a randomly chosen packet that recently preceded it into the buffer. When the two packets are of the same flow, we declare a *hit*. The sequence of hits is used in two ways, and with two different objectives in mind:

- To estimate the number of active flows
- To find candidates for misbehaving flow

The definition of *hit* can be flexible. The strongest plausible requirement is to declare a hit only when the two packets indeed are of the same flow: same destination and source addresses, same destination and source port numbers, and same protocol identifiers. Alternatively, we might use a more lax definition of *hit*—for example, only the same source address. We might also choose not to check for a hit for every arriving packet. Instead, we could test for hits for only a random or deterministic subsequence of arriving packets. Also, we can compare the arriving packet with not one, but with some $K > 1$ randomly chosen packets from the recent past. That would give information of the type "J out of K hits," which could be used to estimate the number of flows more accurately.

Rather than maintaining per-flow state, a small cache is used to store a list of M recently seen flows, with the following extra information for each flow in the list: a *count* and a *timestamp*. The list is called the *zombie list* according to [22], and the flows in the list *zombies*.

The zombie list starts out empty. As packets arrive, as long as the list is not full, for every arriving packet, the packet flow identifier (source address, destination address, etc.) is added to the list, the count of that zombie is set to zero, and its timestamp is set to the arrival time of the packet.

Once the zombie list is full, it works as follows: Whenever a packet arrives, it is compared with a randomly chosen zombie in the zombie list.

Hit: If the arriving packet's flow matches the zombie, we declare a *hit*. In that case, the count of the zombie is increased by one, and the timestamp is reset to the arrival time of the packet in the buffer.

No Hit: If the two are not of the same flow, we declare *no hit*. In that case, with probability p the flow identifier of the packet is overwritten over the zombie chosen for comparison. The count of the zombie is set to zero, and the timestamp is set to the arrival time at the buffer. With probability $1 - p$ there is no change to the zombie list.

Irrespective of whether there was a hit or not, the packet may be dropped if the buffer occupancy is such that the system is in random drop mode. The drop probability may depend on whether there was a hit or not.

Define $P(t)$ to be an estimate for the hit frequency around the time of the arrival of the tth packet at the buffer. For the tth packet, let

$$h(t) = \begin{cases} 0 & \text{if no hit,} \\ 1 & \text{if hit,} \end{cases} \tag{6.4}$$

and let

$$P(t) = (1 - \alpha)P(t - 1) + \alpha h(t), \tag{6.5}$$

with $0 < \alpha < 1$. It has been shown in [22] that $P(t)^{-1}$ is a good estimate for the effective number of active flows in the time shortly before the arrival of packet t.

To reduce comparison overhead, it is allowable to update $P(t)$ not after every packet, but (say) after every L packets or at predetermined epochs. If H hits are got out of L packets, a possible update rule is

$$P(\text{new}) = (1 - L\alpha)P(\text{old}) + \alpha H. \tag{6.6}$$

As long as $0 \leq L\alpha \ll 1$, this has practically the same effect as updating after every packet [22].

6.2.6.2 *Mechanism*

Let us denote the packet drop probability function by p_z. According to [22], a function $q(x)$ is defined as follows:

$$q(x) = \begin{cases} p_{\max} & \text{if } B/3 \leq x < B, \\ p_{\max}/4 & \text{if } B/6 \leq x < B/3, \\ 0 & \text{if } 0 \leq x < B/6, \end{cases} \tag{6.7}$$

where B is the buffer size, x is the backlog, and p_{max} is a parameter ranging between 0 and 1. The range of interest for p_{max} is $(0.09, 0.15)$ according to [22]. Higher values for p_{max} merely drive too many TCP flows into time-out, while much lower values allow relatively large congestion windows.

When packet t arrives at the buffer, SRED first updates $P(t)$ from (6.5). If at the arrival instant the buffer contains x bytes, SRED drops the packet with a probability p_z that equals

$$p_z = q(x) \times \min\left\{1, \frac{1}{[\beta P(t)]^2}\right\} \times \left[1 + \frac{h(t)}{P(t)}\right], \qquad (6.8)$$

where β is a parameter with the suggested value 256 according to [22].

Note that, unlike RED, SRED doesn't compute the average queue length (this operation can easily be added if needed). In [22], it has been shown that using an averaged buffer occupation doesn't improve performance. The motivation for choosing (6.7) and (6.8) is manifold. First, the buffer occupancy can vary considerably because of widely varying round-trip times, flows using different maximum segment sizes (MSSs), and transients caused by new flows before they reach the equilibrium of TCP flow and congestion control window. By making the drop probability depend on the buffer occupancy, which is the role of q in (6.8), SRED ensures that the drop probability increases when the buffer occupancy increases, even when the estimate $P(t)$ remains the same.

The ratio 4 in (6.7) was chosen so that TCP connections would reach the new equilibrium after a single packet loss. When $0 \le P(t) < 1/\beta$, SRED uses $p_z = q$ according to (6.8). This is for two reasons. First, if the drop probability becomes too large, TCP flows spend much or most of their time in time-out. So further increasing p_z is not sensible. Second, when $P(t)$ becomes small (when hits are rare), estimating $P(t)$ becomes unreliable.

SRED uses hits directly in the dropping probabilities, as indicated by the last factor $[1 + h(t)/P(t)]$ in (6.8). This is based on the idea that misbehaving flows are likely to generate more hits. There are two reasons. First, misbehaving flows by definition have more packet arrivals than other flows and so trigger more comparisons. Second, they are more likely to be present in the zombie list. This increases the drop probability for overactive flows and can also reduce TCP's bias in favor of flows with short round-trip times (RTTs). Interested readers are referred to [22] for further details on the SRED algorithm.

6.2.7 Longest Queue Drop (LQD)

6.2.7.1 *Concept* Motivated by the fact that if connections are given equal weights, then connections that use the link more (get a higher share of the bandwidth unused by other connections) tend to have longer queues, the longest queue drop (LQD) approach was proposed in [38] for buffer management. Biasing the packet drops so that connections with longer queue have higher drop rates should make the bandwidth sharing more fair. In addition, the LQD policy offers some flow isolation and protection, since if one connection misbehaves consistently, only that connection experiences an increased loss rate.

The LQD algorithm requires searching through the backlogged queues to determine which is the longest queue. [38] proposes a variant of LQD that is particularly easy to implement, *approximated longest queue drop* (ALQD). A register holds the length and identity of the longest queue as determined at the previous queuing operation (queue, dequeue, drop). On every queuing event (including enqueuing and dequeuing), the current queue length is compared with the longest queue identified in the register. If it is the same queue, the queue length in the register is adjusted. If the current queue is longer, its identity and length are now stored in the register. A similar scheme called *quasi-pushout* is proposed in [39]. Below we briefly introduce the ALQD algorithm and present an implementation architecture.

6.2.7.2 Approximated LQD (ALQD)   Instead of sorting out the real longest queue among N flow queues, the ALQD algorithm [38] tracks the quasi-longest queue (Qmax) by using three comparisons only. One is on the arrival of a packet: the queue length of the destined queue is increased and compared with that of the Qmax. The other two are on the departure or dropping of a packet: the queue length of the selected queue is decreased and compared with that of the Qmax. If the new length of the destined or selected queue is greater than that of the Qmax, the flow identifier (FID) and length of the Qmax are replaced with that of the particular queue. That is, the destined or selected queue becomes the new Qmax.

The ALQD drops the HOL packet of the currently longest queue when the data memory is full, because dropping from the front can trigger TCP's fast retransmit–recovery feature faster and, hence, increase throughput [38]. Figure 6.15 shows the ALQD algorithm.

Since packets are of variable length, they are divided into a number of fixed-length segments (or cells) to fully use the data memory in routers. The above algorithm actually checks on each packet arrival whether the data memory is full. Therefore, dropping a packet will guarantee sufficient memory space for the incoming cell. It is not necessary to double-check the memory status after dropping a packet. A *packet arrival event* is defined as when its last cell has arrived at the system. Similarly, a *packet departure event* is when its last cell has left the system. The queue length updates are performed on packet arrival, departure, and dropping events.

ALQD requires only $O(1)$ complexity in time and space. However, its state does not reflect exactly the state of the system. So optimal behavior at all times cannot be ensured, especially when scheduling weights vary over a very wide range. A scenario can be constructed where ALQD cannot free enough

```
for i = 0 to N − 1
/* N is the number of queues. Each queue may correspond
   to each port of a switch */
{
    if (queue i has an arriving packet) {
        if (data memory full) {
            push out HOL packet of current Qmax;
            decrement length of current Qmax by length
             of the dropped packet;
        }
        store cell by cell the incoming packet;
        increment length of queue i by length
          of the incoming packet;
        if (current Qmax is shorter than queue i)
          assign queue i as the new Qmax;
    }
    if (queue j has a departing packet) {
        send out the packet;
        decrement its queue length by length
          of the outgoing packet;
    }
    if (current Qmax is shorter than queue j)
      assign queue j as the new Qmax;
}
```

Fig. 6.15 The ALQD scheme.

memory and some incoming packets have to be dropped, thereby temporarily breaching the strict flow isolation property of LQD. However, the degradation is such that it makes the complexity–performance tradeoff worthwhile [38].

6.2.7.3 *Architectural Design* A VHDL design and implementation of the ALQD with a PFQ scheduler for IP routers are presented in [40]. Figure 6.16 shows the system architecture with the data memory, where there is a *packet segmentation unit* (PSU) before the input interface, which chops each packet into a number of fixed-length segments (also called cells) before sending them to the input interface. There is also a packet reassembly unit (PRU) after the output interface, which reassembles the segments of each packet into a packet before forwarding it to the next processing unit.

To facilitate the queue management (described later) and packet reassembly operations in the PRU, both the PSU and the queue management system are assumed to transmit cells without interleaving the cells from different packets. Both input and output interfaces are assumed to have one-cell buffer space.

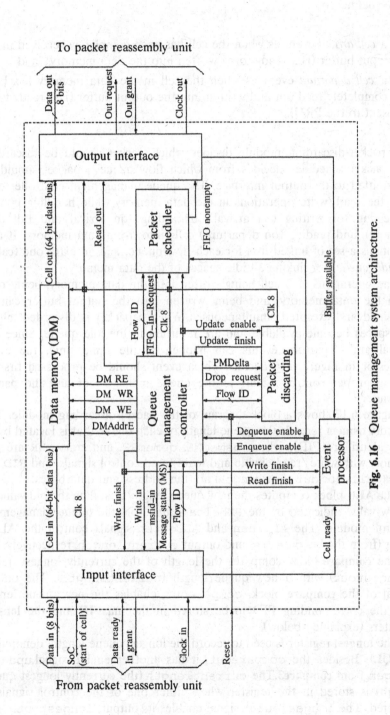

Fig. 6.16 Queue management system architecture.

We define:

- a *cell arrival* event as when the cell has been completely received in the input buffer (i.e., ready to be written into the data memory), and
- a *cell departure* event as when the cell in the data memory has been completely read out and written into the output buffer (i.e., ready to be sent to the PRU).

The packet-discarding module decides which packet should be discarded. The packet scheduler decides from which flow queue a packet should be transmitted to the output interface. The queue management controller controls the read/write operations in the data memory while handling pointer processing for writing (on arrival), discarding (on arrival and full data memory), and reading (on departure) cells into/from data memory. It also maintains a set of linked lists for each flow queue, and an extra one (called the *idle queue*) for linking all idle spaces of the data memory.

The operations of a cell being written into the data memory, being read from the data memory, and being written into the output buffer can be pipelined and executed simultaneously. When a packet is discarded, all its cell spaces become available and are returned to the idle queue. Since cell arrival and departure events can happen at the same time, the event processor in Figure 6.16 decides which event should be processed first by issuing proper control signals to the queue management and the packet scheduler.

Figure 6.17 shows a block diagram of the packet-discarding module. The memory **qsmem** records the queue length of each queue that is located by its flow identifier (FID), where qsmemWE, qsmemRE, and qsmemWR are the memory with write/read enable and write/read control signals, and FID In and #cells (or currentlength) are the address and data buses.

The **ALU** block computes the new queue length by adding the old value to a new value indicated by the PMDelta signal from the queue management control module. The ALUinen and ALUouten signals control the ALU's input (from the bus #cells) and output currentlength, respectively.

The **compare** block compares the length of the currently longest queue (longestout) with a new queue length (currentlength). The output signal of the **compare** block, compareout, enables the new queue length and the corresponding FID to be stored in the **maxFID** and the **longest** registers (explained below).

The **longest** register is used to record the longest queue length, denoted by **reg(QL)**. Besides the compareout, it has another enable signal, updatelongen from **compare3**. The currentlength (the currently longest queue length) is stored in the register when either one of the control signals is asserted. The longestouten signal enables its output, longestout.

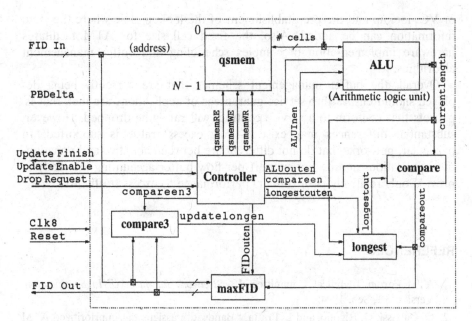

Fig. 6.17   Functional blocks of the packet-discarding module.

The **maxFID** register is used to record the FID of the longest queue, denoted by **reg(FID)**. Its output enable signal is FIDouten and its output signal is FID Out.

The **compare3** block compares the incoming FID and the FID stored in **maxFID**. The comparison is enabled by compareen3. Its output, update-longen, indicates whether both FIDs are the same. The control signal, updatelongen, allows the **longest** register to be updated with the value of the **reg(QL)** without a valid compareout signal. This can happen when a packet is discarded or sent out from the currently longest queue, and only its queue length should be updated. The compareout signal won't be valid, because there is no need to compare the queue length and FID. This block can help identify this particular case when the incoming FID is the same as the FID in the register **reg(FID)**.

The **controller** block generates all the internal control signals as well as the output signal, Update Finish. Further details about this module and the other designs are in [40].

6.3 SUMMARY

IP networks tend not to keep per-flow state information to reduce implementation cost. But in some environments, notably ATM, it is quite natural to maintain state information for each flow or virtual circuit because of the VPI or VCI lookup and the preestablishment of the lookup table where the state information can be added. Also, the fixed cell size for ATM facilitates hardware implementation of complex scheduling and buffer management algorithms.

Overall, the buffer management schemes that are currently being deployed should ensure that service guarantees of applications are met, that is, packets that conform to a service contract will rarely be dropped. However, substantial differences may exist in how excess traffic is supported. In particular, networks that do not differentiate between adaptive and nonadaptive applications and do not support per-flow buffer accounting are likely to provide only relatively poor (unfair) performance to excess traffic [41].

REFERENCES

1. ATM Forum Technical Committee, *User-Network Interface (UNI) Specification*, Version 3.1, Sep. 1994.
2. G. Gallassi, G. Rigolio, and L. Fratta, "Bandwidth assignment in prioritized ATM networks," *Proc. IEEE GLOBECOM*, 1990.
3. A. Y. Lin and J. A. Silvester, "Priority queueing strategies and buffer allocation protocols for traffic control at an ATM integrated broadband switching system," *IEEE J. Select. Areas Commun.*, vol. 9, no. 9, pp. 1524–1536, Dec. 1991.
4. D. W. Petr and V. S. Frost, "Nested threshold cell discarding for ATM overload control: optimization under cell loss constraints," *Proc. IEEE INFOCOM'91*, Bar Harbour, FL, Apr. 1991.
5. L. Tassiulas, Y. C. Hung, and S. S. Panwar, "Optimal buffer control in an ATM network node," *IEEE/ACM Trans. Networking*, vol. 2, no. 4, pp. 374–386, Aug. 1994; also appears in part in *Proc. IEEE INFOCOM'93*).
6. H. J. Chao and H. Cheng, "A new QoS-guaranteed cell discarding strategy: self-calibrating pushout," *IEEE GLOBECOM'94*, San Francisco, Nov. 1994.
7. J. Y. Shue, "VHDL design of self-calibrating pushout queue manager," Master's thesis, Polytechnic University, Brooklyn, NY, Dec. 1995.
8. H. J. Chao, H. Cheng, Y. Jenq, and D. Jeong, "Design of a generalized priority queue manager for ATM switches," *IEEE J. Select. Areas Commun.*, Jun. 1997.
9. "Broadband ISDN switching system generic requirements," TA-NWT-001110, Issue 1, Bellcore, Aug. 1992.
10. A. K. Parekh, "A generalized processor sharing approach to flow control in integrated services network," Ph.D. thesis, MIT, Feb. 1992.
11. ATM Forum, *Traffic Management Specification Version 4.0*, Apr. 1996.
12. G. Armitage and K. Adams, "Packet reassembly during cell loss," *IEEE Net.*, vol. 7, no. 5, 26–34, Sep. 1993.
13. A. Romanow and R. Oskouy, "A performance enhancement for packetized ABR and VBR + data," ATM Forum Contribution 94-0295, Mar. 1994.

14. K. Cheon, "Intelligent cell discarding policies for TCP traffic over ATM–UBR," Ph.D. dissertation, Department of Electrical Engineering, Polytechnic University, Brooklyn, NY, Dec. 1997.

15. A. K. Choudhury and E. L. Hahne, "Dynamic queue length thresholds for shared-memory packet switches," *IEEE/ACM Trans. Net.*, vol. 6, pp. 130–140, Apr. 1998.

16. I. Cidon, R. Guerin, and A. Khamisy, "Protective buffer management policies," *IEEE/ACM Trans. Netw.*, vol. 2, no. 3, pp. 240–246, Jun. 1994.

17. S. Floyd and V. Jacobson, "On traffic phase effects in packet-switched gateways," *Internetw.: Res. Experience*, vol. 3, no. 3, pp. 115–156, Sep. 1992.

18. S. Floyd and V. Jacobson, "Random early detection gateways for congestion avoidance," *IEEE/ACM Trans. Netw.*, vol. 1, no. 4, pp. 397–413, Aug. 1993.

19. H. Kroner, G. Hebuterne, P. Boyer, and A. Gravey, "Priority management in ATM switching nodes," *IEEE Trans. Commun.*, vol. COM-9, no. 3, pp. 418–427, Apr. 1991.

20. D. Lin and R. Morris, "Dynamics of random early detection," *Proc. ACM SIGCOMM*, pp. 127–137, Sep. 1997.

21. "Quality of Service (QoS) Networking," White paper, Cisco Systems, Jun. 1999.

22. T. J. Ott, T. V. Lakshman, and L. H. Wong, "SRED: stabilized RED," *Proc. IEEE INFOCOM*, Mar. 1999.

23. A. Romanow and S. Floyd, "Dynamics of TCP traffic over ATM networks," *IEEE J. Select. Areas Commun.*, vol. 13, no. 4, pp. 633–641, May 1995.

24. J. Turner, "Maintaining high throughput during overload in ATM switches," *Proc. IEEE INFOCOM*, San Francisco, pp. 287–295, Apr. 1996.

25. J. Nagle, "Congestion control in IP/TCP," RFC 896, Internet Engineering Task Force (IETF), Jan. 1984.

26. V. Jacobson, "Congestion avoidance and control," *ACM SIGCOMM'88*, Aug. 1988.

27. R. Braden, Ed., "Requirements for Internet hosts—communication layers," RFC 1122, Internet Engineering Task Force (IETF), Oct. 1989.

28. W. Stevens, "TCP slow start, congestion avoidance, fast retransmit, and fast recovery algorithms," RFC 2001, Internet Engineering Task Force (IETF), Jan. 1997.

29. B. Braden, D. Clark, J. Crowcroft, B. Davie, S. Deering, D. Estrin, S. Floyd, V. Jacobson, G. Minshall, C. Partridge, L. Peterson, K. Ramakrishnan, S. Shenker, J. Wroclawski, and L. Zhang, "Recommendations on queue management and congestion avoidance in the Internet," RFC 2309, Internet Engineering Task Force (IETF), Apr. 1998.

30. W. Leland, M. Taqqu, W. Willinger, and D. Wilson, "On the self-similar nature of Ethernet traffic (extended version)," *IEEE/ACM Trans. Netw.*, vol. 2, no. 1, pp. 1–15, Feb. 1994.

31. T. V. Lakshman, A. Neidhardt, and T. Ott, "The drop from front strategy in TCP over ATM and its interworking with other control features," *IEEE INFOCOM*, MA28.1, 1996.

32. S. Floyd, "RED: discussions of byte and packet modes," Mar. 1997. http://www-nrg.ee.lbl.gov/floyd/REDaveraging.txt.

33. C. Villamizar and C. Song, "High performance TCP in ANSNET," *Comp. Commun. Rev.*, vol. 24, no. 5, pp. 45–60, Oct. 1994.

34. M. Gaynor, "Proactive packet dropping methods for TCP gateways," Oct. 1996, http://www.eecs.harvard.edu/ ~ gaynor/final.ps.

35. D. Clark and J. Wroclawski, "An approach to service allocation in the Internet," Internet Draft, Jul. 1997, draft-diff-svc-alloc-00.txt.

36. K. Nichols, V. Jacobson, and L. Zhang, "A two-bit differentiated services architecture for the Internet," Internet Draft, Nov. 1997, draft-nichols-diff-svc-arch-00.txt.

37. S. Blake, D. Black, M. Carlson, E. Davies, Z. Wang, and W. Weiss, "An architecture for differentiated services," RFC 2475, Internet Engineering Task Force (IETF), Dec. 1998.

38. B. Suter, T. V. Lakshman, D. Stiliadis, and A. K. Choudhury, "Design considerations for supporting TCP with per-flow queueing," *IEEE J. Select. Areas Commun.*, Jun. 1999; *IEEE INFOCOM'98*.

39. Y. S. Lin and C. B. Shung, Quasi-pushout cell discarding, *IEEE Commun. Lett.*, vol. 1, no. 5, pp. 146–148, Sep. 1997.

40. M. Feng, "Design of per flow queuing buffer management and scheduling for IP routers," Master's thesis, Department of Electrical Engineering, Polytechnic University, Brooklyn, NY, Dec. 1998.

41. R. Guerin and V. Peris, "Quality-of-service in packet networks: basic mechanisms and directions," *Comput. Netw.*, vol. 31, no. 3, pp. 169–179, Feb. 1999.

CHAPTER 7

FLOW AND CONGESTION CONTROL

Heavy congestion leads to severe delay and loss problems in packet networks. The basic objective of flow or congestion control in a network is to efficiently regulate the transient streams of sessions, in a decentralized manner, in order to avoid congestion in the network. The term "session" is used somewhat loosely to refer to any communication process to which flow control is applied. That is, a session could be a virtual channel (VC) or a virtual path (VP) (in ATM networks), or the entire data flow between two nodes. The corresponding regulating strategies are called *flow control algorithms* or *protocols*. Section 7.1 reviews the flow control and congestion control mechanisms used in most networks. Section 7.2 describes several flow control mechanisms for ATM networks. The flow and congestion control mechanisms for packet networks are presented in Section 7.3. Section 7.4 describes a link-by-link rate-based flow control scheme called EASY.

7.1 OVERVIEW

In most networks, there are circumstances in which the externally offered load is larger than can be handled. Then, if no measures are taken to restrict the entrance of traffic into the network, queue sizes at bottleneck links will grow and packet delays will increase. In the worst case, this may cause packet loss when buffers overflow, possibly violating maximum delay/loss specifications. One of the solutions is to control the amount of traffic that flows over the network within a time interval. However, if the traffic is over controlled, the throughput will decrease and the data transfer delay will become excessive. In general, it is essential for flow control to rather strike a good compromise between high throughput, low tolerable delay, and low degree of network congestion, than recover the network operations from severe congestion. Henceforth, we will make no distinction between the terms "congestion control" and "flow control."

Usually, flow control is applied independently to each individual session. However, there is a nonnegligible interaction between different sessions, because they may share the same buffer and bandwidth resources at a particular node. Flow control schemes should also ensure fairness while satisfying quality of service (QoS) requirement of each session.

There are two classes of flow control schemes, closed-loop and open-loop. In closed-loop flow control, a delicate handshaking algorithm is necessary between the session's transmitter and receiver in order to prevent network congestion yet achieve high utilization of link capacity. Since the feedback information can be corrupted due to large link propagation delay, an additional predictive mechanism may be needed at the transmitter to issue control messages to avoid buffer overflow or underflow at the receiver, thus adding to the cost of protocol operations. In contrast, open-loop flow control can be much simpler to implement. For example, if there is a minimum data rate negotiated between the two ends, the transmitter can send data at this rate, without any data loss, regardless of the traffic of other sessions. However, it may not fully utilize the network resources even if there are excess link capacity and empty buffers available at the receiver.

Basically, we can control the sessions' data flows by either limiting directly their data populations in the network, or limiting their transmission rates. This yields window-based and rate-based flow control schemes, respectively, as introduced below.

7.1.1 Window-Based Flow Control

Window-based flow control is a kind of closed-loop control in which the receiver keeps a buffer of size W data units for a particular session, where W is called the *window size*. In addition, there are W logical credits, or permits, associated with the session. Each data unit waiting in the buffer of the receiver must hold a credit. When a data unit is dispatched from the receiver, it relinquishes the credit, and this credit is sent back to the transmitter instantaneously. Meanwhile, the transmitter starts to send data when it gets empty credits; otherwise it stops until there is a credit released from the receiver. Credits that are not currently held by data are either stored at the transmitter or on their way back to the transmitter, so the buffer of the receiver will never overflow. In equilibrium, throughput = (window size)/(round-trip delay), where the round-trip delay is the overall time for a credit to be transferred from the transmitter to the receiver, wait at the receiver, and return to the transmitter.

The control loop of a window-based flow control scheme can be end-to-end or hop-by-hop. The end-to-end window scheme is simpler to implement. However, buffer congestion at the intermediate nodes along the path may cause a large end-to-end credit delay, thus slowing down the return of credits to the source, and consequently reducing the rate of fresh data input to the network.

To increase network utilization, the end-to-end window size can be enlarged when the network is lightly loaded, and reduced otherwise. Since queue lengths tend to grow exponentially with congestion, the window must be squeezed at least at the same speed for stability [7, 11]. One successful design is the linear-increase–exponential-decrease window adaptation that has been commonly used by the Transmission Control Protocol (TCP) [7] in ARPANET and by the DECbit scheme [14] in DECNET.

TCP uses round-trip delay to indicate the status of network congestion. When there is packet loss, the window size is slowly increased to reduce the oscillations of the window. DECbit instead uses congestion indication bits encoded in each packet header; such a bit represents the status of a switching node along the session's traversed path. For example, when a switching node detects that a session's data stream is in a congestive state with queue length greater than the threshold value, the node sets the corresponding congestion indication bit to 1. To avoid premature window adjustments, the window size is reduced only if at least half of the congestion bits are set. The short-term fluctuations of queue length are filtered out by averaging the queue length over an interval.

Larger end-to-end window size allows more data units of a session to access the network within a time interval. However, it also indicates a larger demand for buffering at each node along the traversed path, because all these data units could get congested at a particular node. It becomes even more demanding in high-speed data networks, due to the large delay–bandwidth product incurred. (The delay–bandwidth product of a session can be used to estimate the number of its in flight packets from source to destination.) To alleviate this problem, Mitra [27, 26] has proposed an optimal window adaptation for maximizing the session power based on an asymptotic analysis of closed queuing networks, together with dynamic buffer sizing.

On the other hand, hop-by-hop windows can reduce the nodal buffering requirement at the cost of using credit processing at each node. This scheme was first used in TYMNET [2]. Since all packets of a session can be evenly distributed along the nodes from source to destination, they will not get congested at a particular node. Besides, hop-by-hop windows can also have a quick response to network congestion.

In ATM local area networks (LANs) [6, 28, 45], the relatively small delay-bandwidth product may still enable hop-by-hop windows. One recent proposal for available-bit-rate (ABR) services [49] is *flow-controlled virtual channel* (FCVC) [38, 46, 51]. A feedback cell is designed for each VC to transfer its credit information through a reserved channel between two successive nodes on its path. A number of VCs passing the same link can share a common feedback cell to save bandwidth. Each node generates a feedback cell once it has dispatched a certain number of data cells for the VC. This number can be determined so as to control the feedback frequency, the consumed bandwidth of the feedback channel, and the processing capacity at both ends.

However, for voice, video, and other data services that demand the controllability of packet transfer delay and data rate, window schemes are not suitable any more. This is because first, a window-based scheme cannot guarantee a minimum communication rate for a session. Second, its inherent start–stop data transmission may create traffic fluctuations as well as variations of data transfer delay, thus necessitating rate-based flow control.

7.1.2 Rate-Based Flow Control

Rate-based flow control can be either closed-loop or open-loop. In a rate-based flow control scheme, each session is allocated a data rate that commensurates to its need throughout the network. This rate should lie within certain limits that depend on the QoS requirement of this session.

A strict implementation of rate-based open-loop flow control over a session with allocated rate ρ packets/s would admit at most one packet every $1/\rho$ seconds. This, however, amounts to a form of time-division multiplexing (TDM) and tends to introduce a large backlog at the transmitter when the session is bursty, leaving excess bandwidth at the receiver untapped. This problem can be alleviated with a leaky bucket [4, 36, 33] shaper, which allows a certain degree of the burstiness of a session to enter the network without violating the purpose of flow control.

A leaky bucket shaper maintains a token bucket in which tokens are generated at the desired session rate ρ as long as the number of tokens in the bucket does not exceed a certain threshold σ. The head-of-line (HOL) packet should obtain a token from the bucket before it is eligible for transmission. One important characteristic is that during the interval $(\tau, t]$, the total amount of traffic departure from a leaky bucket shaper is always upper bounded by $\sigma + \rho(t - \tau)$. Moreover, as indicated by Cruz [17, 18], if such a linear constraint can be preserved throughout a network, an upper bound on network delay can be warranted.

Parekh and Gallager [39, 37] have proposed an open-loop rate control strategy for high-speed data networks based on the combination of leaky bucket access control and generalized processor sharing (GPS) service discipline, which we call GPS flow control. Since the service policy can provide a minimum bandwidth ρ, the worst-case delay at each node is bounded by σ/ρ. However, this scheme is still close to TDM, except that the logical pipeline channel across the network becomes wider with $\sigma > 1$, allowing some bursty traffic. Therefore, the disadvantage of this scheme is underutilizing the network resources when the network is not that congested. Similar proposals include stop-and-go queuing [24] and virtual clock [20], because they both require leaky-bucket-like access control to enforce the traffic smoothness property of a session, which is characterized by a transmission rate over a constant interval.

The preceding discussions lead to the idea of handshaking, i.e., closed-loop control, between the network and users to increase the network utilization. Mitra's optimal-window scheme uses adaptive leaky bucket control [26] to enforce the expected window throughput, where the updated window size divided by the round-trip delay is set to be the new token generated rate.

Gersht and Lee [25] have proposed another rate-based feedback congestion control scheme for best-effort traffic in ATM networks. In this scheme, each first-class VC is characterized by two negotiated parameters: peak rate and guaranteed rate. On the other hand, each node recognizes two buffer threshold values for the first-class cells: high and low. Once the backlog in the buffer exceeds the high threshold value, the node sends choking signals to all related first-class VCs and allows them to send their cells at guaranteed rates only. Thereafter, when the backlog drops below the low threshold value, the congestion is considered to be abated and the node sends relieving signals so that the VCs may again send their cells at peak rates. Elegant buffer sizing based on the rate agreement can prevent buffer overflow.

Gong and Akyildiz [44] have studied a dynamic leaky bucket scheme for ATM networks in which the token rate of a VC is adjusted to meet the maximum switch utilization with a negotiated target value. This information is carried by a type of specific control cells sent by and then returned to the leaky bucket shaper after traveling through the source–destination–source round-trip path. Upon the return of each control cell, the new token rate is determined according to some heuristic rules.

The ABR traffic management is a kind of end-to-end rate control scheme [49, 50, 52, 47, 53]. In ABR flow control, each source adapts its rate to the change of the network conditions in a linear-increase–exponential-decrease manner [11]. Information concerning the state of the network, such as bandwidth availability, current and/or impending congestion, and so on, is conveyed to the source with special control cells.

Some feedback schemes use congestion bits, instead of control packets, in the header of each packet to take down the feedback information, like Newman's backward explicit congestion notification (BECN) [45] and Hluchyj and Yin's end-to-end closed-loop rate control [43].

7.1.3 Predictive Control Mechanism

The effectiveness of feedback control may be diminished when the propagation delay is large compared with the packet transmission time in high-speed data networks [21, 55]. Recent studies [13, 22, 34] have shown that feedback control exhibits oscillations of queue length at a bottleneck node. To ease this problem, we can use some predictive mechanisms to issue control messages supplementally, or divide a long feedback loop into shorter ones.

Mishra and Kanakia [29, 59] have developed a *hop-by-hop rate-based congestion control* (HRCC) strategy combined with a model-based predictive control mechanism. In HRCC, each node informs its upstream neighbor periodically about the session's backlog and transmission rate at the node. Meanwhile it also revises this transmission rate on receiving feedback information from its downstream neighbor. Between two successive feedback signals, a simulation model is constructed at each node, using the latest feedback information, to predict the queue evolution and session rate at its downstream neighbor. It has been shown that the HRCC scheme can achieve lower oscillation of queue length, less packet loss at the bottleneck node, and higher throughput than any end-to-end rate-based feedback control algorithm.

The model-based predictive control has been widely used in the field of process control [9] to improve the performance of time-delayed control systems. In such control schemes, feedback messages are used to compensate for the prediction inaccuracies so as to minimize the drift of the fictitious model from the actual state of the system.

Another example is the *backlog balancing flow control* (BBFC) scheme [57, 63]. The idea is to evenly distribute all the backlogged data over the entire network from time to time so that each node is congestion-free. A binary feedback protocol is proposed to carry out the flow control while a simulation model is constructed at each node to evaluate the queue evolution of its downstream neighbor. Assume that the service rate at each node is adjusted periodically. A rate update equation is derived from a linear projection model of the flow dynamics. BBFC is the first scheme that identifies an *explicit* objective function of flow control that achieves efficient bandwidth and buffer sharing while preventing network congestion. Further details of the BBFC scheme will be discussed in Section [7.2.2].

7.2 ATM NETWORKS

ATM networks are connection-oriented [36, 49, 48, 33]. A virtual circuit (VC) is established between each source–destination pair. The ATM adaptation layer (AAL), located at a customer premises node (CPN) [48], converts the user data into ATM cells for transmission, and vice versa when it receives ATM cells from the other end. Between any two neighboring nodes, the group of VCs can be bundled up to form a virtual path (VP).

7.2.1 ATM Service Categories

There are six ATM service categories, namely constant bit rate (CBR), real-time variable bit rate (rt-VBR), non-real-time variable bit rate (nrt-VBR), available bit rate (ABR), unspecified bit rate (UBR), and guaranteed frame rate (GFR).

The QoS specifications of CBR, rt-VBR, and nrt-VBR are based on (1) a traffic contract that specifies the characteristics of the cell flow, such as SCR and PCR, and (2) usage parameter control (UPC) performed by the network to enforce the flow. During the connection admission process, the network uses the proposed traffic contract to determine if there are sufficient network resources for this new connection. Once the connection is established, UPC may discard or tag as lower-priority any cell that exceeds the parameters of the traffic contract. There is no feedback congestion control for these three service catagories.

The ABR allows users to declare a minimum cell rate (MCR) that is guaranteed by the network. When additional capacity is available, the user may burst above the MCR. However, the user is required to control its rate, depending on the congestion state of the network. Although the cell transfer delay and cell loss ratio are not guaranteed, it is desirable for the network to minimize delay and loss.

The UBR is a best-effort service. It is designed for those data applications that want to use any left-over capacity and are not sensitive to cell loss or delay. Such connections are not rejected on the basis of bandwidth shortage (no connection admission control) and are not policed for their usage behavior. No amount of capacity is guaranteed, and during congestion, the cells are lost but the sources are not expected to reduce their cell rate. Instead, these applications may have their own higher-level flow control and congestion control mechanisms, such as TCP. Some buffer management schemes such as EPD are required to facilitate this kind of congestion control.

The GFR is the newest of the service classes defined by the ATM Forum (but not completed yet). The GFR user must specify an MCR that is guaranteed by the network. Frame-based traffic shaping is performed by the network to enforce the cell flow. All frames above MCR are given best-effort service. Complete frames are accepted or discarded in the switch. If the GFR traffic is within the MCR limitation, the user can expect the packet loss rate to be very low. The GFR can be regarded as UBR with MCR, and it was originally introduced by Guerin and Heinanen [66] as a modification of UBR called UBR$^+$. The purpose is simple and clear: users, particularly data users, generally have no idea of the detailed characteristics of their traffic; however, they still view it as desirable to have some form of QoS guarantee, as opposed to using UBR service, which gives no guarantees whatsoever. GFR is an admission, to some extent, that ABR is a complex design to implement and that it will not be fully deployed in the near term by all equipment.

Among all six service classes mentioned above, only ABR traffic responds to congestion feedback from the network. The rest of this section presents the backlog balancing flow control [57, 63], and then moves on to ABR traffic.

7.2.2   Backlog Balancing Flow Control

7.2.2.1   *Principles of Backlog Balancing*   This section first describes the flow dynamics of a VC, the concept of backlog balancing, and its implication in the flow control of a VC.

7.2.2.1.1   *Flow Dynamics of a VC*   Flow dynamics of a VC can be characterized by backlog in the buffer of each node on its path and data in transit at the input and output of this node. Consider three consecutive nodes on a VC, as shown in Figure 7.1. Let B_k be the buffer reserved for the VC at node k, and $\mu_k(t)$ be the service rate of node k for the VC at time t; $\mu_k(t) = 0$ for $t \le 0$. Assume that the link propagation delay between any two consecutive nodes is a constant τ. As shown in Figure 7.2, define the *departure process* of

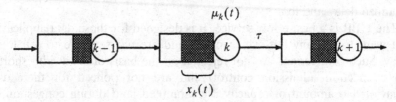

Fig. 7.1   Three consecutive nodes on a VC.

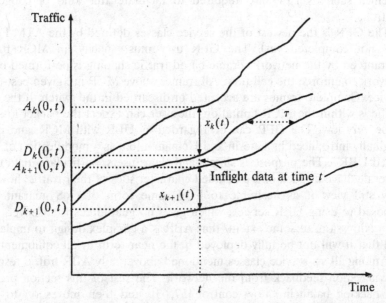

Fig. 7.2   Arrival and departure processes at nodes k and $k + 1$. (©1995 IEEE.)

node k at time t as the total data it sent in $(0, t]$,

$$D_k(0, t) = \int_0^t \mu_k(s) \, ds, \qquad (7.1)$$

and the *arrival process* of node $k + 1$ at time t as the total data it received up to time t, which is also the data node k had sent up to time $t - \tau$, due to the link propagation delay:

$$A_{k+1}(0, t) = D_k(0, t - \tau). \qquad (7.2)$$

The difference between the arrival and departure processes of node k at time t, i.e., the vertical distance between curves $A_k(0, t)$ and $D_k(0, t)$ in Figure 7.2, is its instant *backlog*

$$x_k(t) = A_k(0, t) - D_k(0, t), \qquad (7.3)$$

and $x_k(t)/B_k$ is the *buffer occupancy* at time t. The difference between the arrival and departure (service) rate of node k at time t, i.e., the first-order derivative of its backlog, is its instant *buffer drift*

$$\dot{x}_k(t) = \mu_{k-1}(t - \tau) - \mu_k(t). \qquad (7.4)$$

The vertical distance between $D_k(0, t)$ and $A_{k+1}(0, t)$ in Figure 7.2 represents the inflight data between nodes k and $k + 1$ at time t, that is, the total data node k sent in $(t - \tau, t]$, while the horizontal distance is the constant delay τ.

The backlog balancing is to ensure equal buffer occupancy between nodes k and $k + 1$ at any time,

$$\frac{x_k(t)}{B_k} = \frac{x_{k+1}(t)}{B_{k+1}}. \qquad (7.5)$$

Since according to (7.1), (7.2), and (7.3) the flow dynamics of a VC are completely determined by its source characteristics and the service rate of each node on its path, a hop-by-hop rate control mechanism is needed to achieve an even backlog distribution over this VC from (7.5).

7.2.2.1.2 Objective of Flow Control

Assume that the service rate at each node will be adjusted periodically with period T. Let $t_n = nT$ be the nth updating epoch. The data received by node k in $(t, t + T]$ and its backlog at time t minus that to be dispatched in the same interval will decide the backlog at time $t + T$:

$$x_k(t + T) = x_k(t) + A_k(t, t + T) - D_k(t, t + T), \qquad (7.6)$$

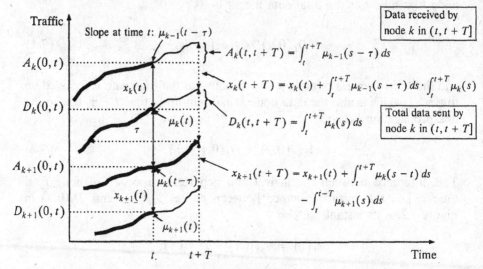

Fig. 7.3 Flow dynamics viewed by node k at time t. (©1995 IEEE.)

which is called the *queue evolution equation*. Let $B_k = B_{k+1}$ for simplicity. Then $x_k(t + T) = x_{k+1}(t + T)$ from (7.5) implies

$$\int_t^{t+T} \mu_k(s) \, ds = \int_t^{t+T} \mu_{k-1}(s - \tau) \, ds + \int_t^{t+T} \mu_{k+1}(s) \, ds$$

$$- \int_t^{t+T} \mu_k(s - \tau) \, ds + [x_k(t) - x_{k+1}(t)], \quad (7.7)$$

where $x_k(t) - x_{k+1}(t)$ is the backlog discrepancy at time t. The point is illustrated in Figure 7.3.

For small T, a linear projection model of the flow dynamics can be constructed by assuming that the arrival rate $\mu_{k-1}(t - \tau)$ and the service rates $\mu_k(t)$ and $\mu_{k+1}(t)$ are constant for $t \in (t_n, t_{n+1}]$, and their values are denoted by $\mu_{k-1}(t_n - \tau)$, $\mu_k(t_n)$, and $\mu_{k+1}(t_n)$, respectively, as shown in Figure 7.4. Therefore, according to (7.7), the expected service rate $\mu_k(t_n)$ must satisfy

$$\mu_k(t_n) = \mu_{k-1}(t_n - \tau) + \mu_{k+1}(t_n) - \frac{1}{T} \int_{t_n}^{t_{n+1}} \mu_k(s - \tau) \, ds$$

$$+ \frac{1}{T} [x_k(t_n) - x_{k+1}(t_n)], \quad (7.8)$$

where the integral term represents those data sent by node k in the interval $(t_n - \tau, t_{n+1} - \tau]$, which can be collected with finite memory by node k at

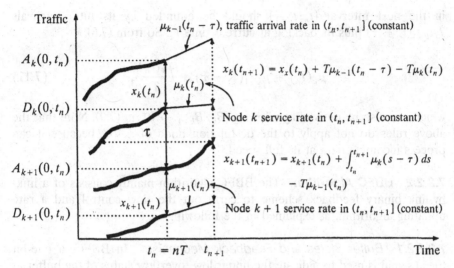

Fig. 7.4   Linear projection of flow dynamics at time t_n $(T \le \tau)$. (©1995 IEEE.)

time t_n for $T \le \tau$. Similarly for $B_k \ne B_{k+1}$,

$$\mu_k(t_n) = \mu_{k-1}(t_n - \tau) + \frac{B_k}{B_{k+1}}\mu_{k+1}(t_n) - \frac{B_k}{TB_{k+1}}A_{k+1}(t_n, t_{n+1})$$

$$+ \frac{1}{T}\left[x_k(t_n) - \frac{B_k}{B_{k+1}}x_{k+1}(t_n)\right]. \tag{7.9}$$

The rate updates for $T > \tau$ are the same as above except that

$$\int_{t_n}^{t_{n+1}}\mu_k(s - \tau)\,ds = \int_{t_n}^{t_n+\tau}\mu_k(s - \tau)\,ds + (T - \tau)\mu_k(t_n)$$

should be used in (7.8) and (7.9), since only $\int_{t_n}^{t_n+\tau}\mu_k(s - \tau)\,ds$ can be collected at time t_n [63]. So, for simplicity, hereinafter the former scenario is chosen for the investigation.

The rate updating (7.9) can be understood by considering the following two extreme cases. First, suppose that node k has a much larger buffer than node $k + 1$ does. According to (7.5), node k should hold proportionally more data in its buffer, with often slow output. If $B_k/B_{k+1} \to \infty$ in (7.5), it follows that $x_k(t) \to \infty$. Since from (7.6)

$$x_k(t_{n+1}) = x_k(t_n) + [\mu_{k-1}(t_n - \tau) - \mu_k(t_n)] \times T \to \infty, \tag{7.10}$$

$\mu_k(t_n)$ should be minus infinity (zero actually).

In the opposite case, since $x_k(t) \to 0^+$ with $B_k \ll B_{k+1}$, node k can deplete its buffer at the highest rate, and the maximum data it can dispatch in the next interval $(t_n, t_{n+1}]$ should be bounded by its future arrivals $T\mu_{k-1}(t_n - \tau)$ plus its backlog at current time t_n. So from (7.6),

$$\mu_k(t_n) = \mu_{k-1}(t_n - \tau) + \frac{x_k(t_n)}{T}, \qquad (7.11)$$

which can also be derived by allowing $B_k/B_{k+1} \to 0^+$ in (7.9). Note that the above rules do not apply to the destination node of a VC, because it can process incoming data at its full speed.

7.2.2.2 BBFC Algorithm   The BBFC algorithm mainly consists of a link-by-link binary feedback scheme to carry out the flow control and a rate updating algorithm, as explained in the following.

7.2.2.2.1 Buffer States and Feedback Mechanism   In BBFC, a one-bit (0–1) signal is used to indicate the underflow–overflow status of the buffer at each node. Consider node $k + 1$ in Figure 7.1. Its buffer is classified into three states with two distinct thresholds B_{k+1}^u and B_{k+1}^o, where $0 \leq B_{k+1}^u < B_{k+1}^o \leq B_{k+1}$. As shown in Figure 7.5(a), the buffer will be in

- overflow state (S_O) if the backlog at time t falls between B_{k+1}^o and B_{k+1}, that is, $B_{k+1} \geq x_{k+1}(t) \geq B_{k+1}^o$;

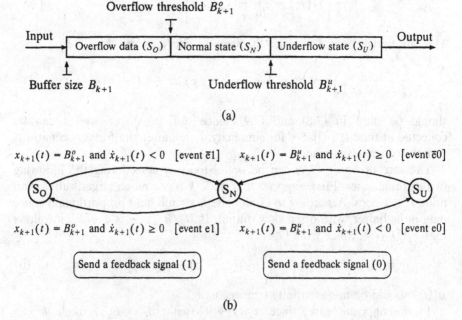

Overflow threshold B_{k+1}^o

Input | Overflow data (S_O) | Normal state (S_N) | Underflow state (S_U) | Output

Buffer size B_{k+1} Underflow threshold B_{k+1}^u

(a)

$x_{k+1}(t) = B_{k+1}^o$ and $\dot{x}_{k+1}(t) < 0$ [event $\bar{e}1$] $x_{k+1}(t) = B_{k+1}^u$ and $\dot{x}_{k+1}(t) \geq 0$ [event $\bar{e}0$]

(S_O) — (S_N) — (S_U)

$x_{k+1}(t) = B_{k+1}^o$ and $\dot{x}_{k+1}(t) \geq 0$ [event e1] $x_{k+1}(t) = B_{k+1}^u$ and $\dot{x}_{k+1}(t) < 0$ [event e0]

Send a feedback signal (1)   Send a feedback signal (0)

(b)

Fig. 7.5   Buffer states and feedback mechanism. (©1995 IEEE.)

- normal state (S_N) if $B_{k+1}^o > x_{k+1}(t) \geq B_{k+1}^u$;
- underflow state (S_U) if $B_{k+1}^u > x_{k+1}(t) \geq 0$.

So B_{k+1}^u and B_{k+1}^o are referred to as the buffer underflow and overflow thresholds, respectively.

There are four different events driving buffer state transitions, as shown in Figure 7.5(b), but only two of them will trigger feedback signals:

(1) The buffer will change from the normal state, S_N, to the overflow state, S_O, at time t if the backlog reaches the threshold B_{k+1}^o at this instant and the data arrival is at least as fast as the departure, indicating a nonnegative buffer drift from (7.4), which is denoted by

$$\text{event e1}: \quad x_{k+1}(t) = B_{k+1}^o \quad \text{and} \quad \dot{x}_{k+1}(t) \geq 0,$$

and node $k + 1$ should send a feedback signal (bit 1) to node k immediately.

(2) The buffer will change from the normal state, S_N, to the underflow state, S_U, at time t if the following event happens:

$$\text{event e0}: \quad x_{k+1}(t) = B_{k+1}^u \quad \text{and} \quad \dot{x}_{k+1}(t) < 0,$$

and node $k + 1$ should send a feedback signal (bit 0) to node k at this moment. (Such a signal will consume negligible bandwidth if it can be piggybacked on a data cell.)

The buffer will return to the normal state, S_N, when the following two events occur, and they will not trigger any feedback signals:

- event $\bar{e}1$: $x_{k+1}(t) = B_{k+1}^o$ and $\dot{x}_{k+1}(t) < 0$;
- event $\bar{e}0$: $x_{k+1}(t) = B_{k+1}^u$ and $\dot{x}_{k+1}(t) \geq 0$.

Operations of such a protocol are based on the following assumptions.

A1. the service rate of each node should be less than or equal to the source's peak data rate S;
A2. each node should provide a guaranteed rate R when it is in the buffer overflow state S_O, where $r < R \leq S$, with r the source's average data rate;
A3. each node should limit its service rate to R, since the latest feedback signal (1) has been received, indicating that its downstream node was in the state S_O.

To prevent buffer overflow, some extra buffers, $B_{k+1} - B_{k+1}^o$, must be reserved to hold the inflight data due to the link propagation delay, as shown

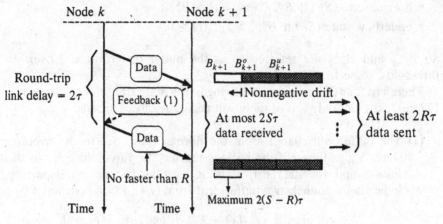

Fig. 7.6 Buffer overflow prevention.

in Figure 7.6. Under the above assumptions, the maximum inflight data would be $2S\tau$ if node k sends data at the peak rate S during the round-trip link propagation delay 2τ. While node $k + 1$ can clear at least $2R\tau$ data, the maximum excess backlog will be $2(S - R)\tau$. So

$$B_{k+1} - B_{k+1}^o = 2(S - R)\tau. \qquad (7.12)$$

To reduce the possibility of buffer underflow, similarly a sufficiently large underflow threshold can be chosen,

$$B_{k+1}^u = 2S\tau. \qquad (7.13)$$

If underflow is not a concern, a small B_{k+1}^u should be used.

To avoid receiving two conflicting feedback signals (01 and 10) in an update period, the period T should be less than their minimal interarrival time, which is the time for the backlog to move between the two thresholds, B_{k+1}^u and B_{k+1}^o, consecutively at the maximum speed S from the above assumptions and (7.4). Thus from the condition on T,

$$T \le \frac{B_{k+1}^o - B_{k+1}^u}{S}. \qquad (7.14)$$

7.2.2.2.2 *Control Structure of the BBFC*   Recall that to determine its new service rate $\mu_k(t_n)$ from (7.9), node k needs to know its arrival rate $\mu_{k-1}(t_n - \tau)$ and the service rate $\mu_{k+1}(t_n)$ in the next time interval $(t_n, t_{n+1}]$, plus the current backlog $x_{k+1}(t_n)$ at node $k + 1$. Let us consider the following approach.

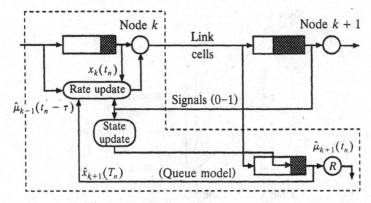

Fig. 7.7   Control structure for the BBFC.

First of all, for small T, $\mu_{k-1}(t_n - \tau)$ can be predicted by the average arrival rate measured in current time interval $(t_{n-1}, t_n]$, that is, the data node k received in $(t_{n-1}, t_n]$ divided by T. Hence,

$$\hat{\mu}_{k-1}(t_n - \tau) = \frac{1}{T} \times A_k(t_{n-1}, t_n), \qquad (7.15)$$

where $\hat{a}$ indicates an estimate of a.

Second, without further information, it can be assumed that $\mu_{k+1}(t_n)$ is equal to the constant rate R, that is,

$$\hat{\mu}_{k+1}(t_n) = R. \qquad (7.16)$$

Accordingly, a simulated model for the real queue of node $k + 1$ can be constructed at node k, as shown in Figure 7.7. It has the same configuration as the real queue of node $k + 1$, except for its constant service rate R. This fictitious model will be used to simulate the queue evolution at node $k + 1$ and thus provide an estimate of the essential input $\hat{x}_{k+1}(t_n)$.

Figure 7.7 also shows the control structure of BBFC. The *state update* is responsible for modifying the status of the queue model according to the latest feedback information; the *rate update* will then choose the new service rate for node k. Both parts are essential to the rate updating algorithm, and each is further explained as follows.

7.2.2.2.3 Rate Updating Algorithm

7.2.2.2.3.1 MODEL STATE UPDATE   The model status will be renewed periodically based on the queue evolution equation (7.6), and thus node k should maintain a running record $D_k(t - 2\tau - T, t)$, for the following reason. Suppose that node k receives a feedback signal (1) at time $s \in (t_{n-1}, t_n]$, as

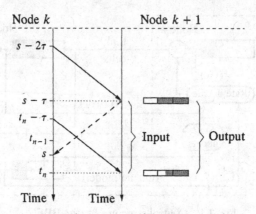

Fig. 7.8   Record $D_k(t - 2\tau - T, t)$ for the model state update.

shown in Figure 7.8, indicating that $x_{k+1}(s - \tau) = B^o_{k+1}$. Only departures from node k after $s - 2\tau$ can affect the status of both the model and the real queue at node $k + 1$ after $s - \tau$ when this signal was generated, and therefore it will be needed for the model state update at time t_n. Since the signal could arrive immediately after t_{n-1}, where $t_n - s \leq T$, the maximum record length should be $2\tau + T$.

The model state update can be outlined as follows:

- As shown in Figure 7.9(a), the feedback signal (1) arriving at time $s \in (t_{n-1}, t_n]$ indicates that $x_{k+1}(s - \tau) = B^o_{k+1}$. The data to be received at node $k + 1$ from $s - \tau$ to t_n, whose number is denoted by $A_{k+1}(s - \tau, t_n)$ [see (7.2)], were previously sent by node k in $(s - 2\tau, t_n - \tau]$, denoted by $D_k(s - 2\tau, t_n - \tau)$ [see (7.1)]. The data dispatched by the model queue according to (7.16) is $R(t_n - s + \tau)$ during $(s - \tau, t_n]$. Thus from the queue evolution equation (7.6),

$$\hat{x}_{k+1}(t_n) = B^o_{k+1} + A_{k+1}(s - \tau, t_n) - R(t_n - s + \tau). \quad (7.17)$$

- If the received signal is bit 0, as shown in Figure 7.9(b), indicating that $x_{k+1}(s - \tau) = B^u_{k+1}$, then

$$\hat{x}_{k+1}(t_n) = B^u_{k+1} + A_{k+1}(s - \tau, t_n) - R(t_n - s + \tau). \quad (7.18)$$

- If no feedback signals have been received within $(t_{n-1}, t_n]$, as shown in Figure 7.9(c), since the total number of data received by the model queue during this interval is $A_{k+1}(t_{n-1}, t_n)$, the number of dispatched data is RT. Using the previous state $\hat{x}_{k+1}(t_{n-1})$, it follows from (7.6) that

$$\hat{x}_{k+1}(t_n) = \hat{x}_{k+1}(t_{n-1}) + A_{k+1}(t_{n-1}, t_n) - RT. \quad (7.19)$$

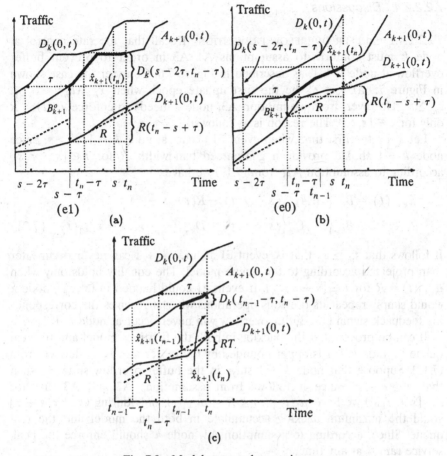

Fig. 7.9 Model state update at time t_n.

7.2.2.2.3.2 SERVICE RATE UPDATE Based on the queue model, node k will choose its new service rate $\mu_k(t_n)$ exactly as it did with the real queue of node $k + 1$, as introduced in Section 7.2.2.3. So

$$\mu_k(t_n) = \hat{\mu}_{k-1}(t_n - \tau) + \frac{B_k}{B_{k+1}}\hat{\mu}_{k+1}(t_n) - \frac{B_k}{TB_{k+1}}A_{k+1}(t_n, t_{n+1})$$

$$+ \frac{1}{T}\left[x_k(t_n) - \frac{B_k}{B_{k+1}}\hat{x}_{k+1}(t_n)\right], \tag{7.20}$$

where $A_{k+1}(t_n, t_{n+1})$ can be extracted from the record $D_k(t_n - 2\tau - T, t_n)$ due to $T \le \tau$. This result can also be obtained from (7.9) with $\mu_{k-1}(t_n - \tau)$ replaced by $\hat{\mu}_{k-1}(t_n - \tau)$ from (7.15), $\mu_{k+1}(t_n)$ by $\hat{\mu}_{k+1}(t_n)$ from (7.16), and $x_{k+1}(t_n)$ by $\hat{x}_{k+1}(t_n)$ from (7.17), (7.18), or (7.19).

7.2.2.2.4 *Discussions*

7.2.2.2.4.1 BUFFER OVERFLOW PREVENTION Recall that the rate control at node k must conform to assumptions A1–A3 in order to prevent buffer overflow at node $k + 1$, as described in Figure 7.6. Consider the case shown in Figure 7.9(a). Let t_m be the first update epoch when $\hat{x}_{k+1}(t_m) < B^o_{k+1}$, $t_m \geq t_n$. However, from assumption A3, node k needs to enforce $\mu_k(t) \leq R$ only for $t \in [s, t_m)$. The reason is as follows.

Let s' be the first time when event $\bar{e}1$ occurs at node $k + 1$, $s' > s$. Since node $k + 1$ should provide a guaranteed bandwidth R for $t \in [s - \tau, s')$ according to assumption A2, from (7.17) we have

$$\hat{x}_{k+1}(t) = B^o_{k+1} + A_{k+1}(s - \tau, t) - R(t - s + \tau)$$

$$\geq B^o_{k+1} + A_{k+1}(s - \tau, t) - D_{k+1}(s - \tau, t) = x_{k+1}(t). \quad (7.21)$$

It follows that $t_m \geq s'$, that is, event $\bar{e}1$ at node $k + 1$ can never occur later than projected according to the queue model. The equality holds only when $\mu_{k+1}(t) = R$ for $t \in [s - \tau, t_m)$. If event $e1$ could happen in $(s', t_m]$, node k would simply repeat its previous operations when it receives the corresponding feedback signal (1). Buffer overflow will never occur at node $k + 1$.

It can be proven that the backlog in both the fictitious model and the real queue of node $k + 1$ is upper bounded by the buffer size B_{k+1} derived from (7.12). Suppose that node $k + 1$ stays in the buffer overflow state S_O such that $s' \geq s + \tau$. Since it follows from assumptions A2 and A3 that for $t \in [s + \tau, s')$ we have $\mu_k(t - \tau) \leq R \leq \mu_{k+1}(t)$, only during $(s - \tau, s + \tau]$ could the maximum backlog accumulate in both the model and the real queue. Since, according to assumption A1, node k should enforce the peak service rate S at any time,

$$A_{k+1}(s - \tau, s + \tau) \leq 2S\tau, \quad (7.22)$$

it follows from (7.21), (7.22), and (7.12) that for $t \in (s - \tau, s + \tau]$,

$$x_{k+1}(t) \leq \hat{x}_{k+1}(t) \leq B^o_{k+1} + 2(S - R)\tau = B_{k+1}. \quad (7.23)$$

This completes the proof.

7.2.2.2.4.2 DATA STARVATION AVOIDANCE There are circumstances in which the buffer at node $k + 1$ stays in the underflow state S_U or even becomes empty due to its service rate being persistently greater than its arrival rate. This is called *data starvation*. It can be even worse if the arrival rate at node $k + 1$ (departure rate of node k) is actually greater than its assumed service rate (which is R) in the fictitious queue model at node k, because without knowing that node $k + 1$ has underflowed, node k will reduce its output to node $k + 1$ as it sees a large backlog accumulating in the fictitious queue model, further extending the duration of starvation at node $k + 1$.

To avoid starvation at node $k + 1$, node k should increase its service rate. From (7.20), the service rate at node k is inversely proportional to the status of its simulating queue model for node $k + 1$; this suggests a heuristic mechanism to decrement the status of the queue model occasionally. Such a mechanism starts at time t_n on the conditions that, first, the last feedback signal is (0), which indicates that the buffer has underflowed previously; second, there is no other feedback signal received since the last signal; and third, the current model status $\hat{x}_{k+1}(t_n)$ exceeds a certain threshold B_{k+1}^{th} ($B_{k+1}^u \leq B_{k+1}^{th} \leq B_{k+1}$). With this mechanism, node k will periodically decrement the model status by a positive constant Unit–Step ($0 <$ Unit–Step $\leq B_{k+1}^{th}$) until it receives new feedback signals, and for $t_h \geq t_n$,

$$\hat{x}_{k+1}(t_h) = \begin{cases} \max\{0, \hat{x}_{k+1}(t_n) - \text{Unit–Step}\} & \text{if } h = n, \\ \max\{0, \hat{x}_{k+1}(t_{h-1}) - \text{Unit–Step}\} & \text{if } h > n, \end{cases} \quad (7.24)$$

where the max function ensures that $\hat{x}_{k+1}(t_h) \geq 0$.

The parameters B_{k+1}^{th} and Unit–Step decide when to start the mechanism and the speed of increasing the service rate, respectively. It would be preferable to choose a small B_{k+1}^{th} and a large Unit–Step for early and fast response in the presence of large link propagation delay. However, it should be noticed that with the peak-rate limitation from assumption A1, a too large Unit–Step may make little difference.

7.2.2.2.4.3 ACCESS FLOW CONTROL Access flow control should be able to ensure high utilization of network resources and yet prevent network congestion. Suppose that node 1 is the CPN for the VC. If $B_1 \gg B_2$, ensuring $x_2(t) = (B_2/B_1)x_1(t)$ from (7.5) and hence $x_1(t) \gg x_2(t)$, the buffer and bandwidth at node 2 may be underutilized, with few incoming data. To avoid this, a positive constant α instead of B_2/B_1 can be chosen:

$$x_2(t) = \alpha x_1(t). \quad (7.25)$$

The same rate update equation as (7.9) can be derived from (7.25), where α can be adjusted for achieving a proper backlog distribution between nodes 1 and 2 and high network utilization [63].

7.2.2.3 Service Policy The next subject is a service policy for the BBFC, which consists of bandwidth allocation, service (time) quantum refreshment, and dequeue processes.

7.2.2.3.1 Bandwidth Allocation Consider N VCs multiplexed at a node. Each of them uses the BBFC and has a logically separate first-in, first-out queue maintained by some buffer management scheme. It is straightforward to use the updated service rate of each VC as its weight factor to decide its share of the link capacity, because this rate indicates its explicit demand for bandwidth.

Let C be the node's outgoing link capacity. For VC i, $i = 1, 2, \ldots, N$, we use $\mu(t, i)$ to denote its expected service rate at time t from the BBFC and $\mu^a(t, i)$ to denote its assigned bandwidth at the same instant. The *capacity constraint* of this node is given by

$$\sum_{i=1}^{N} R(i) \le C, \qquad (7.26)$$

because under assumption A2 of the protocol operations, the node should provide a minimum bandwidth $R(i)$ for VC i when the buffer of this VC is in the overflow state S_O. Note that if this node is the first node over VC i where the above equality holds, its outgoing link is called a *bottleneck link* of this VC.

The bandwidth allocation $\mu^a(t, i)$ should conform to assumptions A1–A3, and normally when the buffers of VC i at both the node and its downstream neighbor over VC i are not in the overflow state, $\mu^a(t, i)$ should be limited by its expected value $\mu(t, i)$,

$$\mu^a(t, i) = \min\left\{ \mu(t, i), \frac{\mu(t, i)}{\sum_{j=1}^{N} \mu(t, j)} \times C \right\}, \qquad (7.27)$$

where $\sum_{j=1}^{N} \mu(t, j) > 0$ is assumed. So, there may be excess bandwidth left after the allocation for every VC if $\sum_{i=1}^{N} \mu^a(t, i) < C$. A dummy VC $N + 1$ can be introduced with its assigned bandwidth equal to the residual link capacity,

$$\mu^a(t, N + 1) = C - \sum_{i=1}^{N} \mu^a(t, i). \qquad (7.28)$$

Figure 7.10 shows the service model, where each VC is served in a cyclic order.

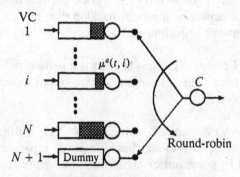

Fig. 7.10 Service model.

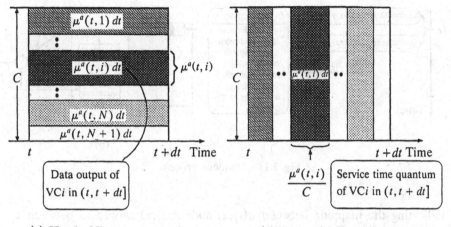

(a) Head-of-line processor sharing   (b) Round-robin emulation

Fig. 7.11   Emulation of head-of-line processor sharing.

7.2.2.3.2 Quantum Refreshment
The quantum refreshment converts the assigned bandwidth of each VC into its service time quantum, in units of the cell time slot in the outgoing link, through a round-robin emulation of HOL processor sharing. Each cell must consume one unit of the quantum before it is dispatched according to the dequeue process (explained next). The process can be detailed as follows.

Consider ideal HOL processor sharing at the node. During a small time interval $(t, t + dt]$, the node will serve a nonempty queue i at a constant rate $\mu^a(t, i)$, dispatching data summed to $\mu^a(t, i)\,dt$ as shown in Figure 7.11(a). $\mu^a(t, N + 1)\,dt$ can be regarded as the fictitious output from the dummy queue. The total departure amounts to $C\,dt$.

Under the round-robin emulation, every queue is served one after another. The service time quantum for queue i in $(t, t + dt]$ is equal to that needed to transmit the same number of data at the link capacity, $\mu^a(t, i)\,dt/C$, as illustrated in Figure 7.11(b), while for the dummy queue, $\mu^a(t, N + 1)\,dt/C$ can be regarded as the node's vacation time.

Assume that bandwidth allocation is exercised periodically with cycle F. Let $t_f = fF$ be the fth allocation epoch. The bandwidth assigned to VC i for $t \in (t_f, t_{f+1}]$ is a constant denoted by $\mu^a(t_f, i)$, $i = 1, 2, \ldots, N + 1$, which is determined at time t_f. Let $q(t_f, i)$ be the service quantum of VC i that should be computed at the same time t_f based on $\mu^a(t_f, i)$ from (7.27) and (7.28). So

$$q(t_f, i) = \frac{\mu^a(t_f, i)}{C} \times F, \qquad (7.29)$$

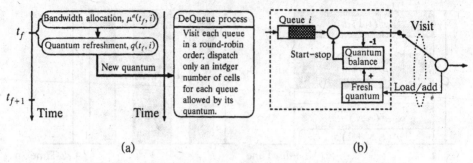

(a)   (b)

Fig. 7.12 Dequeue process.

indicating the mapping between $q(t_f, i)$ and $\mu^a(t_f, i)$ as well as between C and F, and from (7.28) and (7.29),

$$F = \sum_{i=1}^{N+1} q(t_f, i). \tag{7.30}$$

Note that to enforce the service rate of VC i in each of its update intervals, the cycle F should be less than or equal to its update period $T(i)$. For every VC, it follows that

$$F \leq \min_i \{T(i) | i = 1, 2, \ldots, N\}. \tag{7.31}$$

7.2.2.3.3 Dequeue Process The dequeue process, running in parallel with the bandwidth allocation and the quantum refreshment as shown in Figure 7.12(a), performs the job of transmitting cells for each VC in a round-robin fashion according to its service quantum. As illustrated in Figure 7.12(b), when the node visits queue i, $i \neq N + 1$, it should first add the fresh quantum for this queue to its counter, which is recording the current quantum balance. Each cell must consume one unit of the quantum before it leaves. No cell can be sent when the quantum balance becomes zero or less than one unit, and only in this case, or if the queue has been depleted, will the node be allowed to serve other queues. The residual quantum for queue i after current service can be saved in the counter for its next turn. This technique, which is used to preserve a good match between the bit-by-bit round-robin emulation and the HOL processor sharing, was first proposed in *deficit round-robin* [40] for realizing the GPS service policy. The dummy queue has no counter, and its fresh quantum will decide the duration of a visit by the node.

7.2.2.4 Simulation Results We compare BBFC with both FCVC [38, 46, 51] using the GPS service policy [40] and EASY (see Section 7.4 for details) by simulation as shown below. We assume that each scheme here can guarantee no data loss in the network.

7.2.2.4.1 Basic Assumptions

With reference to Figure 7.13, we consider two different network models in our simulation. Each network model is composed of ten hosts and four network nodes. Without loss of generality, host i is the source of VC i and consists of its CPN, for $1 \leq i \leq 10$. Each VC passes node 4, as indicated by the arrows in Figure 7.13. The shaded nodes (node 1 in the first model and node 4 in the second) have the common bottleneck link for every VC in the respective models. Such models can be used to investigate, in an analogy, how the flooding water in converse situations can be controlled across a series of dams under various flow control strategies.

Assume that all links have the identical capacity $C = 1$ cell per unit time. For simplicity, let the cycle $F = 1$. The simulation time is 10,000,000. An on–off data source [41] characterized by (S, r, L) is used, where S is the peak rate, r the average rate, and L the mean burst length of the VC. We also assume that $S = C$ for each VC. The throughput per VC is measured at node 4 as the total number of cells it has dispatched during the simulation time.

On the other hand, each host and node has an infinite buffer, so there is no data loss from end to end. Moreover, there is also an infinite buffer at the downstream neighbor of node 4, so that node 4 can send data as fast as possible under various flow control strategies. Identical nodal buffer allocation is used for each scheme. The feedback information will not experience waiting delay, but only the link propagation delay.

For comparison, let us choose the minimum bandwidth provided by the GPS service policy to be the agreed constant rate specified in both the BBFC and the EASY. For the FCVC, assume that each node will send a feedback cell to its upstream node once it dispatches a data cell for each VC. For the

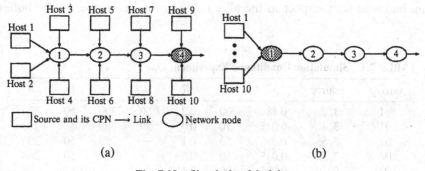

(a)   (b)

Fig. 7.13 Simulation Models.

BBFC, we simply choose $\alpha = 1$ for its access control, and $B_{k+1}^{th} = B_{k+1}^{o}$, Unit–Step $= RT$ for the starvation avoidance.

7.2.2.4.2 *Experiment One*

This experiment is based on the first model in Figure 7.13(a). Assume that the propagation delay of the outgoing link at each CPN is 10, and at node 1 (2, 3) it is 100 (10, 100). (As an example, if $C = 155$ Mbit/s, the propagation delay is 5 μs/km [48], and the length of a link is 54.7 km with a delay of 100.) The ten sources are divided into five distinct groups in Table 7.1. Each group consists of two homogeneous sources with identical end-to-end link propagation delays, so it can be confirmed whether each of them will accept the same service fairly under every scheme. The buffer underflow threshold at node 2 is chosen to be 50 cells for VCs 3, 4 instead of 200 cells from (7.13), and similarly at node 4, 50 cells for VCs 3, 4 and VCs 7, 8, and 100 cells for VCs 5, 6. For simplicity, let $T = 10$ for a VC at each node on its path, including its CPN.

The performance of each scheme is measured in terms of the backlog distribution of a VC, throughput, nodal buffering, and end-to-end and network cell transfer delays (ED and ND), where the ND is equal to the ED minus the cell sojourn time at the CPN.

Consider VC 1. Figure 7.14 shows that the BBFC has tried to ensure an even backlog distribution over the VC. Since node 4 has the common bottleneck link for each VC, the BBFC intentionally reduces its backlog, while the FCVC and the EASY always intend to send data as fast as possible, as indicated by their increasing backlog distributions. This indeed validates the uniqueness of the BBFC in preventing network congestion. Also observed is that each node under the BBFC has desirably fewer queue fluctuations.

Figure 7.15(a) confirms that each scheme can achieve similar throughput in general, because there is no data loss across the network. Besides, each scheme can provide fair service for different VCs in this regard. Defining the sum of buffer allocation per VC at a node to be its *expected total buffer usage* (ETBU), and the maximum aggregate backlog of all the VCs at this node to be its *actual total buffer usage* (ATBU), their ratio can be used to examine the buffer sharing. The smaller the ratio ATBU/ETBU, the more buffer space can be saved with respect to the allocation. In this sense, we say the buffer

TABLE 7.1   Simulation Conditions: Experiment One

Group	Source	r	L	R	B_1	B_2	B_3	B_4
I	1, 2	0.18	200	0.2	60	390	60	390
II	3, 4	0.045	10	0.05	50	260	50	260
III	5, 6	0.095	50	0.1	—	50	50	300
IV	7, 8	0.045	50	0.05	—	—	50	260
V	9, 10	0.095	100	0.1	—	—	—	50

sharing can be more efficient. In this regard, Figure 7.15(b) shows that the BBFC can save more buffer space than the EASY and yet achieves the same throughput performance, while the FCVC has the greatest nodal buffer usage except for node 1 in this case, where the ETBU of node (2, 3, 4) is 220 (1400, 420, 2520) cells.

Figure 7.15(c) shows the end-to-end delay (ED) for each VC. Consider VCs 3, 4 and VCs 7, 8; each of them has identical source loading, but the former pair has a longer end-to-end path. Intuitively VCs 3, 4 should experience greater ED, as is the case for VCs 5, 6 and VCs 9, 10. The BBFC, as proven, can achieve this and thus display its fairness property. The FCVC seems the worst, while the EASY fails in the first case. However, it should be noted that with the state-dependent service policy introduced in Section 7.4, the EASY can achieve overall better ED (small mean and variance) for every VC.

On the other hand, Figure 7.15(d) shows that the BBFC outperforms the other two in terms of the network delay (ND), while the FCVC is still the worst. It may be conceived that by choosing a proper α instead of 1 for the access control of the BBFC, the ED performance can be improved for each VC.

7.2.2.4.3 Experiment Two Our second experiment is based on the network model shown in Figure 7.13(b). Let all input links to node 1 have the same delay of 10. Table 7.2 lists some conditions. The other assumptions are the same as in Experiment One.

Figure 7.16 shows that since most of the flooding water has been regulated by the first dam (node 1), its downstream nodes, as observed, have often stayed in the buffer underflow state during the simulation. The EASY can quickly adapt to this situation by enabling each of these nodes to send data as fast as possible over this VC, causing the least backlog at each of them. It is similar but not so significant for the FCVC, because its credit control means that it cannot take advantage of the minimum bandwidth provided by the GPS service policy.

It is different for the BBFC. Since cell departures at every node are subject to the rate control for the backlog balancing all the time, it would take a longer step to adapt to the above situation, as indicated by a decreasing backlog distribution over the VC. It may be understood that without dispatching data fast like the EASY, each node could intentionally suffer a certain loss in taking advantage of the statistical multiplexing at its downstream neighbor and, as a consequence, might raise its cell transfer delay.

However, the BBFC would be desirable for preventing congestion in a large complex network when every VC may have its bottleneck link at different spots, as seen in Figure 7.14 and Figure 7.16. Also observed in Figure 7.16 is that the BBFC still enables node 1 to have fewer queue fluctuations.

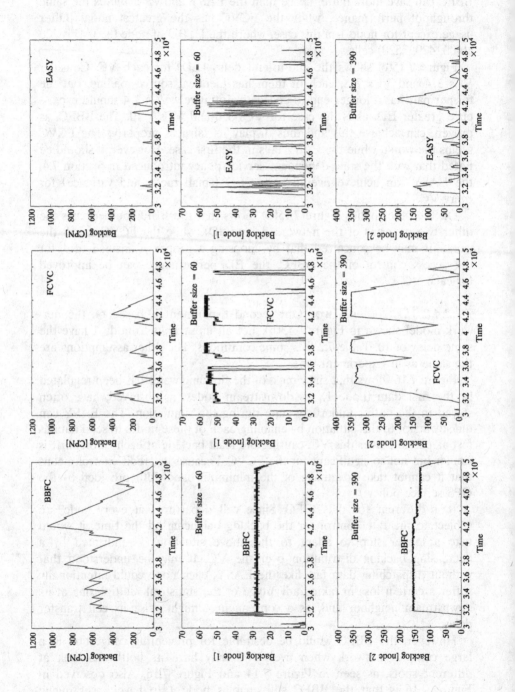

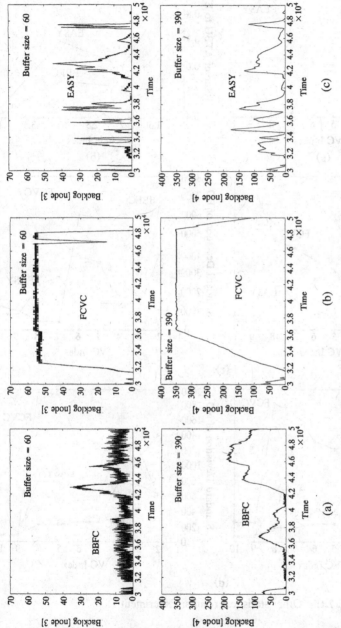

Fig. 7.14 Backlog distribution of VC 1: Experiment One.

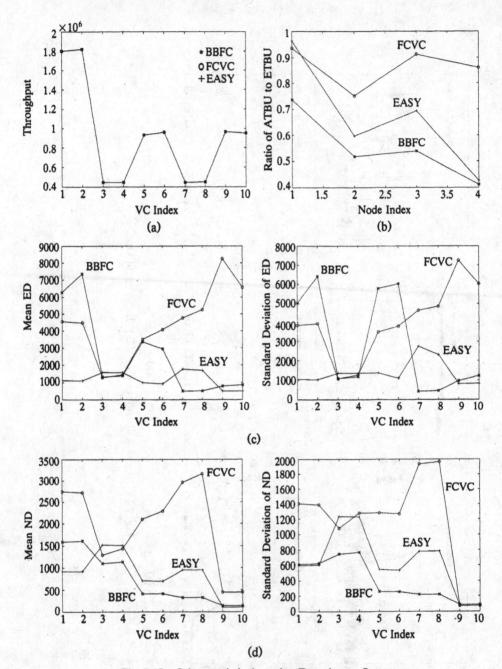

Fig. 7.15 Other statistical results: Experiment One.

TABLE 7.2   Simulation Conditions: Experiment Two

Group	Source	r	L	R	B_1	B_2	B_3	B_4
I	1, 2	0.18	200	0.2	60	390	60	390
II	3, 4	0.045	10	0.05	50	260	50	260
III	5, 6	0.095	50	0.1	50	300	50	300
IV	7, 8	0.045	50	0.05	50	260	50	260
V	9, 10	0.095	100	0.1	50	300	50	300

Figure 7.17(a) validates each scheme's fairness in terms of throughput per VC. Figure 7.17(b) shows, as expected, that it is the EASY that has the smallest ratio ATBU/ETBU at nodes 2, 3, and 4, but it is still the BBFC at node 1. The FCVC always needs the greatest nodal buffering: the ETBU for node 1 (2, 3, 4) is 520 (3020, 520, 3020) cells.

Figure 7.17(c) shows that the BBFC and the EASY are comparable with regard to the end-to-end delay (ED), while the FCVC not only has the worst ED performance but apparently also fails to provide fair service in this regard. The remark can be further justified with the network delay (ND) performance shown in Figure 7.17(d). So it may be concluded at this point that window flow control like the FCVC does have poor controllability of cell transfer delay in terms of mean and variance, due to its inherent start–stop data transmission; however, it could be improved with some service discipline, such as that for the EASY [63].

7.2.2.5 Summary   This section has presented a rate-based flow control algorithm (BBFC) for ATM networks. The basic idea of the algorithm is to adjust the service rate of each node along a VC according to backlog discrepancies between neighboring nodes. The handshaking procedure between any two consecutive nodes is carried out by a link-by-link binary feedback protocol. Each node updates its service rate periodically based on a linear projection model of the flow dynamics. The updated service rate per VC at a node indicates its explicit demand for bandwidth, so a service policy implementing dynamic bandwidth allocation is introduced to enforce such demands.

Three main conclusions are borne out by the investigation. First, the rate update algorithm subject to the backlog balancing always serves to reduce the backlog discrepancy between neighboring nodes. Although it is not feasible to devise any control mechanism to ensure an even backlog distribution across the network all the time, it is possible to develop some algorithms to approach that, and this remark has been substantiated by the BBFC.

Second, when every VC in a network complies with the same set of rules from the BBFC, the maximum backlog at each node can be reduced. The possibility of network congestion is also reduced, while acceptable throughput and delay for each VC can still be achieved (i.e., *efficient bandwidth*

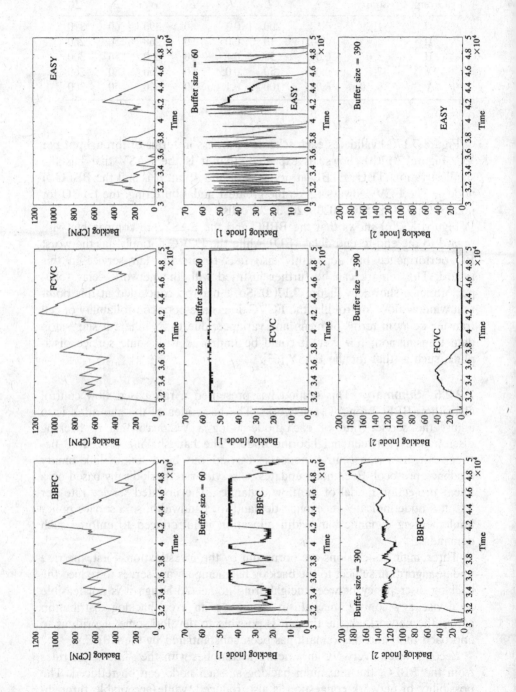

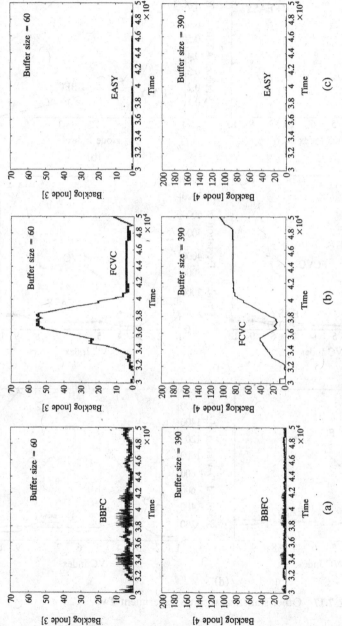

Fig. 7.16 Backlog distribution of VC 1: Experiment Two.

265

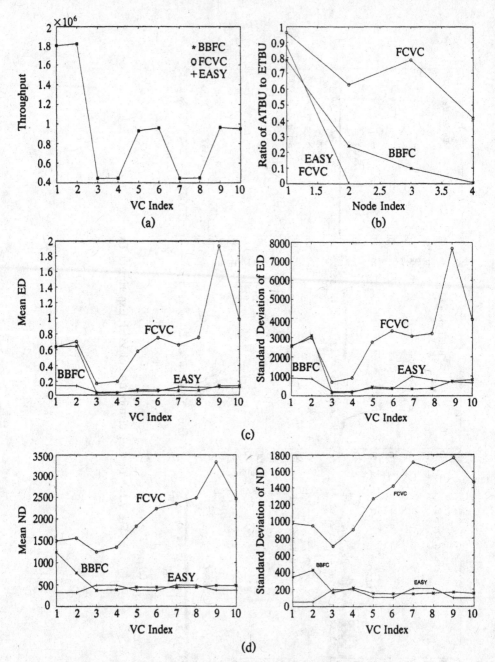

Fig. 7.17 Other statistical results: Experiment Two.

sharing) and nodal buffering can be reduced (*efficient buffer sharing*). Besides, each VC can have small variations of the end-to-end and network cell transfer delay (*traffic smoothing*). It has also been validated that the BBFC can provide fair service for each VC with respect to both throughput and delay performance.

Third, the EASY as described in Section 7.4 can compete with the BBFC in many aspects, except for the nodal buffering. The reduction of its nodal buffering and delay variance should be mainly attributed to the state-dependent service policy introduced in Section 7.4. Therefore, it would be reasonable to conclude that the BBFC is so far a unique flow control algorithm that can achieve efficient bandwidth and buffer sharing, and yet prevent network congestion at the same time. It also provides desirable traffic smoothing. However, the implementation cost of the BBFC could be very high, if not prohibitive, in its current form. Next, we present the ABR flow control as recommended by the ATM Forum.

7.2.3 ABR Flow Control

The ABR flow control is a kind of closed-loop control. With reference to Figure 7.18, a special type of cells, called *resource management* (RM) cells, are used to handle the feedback signals from the network as follows. An ABR source sends RM cells, interleaved with the data cells, to its destination. When these RM cells reach the destination, they are turned around and sent back to the source. Each RM cell can be modified at any network node during its round-trip journey in order to take down the state information of the network. Based on the feedback information from the control loop, the ABR source can adapt its cell transmission rate to achieve high throughput while preventing network congestion and cell loss.

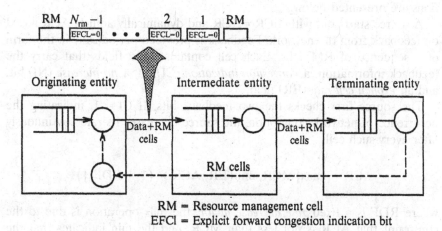

RM = Resource management cell
EFCl = Explicit forward congestion indication bit

Fig. 7.18   ABR rate-based flow control scheme.

The following are the chief characteristics of the ABR service [60].

- ABR connections share the available capacity that is not used by CBR and VBR connections. Thus, ABR can increase network utilization without affecting the QoS of CBR and VBR connections.
- The bandwidth allocated to a single ABR connection is dynamic and varies between its MCR and PCR. A particular connection may ask for an MCR of zero.
- The network provides feedback signals to ABR sources so that ABR flow is limited to the available capacity. Time delay inherent in providing feedback signals dictates the use of buffers along the traverse path. These buffers absorb excess traffic generated prior to the arrival of the feedback signal at the source. Accordingly, the ABR service is appropriate for applications that can tolerate adjustments to their transmission rates and unpredictable cell delays.
- ABR sources adapt their transmission rates dynamically according to feedback signals; thus, a low cell loss rate is guaranteed. This is a major distinction between ABR and UBR.

7.2.3.1 *Source Behavior* The transmission rate of cells from a ABR source is characterized by four parameters: allowed cell rate (ACR), initial cell rate (ICR), MCR, and PCR. ACR is the current rate at which the source is permitted to transmit cells. ICR is the initial value assigned to ACR. During call setup, the network guarantees a source that its ACR will never be smaller than MCR. A source informs the network that its transmission rate will never be higher than PCR. Thus, ACR varies between MCR and PCR. The source may transmit at any rate between zero and ACR. Moreover, immediately after call setup, the network allows the source to send data at ICR. Figure 7.19 shows the parameters used in a call setup. Further explanations are presented below.

A source starts out with ACR = ICR and dynamically adjusts ACR based on feedback from the network. Feedback is provided periodically in the form of a sequence of RM cells. Each cell contains three fields that carry the feedback information: a *congestion indication* (CI) bit, a *no increase* (NI) bit, and an *explicit cell rate* (ER) field.

The source first checks the two feedback bits. If CI = 1, indicating the occurrence of network congestion, the source decreases its rate continuously after every such cell:

$$ACR \leftarrow \max\{MCR, \min\{ER, ACR \times (1 - RDF)\}\},$$

where RDF is a fixed *rate decrease factor*, the max operation is due to the constraint that ACR is not less than MCR, and the min indicates that the source reduces ACR to ER if ER is smaller than ACR. The amount of

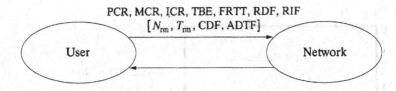

PCR, MCR, ICR, TBE, FRTT, RDF, RIF
$[N_{rm}, T_{rm}, CDF, ADTF]$

Name	Full Name	Units and Range
PCR	Peak cell rate	cells/s, 16-bit floating point format
MCR	Minimum cell rate	cells/s, 16-bit floating point format
ICR	Initial cell rate	cells/s, 16-bit floating point format
TBE	Transient buffer exposure	cells 0 to 16, 777, , 215
FRTT	Fixed round-trip time	μs
RDF	Rate decrease factor	1/32768 to 1
RIF	Rate increase factor	1/32768 to 1
N_{rm}		cells
T_{rm}		ms
CDF	Cutoff decrease factor	none, 1/64 to 1
ADTF	ACR decrease time factor	s, 0.01 to 10.23

Fig. 7.19 Call setup for an ABR connection.

decrease is proportional to the current value of ACR but not less than MCR. The default value for RDF is $\frac{1}{16}$.

If CI = 0, different rules apply depending on the NI bit. If NI = 0, an increase is called for. The source increases its rate by an amount proportional to PCR but no more than PCR, namely, RIF × PCR, and

$$ACR \leftarrow \max\{MCR, \min\{ER, PCR, ACR + RIF \times PCR\}\},$$

where RIF is a fixed *rate increase factor* and its default value is also $\frac{1}{16}$. If NI = 1, no increase is allowed. The source simply makes sure ACR is no less than MCR but no more than ER,

$$ACR \leftarrow \max\{MCR, \min\{ER, ACR\}\}.$$

Therefore, the source will slowly increase its rate when there is no evidence of congestion (i.e., CI = 0).

There are three types of cells associated with an ABR connection: data cell, forward RM (FRM) cell, and backward RM (BRM) cell. Data cells carry user data. RM cells are initiated by the source, which sends one FRM cell for every N_{rm} − 1 data cells, where N_{rm} is a preset parameter (default value 32) and is introduced to control the bandwidth used by RM cells. FRM cells are turned around by the destination and are sent back to the source as BRM cells to convey the feedback information to the source. Each FRM cell contains CI, NI, and ER fields. The source typically sets CI = 0 (no congestion), NI = 0 or 1, and ER equal to some desired rate between ICR and

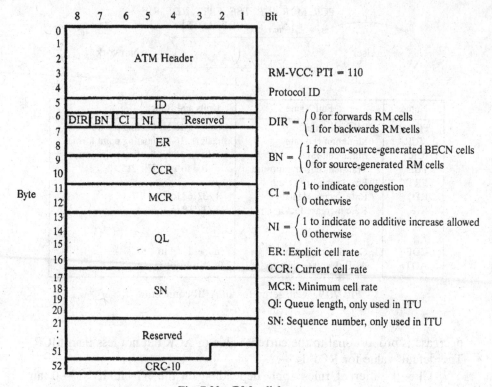

Fig. 7.20 RM cell format.

PCR. Any of these fields in FRM and BRM cells are subject to change by ATM switches or the destination. Figure 7.20 shows the format of a RM cell.

There are two additional details to make the flow control more robust. First, if a source is idle for, say, 500 ms (a different value can be negotiated during call establishment), its ACR is reset to ICR, preventing it from injecting traffic too fast into the network if the previous value of ACR is much higher. Second, RM cells could be lost, so the source should keep track of whether it receives an RM cell back. If it does not receive an RM cells for a prescribed interval, it continually decreases its sending rate. Thus, when a link goes down, all affected sources eventually stop transmission. In other words, the source only increases its sending rate for a connection when given an explicit positive indication (through RM cells) to do so. Figure 7.21 explains the rate representation for an ABR connection.

7.2.3.2 Switch Behavior To support ABR services, ATM switches must monitor queue lengths and begin to throttle back rates as the buffer becomes full. On the other hand, an ATM switch should allocate a fair share of its capacity among all connections that pass through it and selectively throttle

Rates are represented in a binary floating-point representation employing a 5-bit exponent e, a 9-bit mantissa m, and a 1-bit n_z, as described below:

$R = [2^e(1 + m/512)]n_z$ cells/s, with

1 bit reserved	Most significant bit of 16-bit field
$n_z = \{0, 1\}$	Next bit
	If $n_z = 0$, the rate is zero. If $n_z = 1$, the rate is as given by the fields e and m.
$0 \leq e \leq 31$	Next 5 bits
$0 \leq m \leq 511$	Remaining 9 bits
$0 \leq R \leq 4,290,722,992$	

The bit positions of a floating-point rate within a 16-bit word are given below:

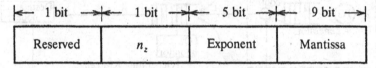

Note: During signaling, cell rates are negotiated with 24-bit integers.

Fig. 7.21   Rate representation.

back on those connections that are using more than their fair share of the capacity on congestion. There are two kinds of methods to provide rate control feedback to a source: binary feedback schemes and explicit rate feedback schemes. Below we give some examples.

7.2.3.2.1 Binary Feedback Schemes Explicit forward congestion indication (EFCI) marking and relative rate marking are two binary feedback schemes. They both have the same structure: the switch monitors its buffer utilization on each output port and, when congestion is pending, performs a binary notification by either approach. The EFCI marking allows the switch to set the EFCI bit in the payload type field of an ATM data cell header as it passes in the forward direction. This will cause the destination end system to set the CI bit in a BRM cell, as illustrated in Figure 7.22.

The relative rate marking allows the switch to directly set the CI or NI bit of a passing FRM cell, which will remain set in the corresponding BRM cells. The switch can also set one of those bits in a passing BRM cell to reduce the feedback delay. The fastest notification, though at the cost of increasing implementation complexity, is to allow the switch to generate a BRM cell with CI or NI bit set rather than waiting for a passing BRM. This special cell can also be regarded as a backward explicit congestion notification (BECN)

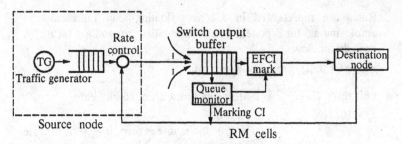

Fig. 7.22 EFCI-based switch behavior.

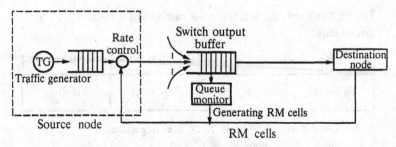

Fig. 7.23 BECN-based switch behavior.

cell with the BN bit set to one, indicating cells initially generated by a switch or destination. (BN = 0 for source.) Figure 7.23 illustrates the BECN-based switch behavior.

There are several approaches to determine which connection to notify first. First, the buffer at each output port is dedicated to a single FIFO queue with one or two thresholds. In the one-threshold case, when buffer occupancy exceeds the threshold, the switch begins to issue binary feedback signals and continues to do so until buffer occupancy falls below the threshold. In the two-threshold case, the binary notifcation begins when buffer occupancy increases and crosses the high threshold; it won't cease until buffer occupancy decreases and crosses the low threshold, thus preventing frequent generation of binary feedback signals.

The second approach is to allocate a separate queue to each VC or to each group of VCs. A separate threshold is used on each queue, so that at any time, binary notification is provided only to VCs with long queues, thus improving the fairness. The third approach is to dynamically allocate a fair share of the capacity, which, for instance, can be the target rate divided by the number of ABR connections. When congestion occurs, the switch marks cells on any VC with its current cell rate (CCR) above the fair share. CCR is set by the source to its current ACR and is carried in RM cells.

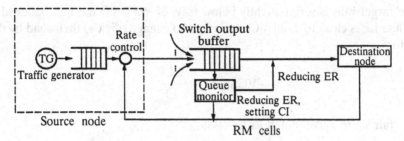

Fig. 7.24 ER-based switch behavior.

7.2.3.2.2 Explicit Rate Feedback Schemes The explicit rate feedback schemes, also called explicit rate marking, allow the switch to reduce the value of the ER field in an FRM or BRM cell. To do that, the switch first computes the fair share for each VC, determines the current load, and then computes a new value of ER for each connection, as illustrated in Figure 7.24. Below are three examples.

In an enhanced proportional rate control algorithm (EPRCA) [67, 68, 49], the switch uses exponential weighted averaging to compute a mean allowed cell rate (MACR) for all connections,

$$\text{MACR} \leftarrow (1 - \alpha) \times \text{MACR} + \alpha \times \text{CCR}.$$

The fair share is set at a fraction of this average,

$$\text{fair share} = \text{DPF} \times \text{MACR},$$

and

$$\text{ER} \leftarrow \min\{\text{ER, fair share}\}.$$

Here, the min operation is due to the constraint that no switch is allowed to increase the ER value in an RM cell. MACR indicates an estimate of the average loading. α is the exponential averaging factor, and DPF (the down pressure factor) is set close to but below 1. The suggested values of α and DPF are $\frac{1}{16}$ and $\frac{7}{8}$, respectively. Because all VCs using more than their fair shares are reduced to the same ER, the throttling is performed fairly.

In a target utilization band algorithm (TUBA) [69, 70], a switch measures its input rate over a fixed "averaging interval" and compares it with its target rate to compute the current load factor (LF):

$$\text{LF} = \frac{\text{input rate}}{\text{target rate}}.$$

The target rate is set at slightly below (say, 85–95% of) the link bandwidth. Unless LF is close to 1, all VCs are asked to change (divide) their load by this factor:

$$\text{VC share} = \frac{\text{CCR}}{\text{LF}}.$$

The fair share is computed as follows:

$$\text{fair share} = \frac{\text{target rate}}{\text{number of connections}}.$$

Thus,

$$\text{ER} \leftarrow \min\{\text{ER}, \max\{\text{fair share}, \text{VC share}\}\}.$$

The interpretation is as follows. Under low load (LF < 1) each VC is assigned an ER above its current CCR. Those VCs whose VC share is less than the fair share receive a proportionately greater increase. Under heavy load (LF > 1), some VCs are assigned an ER greater than their current CCR, and some are assigned a lower ER, with the intention of benefiting those VCs with lesser shares [61].

The TUBA has three attractive features. First, it is a congestion avoidance scheme. By keeping the target rate slightly below the capacity, the algorithm ensures that the queues are very small, typically close to 1, resulting in low delay. Second, the switch has very few parameters compared to EPRCA and is easy to set. Third, the time to reach the steady state is very small compared to EPRCA.

The congestion avoidance using proportional control (CAPC) algorithm [71, 72] also uses the load factor LF to determine ER. The fair share for each VC is initialized as that in [61]. Then, with each arriving RM cell, the fair share is updated as follows. During underload (LF < 1),

$$\text{fair share} \leftarrow (\text{fair share}) \times \min\{\text{ERU}, 1 + (1 - \text{LF}) \times R_{up}\}.$$

Here, R_{up} is a slope parameter in the range 0.025 to 0.1. ERU is the maximum increase allowed and is typically set to 1.5. During overload (LF > 1), the fair share is decreased by

$$\text{fair share} \leftarrow (\text{fair share}) \times \max\{\text{ERF}, 1 - (\text{LF} - 1) \times R_{dn}\}.$$

Here, R_{dn} is also a slope parameter, in the range 0.2 to 0.8. ERF is the maximum decrease allowed and is typically set to 0.5. Thus,

$$\text{ER} \leftarrow \min\{\text{ER}, \text{fair share}\}.$$

This scheme is simple. However, it has been shown to exhibit very large rate oscillations if RIF is set too high, and can sometimes lead to unfairness [62].

7.2.3.3 · *Destination Behaviors*

When a data cell is received, the destination will save its EFCI indicator as the EFCI state of the connection.

Destinations are required to turn the forward RM cells around with minimal modifications as follows: The DIR bit is set to "backward" to indicate that the cell is a BRM cell; the BN bit is set to zero to indicate that the cell was not generated by a switch; the CCR and MCR fields should not be changed. If the last cell has the EFCI bit set, the CI bit in the next BRM is set and the stored EFCI state is cleared.

Destinations are also required to turn around RM cells as quickly as possible; however, an RM cell may be delayed by many factors. If the destination is internally congested, it may reduce the ER or set the CI or NI bits just like a switch. When the destination is too congested and wants the source to reduce its rate immediately without having to wait for the next RM cell, the destinations are allowed to generate BECN RM cells. If the reverse ACR is low, the destinations are allowed a number of options to do it. The implication of these various options is that old, out-of-rate information can be discarded.

It is recommended to turn around as many RM-cells as possible to minimize turnaround delay, first by using in-rate opportunities and then by using out-of-rate opportunities as available.

7.2.3.4 *Virtual Source and Destination*

The round-trip delay of end-to-end rate control can be very large. To reduce the size of the feedback loop, the network can be segmented into smaller pieces and the switches on the path of a connection can act as *virtual source* and/or *virtual destination*. In particular, each ABR control segment, except the first, is sourced by a virtual source, which assumes the behavior of an ABR source end point. Each ABR control segment, except the last, is terminated by a virtual destination, which assumes the behavior of an ABR destination end point, as illustrated in Figure 7.25.

Also, the intermediate segments can use any proprietary congestion control scheme. This allows public telecommunication carriers to follow the

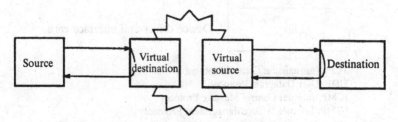

Fig. 7.25 Virtual source and destination.

standard interface only at entry and exit switches. More importantly, virtual source–destination provides a more robust interface to a public network in the sense that the resources inside the network do not have to rely on user compliance. Misbehaving users will be isolated in the first control loop. The users here include private networks with switches that may or may not be compliant.

There is no limit on the number of segments that can be created. In the extreme case, every switch could act as a virtual source–destination and one would get hop-by-hop rate control. Notice that the virtual sources and destinations need to maintain per-VC queuing and may therefore be quite expensive.

7.3 TCP/IP NETWORKS

Transmission Control Protocol (TCP) [91, 92, 72] is used on top of a network level protocol called Internet Protocol (IP), as shown in Figure 7.26, and performs the functions of detecting and recovering lost or corrupted packets, flow control, and multiplexing. TCP uses sequence numbers, cumulative acknowledgment, windows, and software checksums to implement these functions.

IP, which is a connectionless or datagram packet delivery protocol, deals with host addressing and routing, but the latter function is almost totally the task of the Internet level packet switch, or routers. IP also provides the ability for packets to be broken into smaller units (fragmented) on passing into a network with a smaller maximum packet size. The IP layer at the receiving end is responsible for reassembling these fragments.

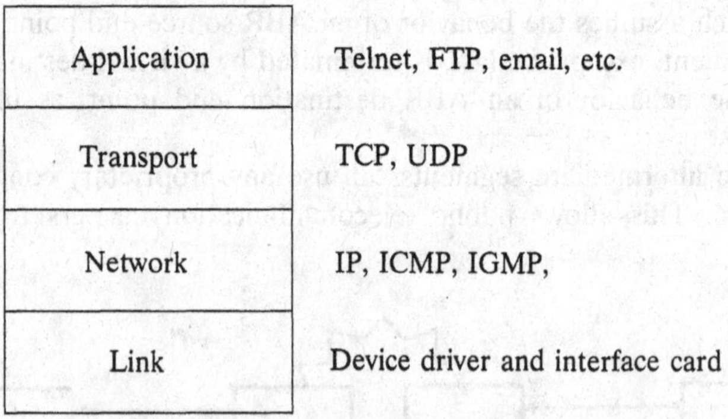

Application	Telnet, FTP, email, etc.
Transport	TCP, UDP
Network	IP, ICMP, IGMP,
Link	Device driver and interface card

TCP: Transmission Control Protocol
UDP: User Datagram Protocol
ICMP: Internet Control Message Protocol
IGMP: Internet Group Management Protocol

Fig. 7.26 TCP/IP protocol suite.

Under IP is the layer dealing with the specific network technology being used. This may be a very simple layer in the case of a local area network such as Ethernet, or a rather complex layer for a network such as X.25. On top of TCP sits one of a number of application protocols, more commonly for remote login, file transfer, or mail. Further explanation of the TCP/IP protocol suite can be found in [91, 92, 72]. Below we focus on TCP only.

7.3.1 TCP Overview

TCP is a connection-oriented, end-to-end reliable protocol designed to fit into a layered hierarchy of protocols that support multinetwork applications. TCP provides for reliable interprocess communication between pairs of processes in host computers attached to distinct but interconnected computer communication networks. Very few assumptions are made as to the reliability of the communication protocols below the TCP layer. TCP assumes it can obtain a simple, potentially unreliable datagram service from the lower-level protocols, such as IP. In principle, the TCP should be able to operate above a wide spectrum of communication systems ranging from hard-wired connections to packet-switched or circuit-switched networks. There have been numerous research efforts, implementation experiments, and papers over a period of decades. Below we briefly describe the operations of TCP. Interested readers are referred to [73-94] for further details.

Simply speaking, TCP data are organized as a stream of bytes, much like a file. Figure 7.27 shows a TCP packet. The basic operations of TCP involve dividing higher-level application data into segments (data portion in Figure 7.27), attaching each segment to a TCP header (thus forming a TCP packet), and forwarding each TCP packet to a lower level, such as IP. In the TCP header, as shown in Figure 7.27, TCP associates port numbers with particular applications and a sequence number with every byte in the data stream. TCP uses a number of control flags to manage the connection, as shown in Figure 7.27. Some of these flags pertain to a single packet, such as the URG flag indicating valid data in the urgent pointer field, but two flags (SYN and FIN) require reliable delivery, as they mark the beginning and end of the data stream. TCP exchanges special segments to start up and close down a data flow between two hosts, and uses acknowledgments (ACKs) and timeouts to ensure the integrity of the data flow.

Sequence numbers are used to coordinate which data have been transmitted and received. TCP will arrange for retransmission if it determines that data have been lost. TCP can dynamically learn the delay characteristics of a network by tracing the round-trip delay of transmitted data and returned ACKs, and can adjust its operation to maximize throughput without overloading the network. TCP also manages data buffers and coordinates traffic so its buffers will never overflow. Fast senders will be stopped periodically to keep up with slower receivers. More details are presented below.

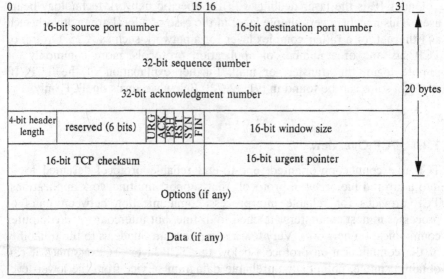

URG: The urgent pointer is valid
ACK: The acknowledgment number is valid
PSH: The receiver sholud pass these data to the application as soon as possible
RST: Reset the connection
SYN: Synchronize sequence numbers to initiate a connection
FIN: The sender is finished sending data

Fig. 7.27 A TCP packet.

7.3.1.1 Full-Duplex Operation

TCP usually operates full duplex. Only during connection start and close sequences can TCP exhibit asymmetric behavior (i.e., data transfer in the forward direction but not in the reverse, or vice versa). A TCP session can be regarded as two independent byte streams, traveling in opposite directions. No TCP mechanism exists to associate data in the forward and reverse byte streams. The algorithms described here operate independently in both directions.

7.3.1.2 Acknowledgment and Sequence Numbers

The acknowledgment mechanism is at the kernel of TCP. Basically, when data arrive at the receiver, the protocol requires that it send back an acknowledgment. TCP uses a 32-bit sequence number that counts bytes in the data stream, as shown in Figure 7.27. The protocol specifies that the bytes of data are sequentially numbered, so that the receiver can acknowledge data by identifying the first byte of data that it has not yet received, which also acknowledges the previous bytes. Forward and reverse sequence numbers are completely independent. Each TCP peer must track both its own sequence numbering and the numbering being used by the remote peer.

Sliding window protocol is performed at the byte level

Advertised window (rwnd)

1 2 |3 4 5 6 7 8| 9 10 11

Sent and acknowledged — Sent but not acknowledged — Can be sent: usable window — Can't be sent

Here sender can transmit sequence numbers 6, 7, 8.

(a)

Transmission of a single byte (with SeqNo = 6) and acknowledgment is received (AckNo = 5, rwnd = 4):

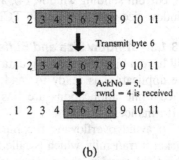

(b)

Acknowledgment is received that enlarges the window to the right (AckNo = 5, rwnd = 6)

1 2 3 4 |5 6 7 8| 9 10 11

↓ AckNo = 5, rwnd = 6 is received

1 2 3 4 |5 6 7 8 9 10| 11

A receiver opens a window when TCP buffer empties (meaning that data are received by the application)

(c)

Acknowledgment is received that reduces the window from the right (AckNo = 5, rwnd = 3)

1 2 3 4 |5 6 7 8| 9 10 11

↓ AckNo = 5, rwnd = 3 is received

1 2 3 4 |5 6 7| 8 9 10 11

Shrinking a window (should not be used)

(d)

Fig. 7.28 Sliding window mechanism: (a) window vs. data; (b) window closes; (c) window opens; (d) window shrinks.

Each TCP packet contains the starting sequence number of the data in that packet, and the sequence number (called the acknowledgment number) of the last byte received from the remote peer. With this information, a sliding window protocol is implemented, as illustrated in Figure 7.28. The sending window, as shown in Figure 7.28(a), which is always upper-bounded by the newly *advertised window* (rwnd) from the receiver, basically consists of sequence numbers for bytes that were sent but haven't been acknowledged and bytes that can be sent immediately. The sequence numbers for the first set of bytes are located at the left side of the window (the sequence numbers are contiguous and increasing, unless wrapped around due to finite-bit overflow), while those for the second set of bytes are at the right side of the window.

The sending window can shrink either from the left side [if, for example, an ACK numbered 5 is received, indicating that two bytes with sequence numbers 3 and 4 have been correctly received, as shown in Fig. 7.28(b)], or from the right side [if an advertised window, e.g., rwnd = 3, is received and has size less than that of the current sending window (4), as shown in Fig.

7.28(d)]. On the other hand, the sending window can be enlarged if an advertised window, e.g., rwnd = 6, is received and has size less than that of the current sending window (4), as shown in Fig. 7.28(c). Next we explain the window size and buffering in TCP.

7.3.1.3 *Window Size and Buffering*
Each end point of a TCP connection will have a buffer for storing data that is transmitted over the network before the application is ready to read the data. This lets network transfers take place while applications are busy with other processing, improving overall performance.

To avoid overflowing the buffer, TCP sets a *window size* field in each packet it transmits, which is called the *advertised window* as mentioned above. This field contains the number of bytes that may be admitted into the buffer or, in other words, the size of the buffer that currently is available for additional data. This number of bytes is the maximum that the remote TCP is permitted to transmit, as illustrated in Figure 7.28.

If the advertised window falls to zero, the remote TCP is required to send a small segment now and then to see if more data are accepted. If the window remains closed at zero for some substantial period with no response from its TCP peer, the remote TCP can conclude that its TCP peer has failed and can close the connection.

7.3.1.4 *Round-Trip Time Estimation*
TCP requires that, when a host transmits a TCP packet to its peer, it must wait a period of time called the *retransmission interval* (rto) for an acknowledgment. If the reply does not come within the expected period, the packet is assumed to have been lost and the data is retransmitted. The value of rto is updated based on a so-called *round-trip time* (RTT) estimation, for which TCP always monitors the normal exchange of data packets. Figure 7.29 illustrates the RTT measurements.

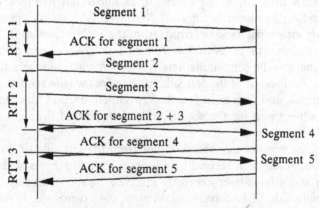

Fig. 7.29 Round-trip time measurements.

RTT estimates are among the most important performance parameters in a TCP exchange, especially for considering the throughput. If the RTT estimate is too low, packets are retransmitted unnecessarily; if too high, the connection can sit idle while the host waits to time out. For example, over an Ethernet, no more than a few microseconds should be needed for a reply. If the traffic must flow over the wide-area Internet, a second or two might be reasonable during peak utilization times. Obviously a robust mechanism is needed to accommodate such diversified networking environments.

As is commonly done today, Jacobson [82] suggests estimating the mean RTT via the low-pass filter

$$a \leftarrow a + g \times (m - a), \tag{7.32}$$

where a is the average RTT estimate, m is a RTT measurement from the most recently ACKed data packet, and g is a filter gain constant with a suggested value of 0.1. The deviation v of the RTT is estimated by

$$v \leftarrow (1 - h)v + h|m - a|, \tag{7.33}$$

where h is also a filter gain constant with a suggested value of 0.25. Once the a and v estimates are updated, the retransmission interval, rto, for the next packet sent is set to

$$\text{rto} = a + 4v. \tag{7.34}$$

After introducing the basic algorithms used in TCP, we discuss TCP congestion control in detail.

7.3.2 TCP Congestion Control

7.3.2.1 Concepts TCP simply runs as follows: a source increases its window size until it detects a packet loss; at this point, the source reduces the window size, and the cycle repeats. The TCP performance doesn't depend on the transfer rate itself, but rather on the product of the transfer rate and the round-trip delay [82, 92, 72]. This *bandwidth–delay product* measures the amount of data that would "fill the pipe" connecting sender and receiver. It is the buffer space required at sender and receiver to obtain maximum throughput on the TCP connection over the path, that is, the amount of unacknowledged data that TCP must handle in order to keep the pipeline full.

According to the packet conservation principle [82], a new packet isn't put into the network until an old packet leaves. The returning ACKs function as pacing signals. In the steady state, the sender's packet rate is equal to the arrival rate of the ACKs, which, in turn, is decided by the slowest link in the round-trip path. Thus, TCP can automatically sense the network bottleneck and regulate its flow accordingly, whether the bottleneck is the receiver or the transit network. This is called the *self-clocking* behavior of TCP.

TCP provides only end-to-end flow control and uses packet loss as an *implicit* signal to detect network congestion. Careful control is necessary to ensure fairness among various users in resource sharing, because this is a distributed algorithm. Congestion control in TCP involves finding places that violate conservation and fixing them, and includes four interwined algorithms: *slow start, congestion avoidance, fast retransmit*, and *fast recovery* [90]. There are two widely used variants of TCP. The *Tahoe* version includes slow start, congestion avoidance, and fast retransmit [82]. The *Reno* version modified the fast retransmission of TCP Tahoe to include fast recovery [83], as explained below.

7.3.2.2 *Slow Start and Congestion Avoidance*   Slow start and congestion avoidance algorithms are used at the beginning of a transfer, or after repairing loss detected by the retransmission timer, to slowly probe a network with unknown conditions to determine the available capacity and to avoid congesting the network with too large a burst of data. To implement them, two variables are added to the TCP per-connection state. The congestion window (cwnd) is a sender-side limit on the amount of data the sender can transmit into the network before receiving an acknowledgment (ACK), while the receiver's advertised window (rwnd) is a receiver-side limit on the amount of outstanding data. The minimum of cwnd and rwnd (i.e., the sending window shown in Figure 7.28) governs data transmission. The initial value of cwnd (IW) must be less than or equal to twice the sender maximum segment size (SMSS) in bytes, and must not be more than two segments [90].

The slow start threshold (ssthresh) is used to determine whether the slow start or the congestion avoidance algorithm is used to control data transmission. The slow start algorithm is used when cwnd < ssthresh, while the congestion avoidance algorithm is used when cwnd > ssthresh. When cwnd and ssthresh are equal, the sender may use either slow start or congestion avoidance. The initial value of ssthresh may be arbitrarily high, but it may be reduced in response to congestion.

During slow start, TCP increments cwnd by at most SMSS bytes for each ACK received that acknowledges new data. Slow start ends when cwnd exceeds or reaches ssthresh as noted above, or when congestion is observed. The slow start window is actually increased in an exponential manner (not slow at all!), as illustrated in Figure 7.30.

During congestion avoidance, cwnd is incremented by one full-size segment per RTT every time a nonduplicate ACK is received; one commonly used formula according to [92, 72] is

$$\text{cwnd} \leftarrow \max\left\{1, \text{cwnd} + \frac{\text{SMSS} \times \text{SMSS}}{\text{cwnd}}\right\}, \qquad (7.35)$$

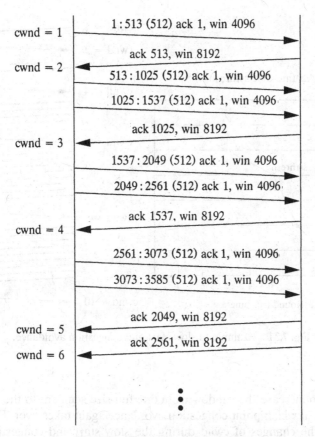

Fig. 7.30 Example of slow start.

where the max function is used to round up the result to 1 byte on account of the integer arithmetic usually used in TCP implementations. The adjustment based on (7.35) will be slightly more aggressive than one segment per RTT for a connection in which the receiver acknowledges every data segment, but will be less aggressive for a receiver acknowledging every other packet [90]. Congestion avoidance continues until congestion is detected.

When the retransmission timer expires (recall that this is when a TCP sender assumes a segment loss as explained in the previous section), the value of ssthresh must be set to no more than the value given by

$$\text{ssthresh} = \max\{\text{DataInFlight}/2, 2 \times \text{SMSS}\}, \qquad (7.36)$$

where DataInFlight is the number of outstanding data in the network that, as illustrated in Figure 7.28, were sent but have not been ACKed yet.

Furthermore, upon a timeout, cwnd must be set to no more than one full-sized segment regardless of its initial value (IW). Therefore, after re-transmitting the dropped segment, the TCP sender uses the slow start

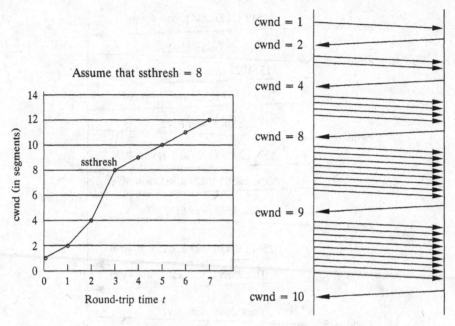

Fig. 7.31 Example of slow start and congestion avoidance.

algorithm to increase the window from one full-size segment to the new value of ssthresh, at which point congestion avoidance again takes over. Figure 7.31 illustrates the changes of cwnd during the slow start and congestion avoidance.

7.3.2.3 *Fast Retransmit and Fast Recovery*

TCP requires a receiver to send an immediate *duplicate* ACK when an out-of-order segment arrives; when the incoming segment fills in all or part of a gap in the sequence space, the TCP receiver should send an immediate ACK. The purpose of the duplicate ACK is to inform the sender that a segment was received out of order and which sequence number is expected.

The fast retransmit algorithm is based on incoming duplicate ACKs to detect and repair loss and does not require any timers. According to [83], however—since duplicate ACKs can be caused by dropped segments, the reordering of data segments by the network, or replication of ACK or data segments by the network—only after receiving three duplicate ACKs (four identical ACKs without the arrival of any other intervening packets) does the algorithm assume that a segment has been lost. After detecting the loss, TCP performs a retransmission of what appears to be the missing segment, without waiting for the retransmission timer to expire, leading to higher channel utilization and connection throughput.

After sending what appears to be the missing segment, the fast recovery algorithm governs the transmission of new data until a nonduplicate ACK arrives. Under this mechanism, the sender can continue to transmit new segments (using a reduced cwnd), instead of following slow start, because (by the rules of generating duplicate ACKs) the receipt of a duplicate ACK also indicates that a segment has left the network, and a new segment can be put into the network according to the packet conservation principle [82].

The fast retransmit and fast recovery algorithms are usually implemented together as follows [72, 83, 92]:

1. When the third duplicate ACK is received, set ssthresh to no more than the value given in (7.36).

2. Retransmit the lost segment and set cwnd to ssthresh plus $3 \times$ SMSS. This artificially *inflates* the congestion window by the number of segments (three) that have left the network and that the receiver has buffered.

3. Increment cwnd by SMSS for each additional duplicate ACK received. This artificially inflates the congestion window in order to reflect the additional segment that has left the network.

4. Transmit a segment, if allowed by the new value of cwnd and the receiver's advertised window rwnd.

5. When the next ACK arrives that acknowledges new data, set cwnd to ssthresh (the value set in step 1). This is termed *deflating* the window. This ACK should be the acknowledgment elicited by the retransmission from step 1, one RTT after the retransmission (though it may arrive sooner in the presence of significant out-of-order delivery of data segments at the receiver).

The recovery period ends when a recovery ACK is received acknowledging all data that were outstanding when fast recovery was entered. During the recovery period, cwnd is held constant. When recovery ends, TCP returns to congestion avoidance as usual. A timeout is forced if it is detected that a retransmitted segment has been lost (again). Under such circumstances, TCP will retransmit the dropped packet and then perform slow start. This condition is included to prevent TCP from being too aggressive in the presence of persistent network congestion.

However, the above algorithms cannot recover very efficiently from multiple losses in a single flight of packets. That is because TCP [91] uses a cumulative acknowledgment scheme in which received segments (that are not at the left edge of the receive window) are not acknowledged. The left edge indicates the highest *in-order* sequence number the receiver has seen so far. This forces the sender to either wait a RTT to find out about each lost packet, or to unnecessarily retransmit segments that have been correctly received [77]. With the cumulative acknowledgment scheme, multiple dropped segments generally cause TCP to lose its ACK-based clock (i.e., utilizing individual segment departures from the network), reducing the overall throughput [77]. One solution to this is discussed in the next section.

7.3.3 *Other TCP Variants* There are a number of other proposals for TCP congestion control algorithms. Below we present a few examples.

7.3.3.1 *New Reno TCP* The *new Reno* TCP [80] includes a small change to the Reno algorithm at the sender that eliminates Reno's wait for a retransmit timer when *multiple* packets are lost from a window. The change concerns the sender's behavior during fast recovery when a *partial* ACK is received that acknowledges some, but not all, of the packets that were outstanding at the start of that fast recovery period. In Reno, partial ACKs take TCP out of fast recovery by deflating the usable window back to the size of the congestion window. In new Reno, partial ACKs received during fast recovery are instead treated as an indication that the packet immediately following the acknowledged packet in the sequence space has been lost, and should be retransmitted. Thus, when multiple packets are lost from a single window of data, new Reno can recover without a retransmission timeout, retransmitting one lost packet per RTT until all of the lost packets from that window have been retransmitted. New Reno remains in fast recovery until all of the data outstanding when fast recovery was initiated has been acknowledged.

7.3.3.2 *SACK TCP* *Selective acknowledgment* (SACK) [77] is a strategy allowing TCP to recover efficiently from multiple losses in a single flight of packets, as mentioned in the fast recovery algorithm in Section 7.3.2.3. With SACK, the data receiver can inform the sender about all segments that have arrived successfully, so the sender needs to retransmit only the segments that have actually been lost. SACK options [87] have been proposed to be added to TCP standards. The main difference between the SACK TCP implementation and the Reno TCP implementation is in the behavior when multiple packets are dropped from one window of data [77].

During fast recovery, SACK maintains a variable called *pipe* that represents the estimated number of packets outstanding in the path, differing from the mechanisms in the Reno implementation. The sender only sends new or retransmitted data when the estimated number of packets in the path is less than the congestion window. The variable *pipe* is incremented by one when the sender either sends a new packet or retransmits an old packet. It is decremented by one when the sender receives a duplicate ACK packet with a SACK option reporting that new data have been received at the receiver.

When receiving the first partial ACK (ACK received during fast recovery that advances the acknowledgment number field of the TCP header, as shown in Figure 7.27, but does not take the sender out of fast recovery), the sender decrements *pipe* by two packets rather than one, counting the original packet (assumed to have been dropped) and the departed packet. For any succeeding partial ACKs, *pipe* is incremented by one when the retransmitted packet enters the pipe, but is never decremented for the original packet, complying with the packet conservation rule [82]. By doing this the sender never recovers more slowly than a slow start [77].

The sender maintains a data structure, *scoreboard*, that remembers acknowledgments from previous SACK options. When the sender is allowed to send a packet, it retransmits the next packet from the list of packets inferred to be missing at the receiver. If there are no such packets and the receiver's advertised window is sufficiently large, the sender sends a new packet.

7.3.3.3 FACK TCP *Forward acknowledgment* (FACK) TCP [85, 86] adds more precise control to the injection of data into a network during recovery. It is able to do this by keeping an explicit measure of the total number of bytes of data outstanding in the network, using the additional information provided by the SACK option. In contrast, Reno and SACK TCP both attempt to estimate this number by assuming that each duplicate ACK received represents one segment that has left the network.

The FACK algorithm introduces two new state variables. The variable sand.fack reflects the correctly received data held by the receiver with the highest sequence number (x), and snd.fack $= x + 1$. It is updated from the acknowledgment number in the TCP header and information contained in TCP SACK options during recovery. Another state variable, retran–data, is used to record the quantity of outstanding retransmitted data in the network. Each time a segment is retransmitted, retran–data is increased by the segment's size. When a retransmitted segment is determined to have left the network, retran–data is decreased by the segment's size.

Let snd.nxt be the sequence number of the first byte of unsent data [91]. Define awnd to be the sender's estimate of the actual quantity of data outstanding in the network,

$$awnd = snd.nxt - snd.fack + retran\text{–}data. \tag{7.37}$$

A sender can continually send segments as long as awnd < cwnd. Thus, FACK TCP can regulate the amount of data outstanding in the network to be within one SMSS of the current cwnd. When awnd > cwnd, the sender can stop transmitting data until enough data have left the network that awnd falls below the current cwnd, and then resumes transmission of data. Discussions on various senders' behavior are in [85, 86].

7.3.3.4 Vegas TCP Vegas TCP [95, 96] is intended to increase TCP throughput and decrease losses by using new retransmission, congestion avoidance, and slow-start mechanisms. First of all, it extends Reno's retransmission mechanisms. In Reno, it is possible to decrease the congestion window more than once for losses that occurred during one RTT interval. In contrast, Vegas only decreases the congestion window if the retransmitted segment was previously sent *after* the last window decrease.

Simply, Vegas reads and records the system clock each time a segment is sent. When an ACK arrives, Vegas reads the clock again and does the RTT calculation using this time and the timestamp recorded for the relevant segment. Based on the more accurate RTT estimate, Vegas decides retransmission as follows. When a duplicate ACK is received, Vegas checks to see if the difference between the current time and the timestamp recorded for the relevant segment is greater than the timeout value. If it is, Vegas assumes that this loss happened before the last window decrease (i.e., beyond the current RTT interval). This does not imply that the network is congested for the *current* congestion window size, and therefore, Vegas retransmits the segment without having to wait for three duplicate ACKs.

When a nonduplicate ACK is received, if it is the first or second one after a retransmission, Vegas again checks to see if the time interval since the segment was sent is larger than the timeout value. If it is, then Vegas retransmits the segment. This will catch any other segment that may have been lost previous to the retransmission without having to wait for a duplicate ACK, thus making loss detection much sooner than Reno.

Second, Vegas TCP performs congestion avoidance based on changes in the estimated amount of extra data in the network in addition to dropped segments. Vegas uses the difference between expected and actual (measured) sending rates, denoted by μ and $\hat{\mu}$, respectively, to estimate the available bandwidth in the network (or the level of network congestion). The expected throughput μ is given by

$$\mu = \frac{\text{cwnd}}{\text{RTT}_{\min}}, \tag{7.38}$$

where $\text{RTT}_{\min}$ is the minimum of all measured RTTs for a connection. Assume that a specific segment, s, was sent at time sent(s), its acknowledgment is received at time ack(s), and thus the RTT for this segment is ack(s) − sent(s). Suppose $N(s)$ is the number of bytes transmitted during this sample RTT. After recording {sent(s), ack(s), $N(s)$}, Vegas calculates the current sending rate by

$$\hat{\mu} = \frac{N(s)}{\text{ack}(s) - \text{sent}(s)}. \tag{7.39}$$

Note that $\mu \geq \hat{\mu}$ by definition.

Based on the difference between the expected and actual sending rates, Vegas then adjusts the window accordingly. That is, when $\mu - \hat{\mu} < \beta_L$, Vegas assumes that the network is not congested because the actual flow rate is close to the expected rate; therefore, Vegas increases the congestion window cwnd linearly during the next RTT. When $\mu - \hat{\mu} > \beta_H$, since the actual rate is smaller than the expected rate, Vegas assumes that the network is over-loaded and will decrease cwnd linearly during the next RTT. When the rate difference falls between β_L and β_H ($\beta_L < \beta_H$), Vegas simply leaves cwnd unchanged. The thresholds β_L and β_H represent the desired range of extra data in transit in the network.

Finally, to find a connection's available bandwidth at the beginning of a transfer without incurring loss, Vegas slightly modifies the slow-start algorithm by incorporating a congestion detection mechanism. It allows exponential growth only every other RTT. In between, the congestion window stays fixed so a valid comparison of the expected and actual rates can be made. When the actual rate falls below the expected rate by one segment size over the current RTT_{min} [95, 96], Vegas changes from slow-start mode to congestion avoidance mode.

It has been shown in [95, 96] that compared with Reno TCP, Vegas TCP does lead to a fair allocation of bandwidth and does not suffer from the delay bias as Reno TCP does. However, when competing with other Reno TCP connections [97], Vegas TCP gets penalized due to the relatively aggressive window control in Reno TCP. Interested readers are referred to [95, 96, 97] for further details.

7.3.4 TCP with Explicit Congestion Notification

Explicit congestion notification (ECN) [100] is a strategy that is used to accommodate applications sensitive to delay or loss of one or more individual packets without using packet drop as an indication of network congestion. In contrast, the TCP variants mentioned before are only appropriate for pure best-effort traffic that has little or no sensitivity to delay or loss of individual packets. ECN has been recently added to IP to facilitate end-to-end congestion control in the Internet.

The ECN mechanism requires an ECN field in the IP header with two bits. The *ECN-capable transport* (ECT) bit is set by the data sender to indicate that the end points of the transport protocol are ECN-capable. The *congestion experienced* (CE) bit is set by the router to indicate congestion to the end nodes (active queue management mechanisms, such as RED, in routers can facilitate early congestion detection before the queue overflows [98, 99]). We use the term *CE packet* to denote a packet that has the CE bit set. Currently, bits 6 and 7 in the IPv4 type of service (TOS) octet [91] are designated as the ECN field. Bit 6 is designated as the ECT bit, and bit 7 is designated as the CE bit, as illustrated in Figure 7.32(b).

ECN-capable TCP reacts to the receipt of a single CE packet the same as it responds to a *single* dropped packet. To avoid reacting multiple times to multiple indications of congestion within a RTT, the sender should react to congestion, at most, once per window of data (i.e., at most once per RTT). In addition, the sender should not decrease the slow-start threshold, ssthresh,

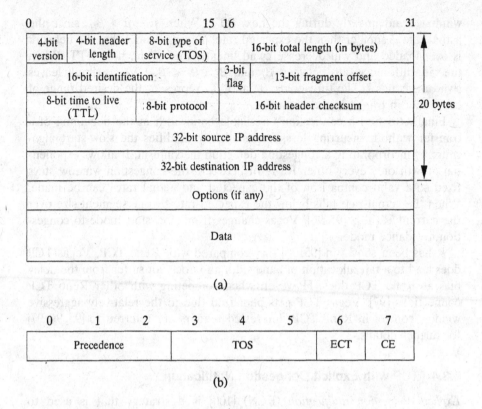

ECN: Explicit congestion notification
ECT: ECN-capable transport
CE: Congestion experienced

Fig. 7.32   (a) IP packet header; (b) IP TOS octet.

if it has been decreased within the last RTT. However, if any retransmitted packets are dropped or have the CE bit set, then this is interpreted by the source TCP as a new instance of congestion. For a router, the CE bit of an ECN-capable packet should only be set if the router would otherwise have dropped the packet as an indication of congestion to the end nodes.

ECN requires the support of three new mechanisms from TCP, in addition to the functionality given by the ECN field in the IP packet header. First, negotiation is used between the end points during setup to determine if they are both ECN-capable. Second, an ECN-echo flag is carried in the TCP header for the receiver to inform the sender when a CE packet has been received. Third, a congestion window reduced (CWR) flag is also carried in the TCP header for the sender to inform the receiver that the, congestion window has been reduced (and the receiver can determine when to stop setting the ECN-echo flag). Bit 9 in the reserved field of the TCP header is designated as the ECN-echo flag. The CWR flag is assigned to bit 8 in the reserved field of the TCP header.

Very simply, for a TCP connection using ECN, data packets are transmitted with the ECT bit set in the IP header (set to a 1). If the sender receives an ECN-echo ACK packet (that is, an ACK packet with the ECN-echo flag set in the TCP header), then the sender knows that congestion was encountered in the network on the path from the sender to the receiver. The TCP sets the CWR flag in the TCP header of the first data packet sent after the window reduction. If that data packet is dropped in the network, then the sender will have to reduce the congestion window again and retransmit the dropped packet. Thus, the CWR message is reliably delivered to the data receiver.

When a CE data packet arrives, the receiver sends an ACK packet with the ECN-echo bit set. Then it should continue to set the ECN-echo flag in ACK packets until it receives a CWR packet (a packet with the CWR flag set). After the receipt of the CWR packet, acknowledgments for subsequent non-CE data packets do not have the ECN-echo flag set. If another CE packet is received by the data receiver, the receiver will once again send ACK packets with the ECN-echo flag set. Interested readers are referred to the ECN Web page

http://www-nrg.ee.lbl.gov/floyd/ecn.html

for further details.

7.4 EASY—ANOTHER RATE-BASED FLOW CONTROL SCHEME

From the common feedback mechanism of the BBFC, as shown in Section 7.2.2.2, another link-by-link rate-based flow control scheme can also be developed, which is called EASY in [63] because node k will simply limit its service rate by S or R on receiving a corresponding feedback signal (0–1), as shown in Figure 7.33.

Unlike the BBFC, the EASY has no explicit rate control mechanism, so a different service policy is needed. In [63], it is suggested that the link capacity

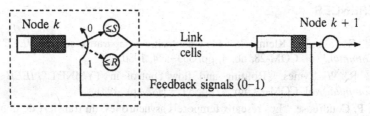

Fig. 7.33 Control structure for the EASY.

be divided among all the VCs at a node, mainly according to their buffer occupancies. The motivation is as follows. First, the backlog of a VC can indicate its short-term input loading as well as its implicit demand for bandwidth. Second, longer queues should intuitively be served faster than the shorter ones so that maximum backlog can be reduced, as can the aggregate backlog and the total buffering at this node [64, 65]. Third, since the node can maintain a certain level of buffer occupancy for each VC by serving it faster in the presence of a longer queue and slower otherwise, the variations of cell transfer delay can be reduced, as illustrated in a way with single-queue studies [48, pp. 216–217; 41].

This state-dependent service policy also consists of bandwidth allocation, quantum refreshment, and dequeue processes [63]. The latter two components are addressed in Section 7.2.2.5. Below we only introduce the bandwidth allocation. Unless otherwise stated, we use the same notation as in Section 7.2.2.5. Let $\phi(t, i)$ be the weight factor for VC i at time t, $i = 1, 2, \ldots, N$. $\phi(t, i)$ can be defined as follows:

$$\phi(t, i) = \frac{x(t, i)}{B(i)} \times \rho(i), \qquad (7.40)$$

where $x(t, i)$ is the backlog of VC i at time t, $B(i)$ is its buffer size, and $\rho(i)$ is a constant parameter that can be independent of or depend on the source characteristics and/or QoS requirement of this VC. As an example, $\rho(i) = R(i)$ is chosen in Section 7.2.2.6. Assuming that $\sum_{j=1}^{N} \phi(t, j) > 0$, the bandwidth allocation is defined as follows:

$$\mu^a(t, i) = \frac{\phi(t, i)}{\sum_{j=1}^{N} \phi(t, j)} \times C. \qquad (7.41)$$

The dummy VC $N + 1$ is still needed because when every backlogged VC i has to send data at the rate $R(i)$ due to its buffer overflow downstream, the sum of these rates may be less than the link capacity [i.e., the inequality (7.26) holds], and (7.28) should be used. The service model is illustrated in Figure 7.10.

REFERENCES

1. M. Gerla and L. Kleinrock, "Flow control: a comparative survey," *IEEE Trans. Commun.*, vol. COM-28, no. 4, pp. 553–574, 1980.

2. L. R. W. Tymes, "Routing and flow control in TYMNET," *IEEE Trans. Commun.*, vol. COM-29, no. 4, pp. 392–398, Apr. 1981.

3. J. P. Coudreuse, "Les réseaux temporels asynchrones: du transfert de données à l'image animée," *Echo Rech.*, no. 112, 1983.

4. J. S. Turner, "New directions in communications (or which way to the information age?)," *IEEE Commun. Mag.*, vol. 24, no. 10, pp. 8–15, Oct. 1986.

5. J. S. Turner, "Design of an integrated services packet network," *IEEE J. Select. Areas Commun.*, vol. SAC-4, no. 8, pp. 1373–1380, Nov. 1986.

6. M. G. H. Katevenis, "Fair switching and fair control of congested flow in broadband networks," *IEEE J. Select. Areas Comm.*, vol. 5, no. 8, pp. 1315–1326, Oct. 1987.

7. V. Jacobson, "Congestion avoidance and control," *Proc. ACM SIGCOMM'88*, pp. 158–181, 1988.

8. J. Y. Hui, "Resource allocation for broadband networks," *IEEE J. Select. Areas Commun.*, vol. 6, no. 9, pp. 1598–1608, Dec. 1988.

9. F. G. Shinskey, *Process Control Systems—Application, Design and Tuning*, McGraw-Hill, 3rd Edition, 1988.

10. I. Cidon and I. S. Gopal, "PARIS: an approach to integrated high-speed private networks," *Int. J. Digital and Analog Cabled Syst.*, vol. 1, pp. 77–85, 1988.

11. D. M. Chiu and R. Jain, "Analysis of the increase and decrease algorithms for congestion avoidance in computer networks," *Comput. Netw. ISDN Syst.*, vol. 17, pp. 1–14, 1989.

12. K. Bala, I. Cidon, and K. Sohraby, "Congestion control for high speed packet switched networks," *IEEE INFOCOM'90*, pp. 520–526, 1990.

13. J. C. Bolot and A. U. Shankar, "Dynamical behavior of rate-based flow control mechanisms," *Comput. Commun. Rev.*, pp. 35–49, Apr. 1990.

14. K. K. Ramakrishnan and R. Jain, "A binary feedback scheme for congestion avoidance in computer networks," *ACM Trans. Comput. Syst.*, vol. 8, no. 2, pp. 158–181, May 1990.

15. S. Shenker, "A theoretical analysis of feedback flow control," *ACM SIGCOM'90*, pp. 156–165, 1990.

16. I. Cidon and I. S. Gopal, "Control mechanisms for high speed networks," *IEEE GLOBECOM'90*, pp. 259–263, 1990.

17. R. L. Cruz, "A calculus for network delay, part I: network elements in isolation," *IEEE Trans. Inf. Theory*, vol. 37, no. 1, pp. 114–131, Jan. 1991.

18. R. L. Cruz, "A calculus for network delay, part II: network analysis," *IEEE Trans. Inform. Theory*, vol. 37, no. 1, pp. 132–141, Jan. 1991.

19. S. Keshav, "A control-theoretic approach to flow control," *ACM SIGCOM'91*, pp. 3–15, 1991.

20. L. Zhang, "Virtual clock: a new traffic control algorithm for packet-switched networks," *ACM Trans. Comput. Syst.*, vol. 9, no. 2, pp. 101–124, May 1991.

21. Y. T. Wang and B. Sengupta, "Performance analysis of a feedback congestion control policy under non-negligible propagation delay," *Proc. ACM SIGCOMM'91*, pp. 149–157, Aug. 1991.

22. A. Mukherjee and J. C. Strikwerda, "Analysis of dynamic congestion control protocols—a Fokker–Planck approximation," *Proc. ACM SIGCOMM'91*, pp. 159–169, Aug. 1991.

23. H. J. Fowler and W. E. Leland, "Local area network traffic characteristics, with implications for broadband network congestion management," *IEEE J. Select. Areas Commun.*, vol. 9, no. 7, pp. 1139–1149, Sep. 1991.

24. S. J. Golestani, "A framing strategy for congestion management," *IEEE J. Select. Areas Commun.*, vol. 9, no. 7, pp. 1064–1077, Sep. 1991.

25. A. Gersht and K. J. Lee, "A congestion control framework for ATM networks," *IEEE J. Select. Areas Commun.*, vol. 9, no. 7, pp. 1119–1130, Sep. 1991.

26. K. W. Fendick, D. Mitra, I. Mitrani, M. A. Rodrigues, J. B. Seery, and A. Weiss, "An approach to high-performance, high-speed data networks," *IEEE Commun. Mag.*, pp. 74–82, Oct. 1991.

27. D. Mitra, "Asymptotically optimal design of congestion control for high-speed data networks," *IEEE Trans. Commun.*, vol. 40, no. 2, pp. 301–311, Feb. 1992.

28. H. T. Kung, "Gigabit local area networks: a systems perspective," *IEEE Commun. Mag.*, pp. 79–89, Apr. 1992.

29. P. P. Mishra and H. Kanakia, "A hop by hop rate-based congestion control scheme," *ACM SIGCOM'92*, pp. 112–123, 1992.

30. C. T. Lea, "What should be the goal for ATM," *IEEE Netw. Mag.*, pp. 60–66, Sep. 1992.

31. J. S. Turner, "Managing bandwidth in ATM networks with bursty traffic," *IEEE Netw. Mag.*, vol. 6, no. 4, pp. 50–58, Sep. 1992.

32. P. E. Boyer and D. P. Tranchier, "A reservation principle with applications to the ATM traffic control," *Comput. Netw. ISDN Syst.*, vol. 24, pp. 321–334, 1992.

33. D. Bertsekas and R. Gallager, *Data Networks*, Prentice-Hall, Englewood Cliffs, NJ, 1992.

34. K. W. Fendick, M. A. Rodrigues, and A. Weiss, "Analysis of a rate-based feedback control strategy for long haul data transport," *Performance Evaluation*, vol. 16, pp. 67–84, 1992.

35. J. Kurose, "Open issues and challenges in providing quality of service guarantees in high-speed networks," *ACM Comp. Commun. Rev.*, pp. 6–15, Jan. 1993.

36. ATM Forum, *ATM User-Network Interface Specification*, Version 3.0, Prentice-Hall, Englewood Cliffs, NJ, 1993.

37. A. K. Parekh and R. G. Gallager, "A generalized processor sharing approach to flow control in integrated services networks: the multiple node case," *IEEE INFOCOM'93*, pp. 521–530, Mar. 1993.

38. H. T. Kung and A. Chapman, "The FCVC (flow controlled virtual channels) proposal for ATM networks" (DRAFT), Jun. 1993.

39. A. K. Parekh and R. G. Gallager, "A Generalized processor sharing approach to flow control in integrated services networks: the single-node case," *IEEE/ACM Trans. Netw.*, vol. 1, no. 3, pp. 344–357, Jun. 1993.

40. M. Shreedhar and G. Varghese, "Efficient fair queuing using deficit round-robin," *IEEE/ACM Trans. Netw.*, vol. 4, no. 3, pp. 375–385, Jun. 1996.

41. C. H. Lam and T. T. Lee, "Fluid flow models with state-dependent service rate," *Stochastic Models*, vol. 13, no. 3, 1997; previous version, *Proc. IEEE GLOBECOM*, 1995.

42. Y. Z. Cho and A. Leon-Garcia, "Performance of burst-level bandwidth reservation in ATM LANs," *IEEE INFOCOM'94*, pp. 812–820, 1994.

43. M. G. Hluchyj and N. Yin, "On close-loop rate control for ATM networks," *IEEE INFOCOM'94*, pp. 99–108, 1994.

44. Y. Gong and I. F. Akyildiz, "Dynamic traffic control using feedback and traffic prediction in ATM networks," *IEEE INFOCOM'94*, pp. 91–98, Jun. 1994.

45. P. Newman,"Traffic management for ATM local area networks," *IEEE Commun. Mag.*, vol. 32, no. 8, pp. 44–50, Aug. 1994.

46. H. T. Kung, T. Blackwell, and A. Chapman, "Credit-based flow control for ATM networks: credit update protocol, adaptive credit allocation, and statistical multiplexing," *ACM SIGCOM'94*, pp. 101–114, 1994.

47. R. J. Simcoe and L. Roberts, "The great debate over ATM congestion control," *Data Commun.*, pp. 75–80, Sep. 1994.

48. A. S. Acampora, *An Introduction to Broadband Networks: LANs, MANs, ATM, B-ISDN, and Optical Networks for Integrated Multimedia Telecommunications*, Plenum Press, New York, 1994.

49. ATM Forum, *Traffic Management Specification*, Version 4.1, AF-TM-0121.000, Mar. 1999.

50. F. Bonomi and K. W. Fendick, "The rate-based flow control framework for the available bit rate ATM service," *IEEE Netw. Mag.*, vol. 9, no. 2, pp. 25–39, Mar./Apr. 1995.

51. H. T. Kung and R. Morris, "Credit-based flow control for ATM networks," *IEEE Netw. Mag.*, vol. 9, no. 2, pp. 40–48, Mar./Apr. 1995.

52. K. K. Ramakrishnan and P. Newman, "Integration of rate and credit schemes for ATM flow control," *IEEE Netw. Mag.*, vol. 9, no. 2, pp. 49–56, Mar./Apr. 1995.

53. K. K. Ramakrishnan and P. Newman, "ATM flow control: inside the great debate," *Data Commun.*, pp. 111–120, Jun. 1995.

54. C. Q. Yang and A. V. S. Reddy, "A taxonomy for congestion control algorithms in packet switching networks," *IEEE Netw. Mag.*, pp. 34–45, Jul./Aug. 1995.

55. R. S. Pazhyannur and R. Agrawal, "Feedback-based flow control of B-ISDN/ATM networks," *IEEE J. Select. Areas Commun.*, vol. 13, no. 7, pp. 1252–1266, Sep. 1995.

56. I. Iliadis, "A new feedback congestion control policy for long propagation delays," *IEEE J. Select. Areas in Commun.*, vol. 13, no. 7, pp. 1284–1295, Sep. 1995.

57. X. Guo and T. T. Lee, "Backlog balancing flow control in high-speed data networks," *IEEE GLOBECOM'95*, pp. 690–695, 1995.

58. X. L. Guo, T. T. Lee, and H. J. Chao, "Concept of backlog balancing and Its application to flow control and congestion control in high-speed networks," *IEICE Trans. Commun.*, vol. E83-B, no. 9, pp. 2100–2116, Sep. 2000.

59. P. P. Mishra, H. Kanakia, and S. K. Tripathi, "On hop-by-hop rate-based congestion control," *IEEE/ACM Trans. Netw.*, vol. 4, no. 2, pp. 224–239, Apr. 1996; previous version, *Proc. ACM SIGCOMM*, 1992.

60. T. Chen, S. Liu, and V. Samalam, "The available bit rate service for data in ATM networks," *IEEE Commun. Mag.*, May 1996.

61. R. Jain, S. Kalyanaraman, R. Goyal, S. Fahmy, and R. Viswanathan, "ERICA switch algorithm: a complete description," ATM Forum Contribution 96-1172, Aug. 1996.

62. A. Arulambalam, X. Chen, and N. Ansari, "Allocating fair rates for available bit rate service in ATM networks," *IEEE Commun. Mag.*, Nov. 1996.

63. X. Guo, "Principles of backlog balancing for rate-based flow control and congestion control in ATM networks," Ph.D. dissertation, Department of Information Engineering, The Chinese University of Hong Kong, Hong Kong, Dec. 1996.

64. H. R. Gail, G. Grover, R. Guérin, S. L. Hantler, Z. Rosberg, and M. Sidi, "Buffer size requirements under longest queue first," *Performance Eval.*, vol. 18, pp. 133–140, 1993.

65. A. Birman, H. R. Gail, and S. L. Hantler, "An optimal service policy for buffer systems," *J. ACM*, vol. 42, no. 3, pp. 641–657, 1995.

66. R. Guerin and J. Heinanen, "UBR + service category," ATM Forum Contribution 96-1598, Dec. 1996.

67. L. Roberts, "Enhanced PRCA (proportional rate-control algorithm)," AF-TM 94-0735R1, ATM Forum, Aug. 1994.

68. L. Roberts, et al., "New pseudocode for explicit rate plus EFCI support," AF-TM 94-0974, ATM Forum, Oct. 1994.

69. R. Jain, S. Kalyanaraman, and R. Viswanathan, "The OSU scheme for congestion avoidance using explicit rate indication," AF-TM 94-0883, ATM Forum, Sep. 1994.

70. R. Jain, S. Kalyanaraman, and R. Viswanathan, "The EPRCA + scheme," AF-TM 94-0988, ATM Forum, Oct. 1994.

71. A. W. Barnhart, "Explicit rate performance evaluation," AF-TM 94-0983R1, ATM Forum, Oct. 1994.

72. W. Stallings, *High-Speed Networks: TCP/IP and ATM Design Principles*, Prentice-Hall, 1998.

73. M. Allman, S. Floyd, and C. Partridge, "Increasing TCP's initial window size," RFC 2414, Internet Engineering Task Force (IETF), Sep. 1998.

74. R. Braden, "Requirements for internet hosts—communication layers," RFC 1122, Internet Engineering Task Force (IETF), Oct. 1989.

75. S. Bradner, "Key words for use in RFCs to indicate requirement levels," RFC 2119, Internet Engineering Task Force (IETF), Mar. 1997.

76. D. Clark, "Window and acknowledgment strategy in TCP," RFC 813, Internet Engineering Task Force (IETF), Jul. 1982.

77. K. Fall and S. Floyd, "Simulation-based comparisons of Tahoe, Reno and SACK TCP," *Comput. Commun. Rev.*, Jul. 1996.

78. S. Floyd and T. Henderson, "The new Reno modification to TCP's fast recovery algorithm," RFC 2582, Internet Engineering Task Force (IETF), Apr. 1999.

79. S. Floyd, "TCP and successive fast retransmits," Technical report, ftp://ftp.ee.lbl.gov/papers/fastretrans.ps, Oct. 1994.

80. J. Hoe, "Improving the start-up behavior of a congestion control scheme for TCP," *Proc. ACM SIGCOMM*, Aug. 1996.

81. A. Hughes, J. Touch, and J. Heidemann, "Issues in TCP slow-start restart after idle," work in progress.

82. V. Jacobson, "Congestion avoidance and control," *Comput. Commun. Rev.*, vol. 18, no. 4, pp. 314–329, Aug. 1988.

83. V. Jacobson, "Modified TCP congestion avoidance algorithm," end2end-interest mailing list, Apr. 30, 1990.

84. J. Mogul and S. Deering, "Path MTU discovery," RFC 1191, Internet Engineering Task Force (IETF), Nov. 1990.

85. M. Mathis and J. Mahdavi, "Forward acknowledgment: refining TCP congestion control," *Proc. SIGCOMM*, Aug. 1996.

86. M. Mathis and J. Mahdavi, "TCP rate-halving with bounding parameters," Technical report. http:/www.psc.edu/networking/papers/FACKnotes/current.

87. M. Mathis, J. Mahdavi, S. Floyd, and A. Romanow, "TCP selective acknowledgement options," RFC 2018, Internet Engineering Task Force (IETF), Oct. 1996.

88. V. Paxson, M. Allman, S. Dawson, W. Fenner, J. Griner, I. Heavens, K. Lahey, J. Semke, and B. Volz, "Known TCP implementation problems," RFC 2525, Internet Engineering Task Force (IETF), Mar. 1999.

89. V. Paxson, "End-to-End Internet packet dynamics," *Proc. ACM SIGCOMM*, Sep. 1997.

90. M. Allman, V. Paxson, and W. Stevens, "TCP congestion control," RFC 2581, Internet Engineering Task Force (IETF), Apr. 1999.

91. J. Postel, "Transmission control protocol," RFC 793, Internet Engineering Task Force (IETF), Sep. 1981.

92. W. Stevens, *TCP/IP Illustrated, Volume* 1: *The Protocols*, Addison-Wesley, 1994.

93. W. Stevens, "TCP slow start, congestion avoidance, fast retransmit, and fast recovery algorithms," RFC 2001, Internet Engineering Task Force (IETF), Jan. 1997.

94. G. Wright and W. Stevens, *TCP/IP Illustrated, Volume* 2: *The Implementation*, Addison-Wesley, 1995.

95. L. S. Brakmo, S. W. O'Malley, and L. L. Peterson, "TCP Vegas: new techniques for congestion detection and avoidance," *ACM SIGCOMM*, pp. 24–35, Aug. 1994.

96. L. S. Brakmo and L. L. Peterson, "TCP Vegas: end to end congestion avoidance on a global internet," *IEEE J. Select. Areas Commun.*, vol. 13, no. 8, pp. 1465–1480, Oct. 1995.

97. J. Mo, R. J. La, V. Anantharam, and J. Walrand, "Analysis and comparison of TCP Reno and Vegas," *IEEE INFOCOM*, Mar. 1999.

98. S. Floyd and V. Jacobson, "Random early detection gateways for congestion avoidance," *IEEE/ACM Trans. Netw.*, vol. 1, no. 4, pp. 397–413, Aug. 1993.

99. B. Braden, D. Clark, J. Crowcroft, B. Davie, S. Deering, D. Estrin, S. Floyd, V. Jacobson, G. Minshall, C. Partridge, L. Peterson, K. Ramakrishnan, S. Shenker, J. Wroclawski, and L. Zhang, "Recommendations on queue management and congestion avoidance in the Internet," RFC 2309, Internet Engineering Task Force (IETF), Apr. 1998.

100. K. Ramakrishnan and S. Floyd, "A proposal to add explicit congestion notification (ECN) to IP," RFC 2481, Internet Engineering Task Force (IETF), Jan. 1999.

CHAPTER 8

QoS ROUTING

In traditional data networks, routing is primarily concerned with connectivity. Routing protocols usually characterize the network with a single metric, such as hop count or delay, and use shortest-path algorithms for path computation, which is typically transparent to any QoS requirements that different packets or flows may have. As a result, routing decisions are made without any awareness of resource availability and requirements. This means that flows are often routed over paths that are unable to support their requirements, while alternate paths with sufficient resources are available. This may result in significant deterioration in performance, in terms of call blocking probability.

To meet the QoS requirements of the applications and improve the network performance, strict resource constraints may have to be imposed on the paths being used. QoS routing is the process of selecting the path to be used by the packets of a flow based on its QoS requirements, such as bandwidth or delay. It refers to a set of routing algorithms that are able to identify a path that has sufficient residual (unused) resources to satisfy the QoS constraints of a given connection [2, 4–18]. Such a path is called a *feasible path*. In addition, most QoS routing algorithms also consider the optimization of resource utilization measured by metrics.

The problem of QoS routing is challenging because selecting paths that meet multiple QoS constraints is a complex algorithmic problem. As current routing protocols are already reaching the limit of feasible complexity, it is important that the complexity introduced by the QoS support should not impair the scalability of routing protocols. Besides, any future integrated services network is likely to carry both QoS and best-effort traffic. It is hard to determine the best operating point for both types of traffic if their distributions are independent. Best-effort flows can experience congestion or even starvation if guaranteed flows are not routed appropriately.

On the other hand, the network state changes dynamically due to transient load fluctuation, connections in and out, and links up and down. The growing network size makes it increasingly difficult to gather up-to-date state information in a dynamic environment. The performance of a QoS routing algorithm can be seriously degraded if the state information used is outdated.

Since many of the properties (e.g., call blocking probability) of resource reservation, admission control, and traffic access control are functions of the path taken through the network topology, ideally route selection and resource reservation should be integrated so that the choice of route can be a function of QoS. The stability of routes can be correlated with the life of the reservation and reservations can be efficiently repaired after topology failures.

Section 8.1 describes QoS signaling and routing in ATM networks. Section 8.2 discusses QoS routing in the Internet.

8.1 ATM SIGNALING AND ROUTING

Signaling in a communication network is the collection of procedures used to dynamically establish, maintain, and terminate connections. Figure 8.1 shows a set of B-ISDN signaling interfaces [1]. The *public UNI* specifies the criteria for connecting customer premises equipment (e.g., ATM end points and private ATM switch) to a public service provider's ATM switch, while the *private UNI* specifies the criteria for connecting user equipment (e.g., workstation, router) to a private (on-premises) ATM switch [1]. The *public NNI* specifies the criteria for connecting two public networks or switching systems, while the *private NNI* specifies the criteria for connecting two private networks or switching systems [2].

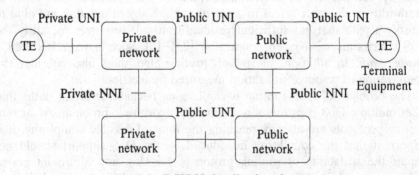

Fig. 8.1 B-ISDN signaling interfaces.

8.1.1 User-to-Network (UNI) Signaling

UNI signaling [1] is used to dynamically establish, maintain, and terminate connections at the edge of the network between an ATM end station and a public or private network. Its capabilities include

- establishment of point-to-point and point-to-multipoint virtual channel connections (VCCs);
- three different ATM private address formats;
- one ATM public address format;
- symmetric and asymmetric QoS connections with a declarable QoS class;
- symmetric and asymmetric bandwidth connections with a declarable bandwidth;
- transport of user-to-user information;
- support of error handling.

Figure 8.2 illustrates the ATM address formats, where the *authority and format identifier* (AFI) and *initial domain identifier* (IDI) constitute the *initial domain part* (IDP) that uniquely specifies an administrative authority, which has the responsibility for allocating and assigning values of the *domain specific part* (DSP).

- Based on ISO NSAP (network service access point)
- Support data: country code, international code designator, and ISDN's E.164

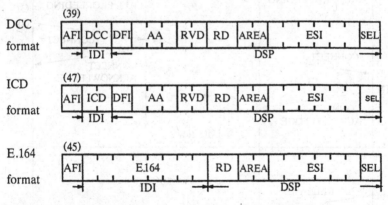

AFI: Authority and format identifier   DSP: Domain-specific part
IDL: Initial domain identifier   DFI: Domain-specific part format ID
DCC: Data country code   AA: Administrative authority
ICD: International code designator   RD: Routing domain
E.164: 15 digits (BCD)   ESI: End system ID (e.g., MAC ID)

Fig. 8.2 ATM address formats.

The AFI identifies the authority allocating the data country code (DCC), international code designator (ICD), or E.164 number; the format of the IDI; and the syntax of the remainder of the address. For example, the DCC ATM format is used when AFI = 39, as illustrated in Figure 8.2.

The DCC specifies the country in which an address is registered. The ICD identifies an international organization. The E.164 specifies Integrated Services Digital Network (ISDN) numbers, which include telephone numbers. Explanation of other fields are in [1].

There are two types of UNI signaling: point-to-point and point-to-multipoint. For point-to-point call/connection control, there is only one signaling endpoint on the user side. A single permanently established point-to-point signaling VC (VPI = 0, VCI = 5) is required

Figure 8.3 shows the UNI signaling flow for a point-to-point connection between two end systems. Suppose end system 1 wants to communicate with end system 2. The signaling proceeds as follows:

(1) End system 1 initiates call establishment by transferring a SET UP message on the signaling virtual channel across the interface, which, if possible, is forwarded through a number of network

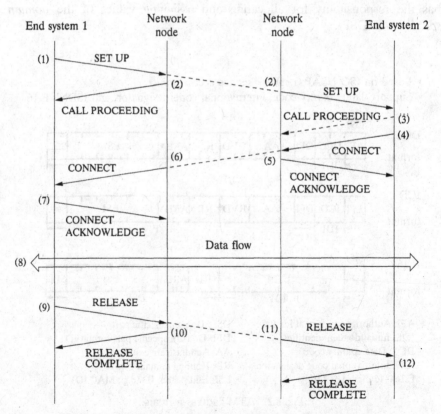

Fig. 8.3   Signaling message flows for point-to-point connections.

nodes (Fig. 8.3 shows two of them) to end system 2. The SET UP message shall contain all the information required by the network to process the call, such as end system address information, traffic descriptor, and QoS parameter information.

(2, 3) Upon receiving the SET UP message, the network (including end system 2) determines whether the access to the requested service is authorized and available. If so, the network may send a CALL PROCEEDING message to indicate that the call is being processed (this is necessary to keep the call timer alive in end system 1).

(4–6) If end system 2 finally determines that the access to the requested service is authorized and available, it may send a CONNECT message to indicate that the call has been accepted. Upon receiving such an indication, the network shall send a CONNECT message across the user-network interface to the calling user (end system 1) and indicate to end system 1 that a connection has been established through the network, as illustrated by (4), (5), and (6) in Figure 8.3.

(7) Upon receipt of the CONNECT message, end system 1 shall send a CONNECT ACKNOWLEDGE message to the network, and the network shall not take any action on receipt of such a message (since it perceives the call to be in the active state). At this point, an end-to-end connection is established.

(8) The data transmission starts at this time.

(9–12) When end system 1 completes the transaction, it can initiate clearing the call by sending a RELEASE message to the network. The network shall enter the *release request* state upon receipt of a RELEASE message. This message prompts the network to disconnect the virtual channel, and to initiate procedures for clearing the network connection to the remote user. Once the virtual channel used for the call has been disconnected, the network shall send a RELEASE COMPLETE message to end system 1, and shall release both the call reference and virtual channel (i.e., the VPI or VCI). End system 1 then disconnects the virtual channel (a timeout method is used when a RELEASE COMPLETE message hasn't been received for a certain amount of time), as illustrated by (9)–(12) in Figure 8.3.

For point-to-multipoint call/connection control, the signaling channel used is the same as the one assigned for point-to-point connections. The signaling specification supports point-to-multipoint calls where information is multicast unidirectionally from one calling user to a set of called users; in that case the calling user is also referred to as the *root* and the called users are also referred to as *leaves*. The setup of the first party of a point-to-multi-

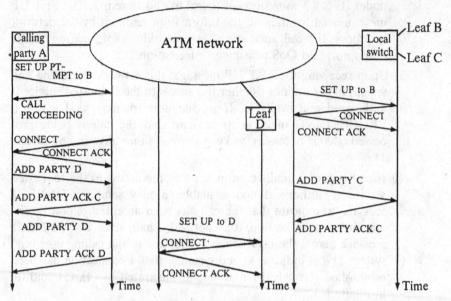

Fig. 8.4   Point-to-multipoint call setup example.

point call is always initiated by the root and follows the same procedure for point-to-point calls as explained above. Then, the root can initiate the addition of a party by transferring an ADD PARTY message to the network.

Figure 8.4 illustrates a point-to-multipoint connection setup, where end system A wants to communicate with B, C, and D. First of all, calling party A sets up a point-to-multipoint switched virtual connection (SVC) to B. After receiving a CONNECT message, end system A initiates the process of adding party C. After it is confirmed that C has been added, another similar process to add party D is initiated. Note that the success of adding a party requires another call setup from some branch node (in a tree based on the root A) to the party, such as C and D, as illustrated in Figure 8.4. Further details can be found in [1].

Figure 8.5 shows the format of a signaling message. The protocol discriminator is used to distinguish messages for UNI signaling from other types of messages or other protocols.

The call reference is used to identify the call at the UNI to which the particular message applies. The call reference is assigned by the originating side of the interface for a call, and remains fixed for the lifetime of the call. The call reference flag is used to identify which end of the signaling originated a call reference when the call references are the same: 0 is outgoing, 1 is incoming. The message type is used to identify the function of the message being sent, as shown in Figure 8.6 and Table 8.1.

Octets	8	7	6	5	4	3	2	1
1	Protocol discriminator							
2	0	0	0	0	Length of call reference value (in octets)			
3	Flag							
4	Call reference value							
5								
6	Message type							
7								
8	Message length							
9								
⋮	Information elements (variable length)							

Fig. 8.5 Signaling message format.

Bits

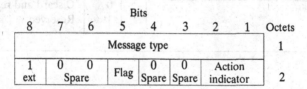

Fig. 8.6 Message type.

Figure 8.7 shows a general information element format. Table 8.2 explains the meaning of the information element identifier as indicated by its possible values. Examples of variable-length parameters include adaptation layer parameters (such as type of AAL, and rate and clock recovery method for AAL 1) and ATM user cell rates (such as peak cell rate, maximum burst size, and sustainable cell rates).

TABLE 8.1   Two Octets of Message Type

Octet 1

Bits	Meaning
000– – – – –	Call establishment messages:
0 0010	CALL PROCEEDING
0 0111	CONNECT
0 1111	CONNECT ACKNOWLEDGE
0 0101	SETUP
010– – – – –	Call clearing messages:
0 1101	RELEASE
1 1010	RELEASE COMPLETE
0 0110	RESTART
0 1110	RESTART ACKNOWLEDGE
011– – – – –	Miscellaneous messages:
1 1101	STATUS
1 0101	STATUS ENQUIRY
110– – – – –	Point-to-multipoint messages:
0 0000	ADD PARTY
0 0001	ADD PARTY ACKNOWLEDGE
0 0011	DROP PARTY
0 0100	DROP PARTY ACKNOWLEDGE

Octet 2

Bit (5)	Meaning	Bit (2, 1)	Meaning
0	Action indicator ignore	0 0	Clear call
1	Follow explicit instruction	0 1	Discard and ignore
		1 0	Discard and report status
		1 1	Reserved

8.1.2 PNNI Signaling

PNNI signaling [2, 3, 4] is used to dynamically establish, maintain, and clear ATM connections at the private network–network or network–node interface (PNNI) between two ATM networks or two ATM network nodes, as shown in Figure 8.8. Figure 8.9 describes the switching system architectural reference model.

PNNI routing is based on the well-known link-state routing technique, similar for example to OSPF. In addition to the basic link-state mechanism, PNNI includes two key extensions that make it suitable for use in today's

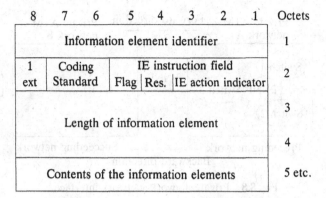

8	7	6	5	4	3	2	1	Octets
Information element identifier								1
1 ext	Coding Standard	IE instruction field						2
		Flag	Res.	IE action indicator				
Length of information element								3
								4
Contents of the information elements								5 etc.

Fig. 8.7   General information element format.

TABLE 8.2   Different Types of Information Element

Bits	Information Element
0000 1000	Cause
0001 0100	Call state
0101 0100	End point reference
0101 0101	End point state
0101 1000	ATM adaptation layer parameters
0101 1001	ATM traffic descriptor
0101 1010	Connection identifier
0101 1100	QoS parameter
0101 1101	Broadband high-layer information
0101 1110	Broadband bearer capability
0101 1111	Broadband low-layer information
0110 0000	Broadband locking shift
0110 0001	Broadband nonlocking shift
0110 0010	Broadband sending complete
0110 0011	Broadband repeat indicator
0110 1100	Calling party number
0110 1101	Calling party subaddress
0111 0000	Called party number
0111 0001	Called party subaddress
0111 1000	Transit network selection
0111 1001	Restart indicator

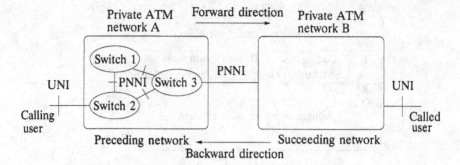

Fig. 8.8 Private network–network interface.

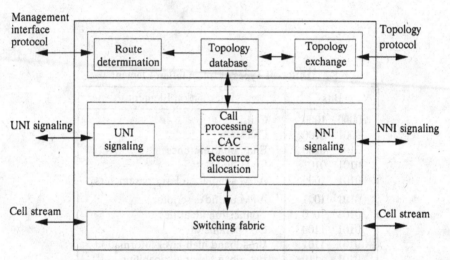

Fig. 8.9 Switching system architectural reference model.

communication environment:

- Support for QoS routing, which is required for applications with real-time requirements.
- A hierarchy mechanism to allow scalability to large world-wide networks. Use of a single routing protocol for the entire network, as opposed to the Internet strategy of using different protocols at different levels (e.g., OSPF, BGP, EGP), is advantageous for supporting end-to-end QoS routing, and also to reduce the configuration associated with multiple levels of routing.

8.1.2.1 Basic Concepts In PNNI, nodes are grouped into clusters called *peer groups* (PGs). The topology database (explained later) contains detailed information about nodes in the same PG and summary information about

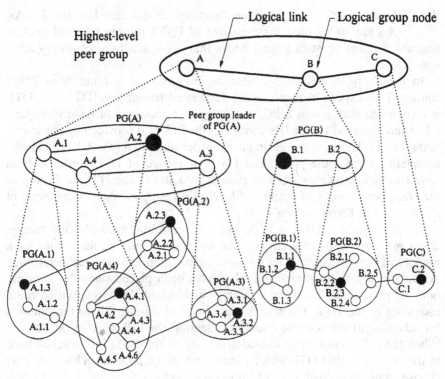

Fig. 8.10 PNNI routing domain hierarchy.

nodes of other PGs. PGs are represented as *logical group nodes* (LGNs) that can be grouped recursively into higher-level PGs, thus forming a hierarchical tree structure. Figure 8.10 illustrates such a PNNI routing domain hierarchy.

One node in a PG is elected as peer group leader. (PGL), as shown with the black nodes in Figure 8.10. PGLs communicate via switched virtual channel connections (SVCCs). A node of a PG that performs the extra work of collecting, aggregating, and building data will be suitable to represent the entire PG as a single node. This representation is made available in the parent node. The PGL floods the higher-level topology to the lower-level nodes in the PG. The PGL participates in the PGL election and spawns or creates the LGN after it has been elected PGL. The logical links between nodes in the same group are called *horizontal links*.

The LGN is an abstract representation of a lower-level PG as a single point for purposes of operating at one level of the PNNI hierarchy, as illustrated in Figure 8.10. It is instantiated by establishment of a routing entity (or node). An LGN establishes and manages an SVCC-based routing control channel (RCC) to other LGNs, summarizes lower-level PG topology, floods summarized PG topology to other LGNs, receives summarized PG topology from other LGNs, passes that summarized topology information to the PGL, and performs *aggregation* functions to describe horizontal links. The RCCs are VCCs used for exchange of PNNI routing protocol packets (explained later) between logical nodes that are logically or physically adjacent [2].

In processing a call, PNNI signaling may request a route from PNNI routing. PNNI specifies routes using designated transit lists (DTLs). A DTL is a complete path across a PG, consisting of a sequence of node identifiers (IDs) and, optionally, port IDs traversing the PG. It is provided by the source node (i.e., DTL originator) or an entry border node to a PG. A hierarchically complete source route is expressed as a sequence of DTLs ordered from lowest to highest PG level and organized as a stack. The DTL at the top of the stack corresponds to the lowest-level PG. Later, we show an example of using DTLs in PNNI routing.

When creating a DTL, a node uses the currently available information about resources and connectivity. That information may be inaccurate for a number of reasons, such as hierarchical aggregation and changes in resource availability due to additional calls that have been placed since the information was produced. Therefore, a call may be blocked along its specified route according to the DTL. Crankback and alternate routing [2] are mechanisms for adapting to this situation short of clearing the call back to the source. When the call cannot be processed according to the DTL, it is cranked back to the creator of that DTL with an indication of the problem. This node may choose an alternate path over which to progress the call or may further crank back the call.

A *border node* is a logical node that is in a specified PG and has at least one link that crosses the PG boundary. For each node in a PG to realize which (border) nodes have connectivity to which higher-level nodes, the border nodes must advertise links to those higher-level nodes. These are called *uplinks*, as illustrated in Figure 8.11. The functions of a border node include: generating an uplink; refraining from normal flooding on that uplink; expanding DTLs by performing *path computation* functions when necessary to provide a path traversing or terminating in the border node's PG; and handling crankback, if possible, to provide alternative routes.

By default, the signaling channel on VPI = 0 controls all of the virtual paths (VPs) on the physical interface. PNNI also supports signaling and routing over multiple virtual path connections (VPCs) to multiple destinations through a single physical interface. In this situation, each VPC configured for use as a logical link has a signaling channel associated with it. VCs within these VPCs are controlled only by the associated signaling channel of that particular VPC, that is, the default signaling channel (on VPI = 0) on the same physical interface does not control VCs within VPCs used as logical links, but does control all the remaining VCs and VPs on the physical link.

A soft permanent virtual path connection (PVPC) or permanent virtual channel connection (PVCC) is established and released between two network interfaces serving the permanent virtual connections as a management ac-

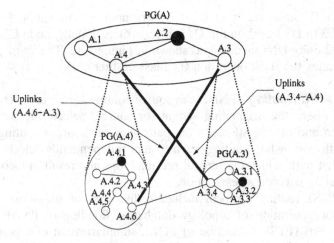

Fig. 8.11 Uplinks in PNNI.

tion. Here, the qualifier "permanent" signifies that it is established administratively (i.e., by network management) rather than on demand (i.e., by the use of signaling across the UNI). The establishment within the network is done by signaling. By configuration, the switching system at one end of the soft PVPC or PVCC initiates the signaling for this.

Figure 8.12 shows an example of PNNI routing, where host H1 wants to communicate with host H2. First, H1 connects to node A1 in PG A. A1 constructs a DTL, {(A1 → A2), B, C}, indicating the connection request should be sent to A2 in the same PG B, and then C before reaching H2. Similarly, the entry node B1 in PG B forwards the request through B2 and

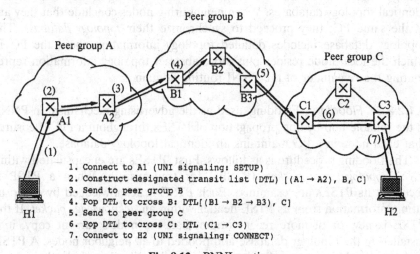

1. Connect to A1 (UNI signaling: SETUP)
2. Construct designated transit list (DTL) [(A1 → A2), B, C]
3. Send to peer group B
4. Pop DTL to cross B: DTL[(B1 → B2 → B3), C]
5. Send to peer group C
6. Pop DTL to cross C: DTL (C1 → C3)
7. Connect to H2 (UNI signaling: CONNECT)

Fig. 8.12 PNNI routing.

B3 to PG C. Since H2 is in C, the entry node C1 forwards the request through C3 to H2 based on the DTL it constructs. Finally, node C3 connects to host H2 using UNI signaling, as shown in Figure 8.12. The word *pop* used here indicates the stack operation (i.e., last in, first out).

8.1.2.2   PNNI Routing   PNNI uses source routing for all connection setup requests, where the originating system (or source) selects the path to the destination and other systems on the path obey the source's routing instructions. With hop-by-hop routing, each system independently selects the next hop for that path, which may create routing loops and result in inconsistency in the routing decision at each node.

The PNNI routing protocol includes: discovery of neighbors and link status, synchronization of topology databases, flooding of PNNI topology state elements (PTSEs), election of PGLs, summarization of topology state information, and construction of the routing hierarchy, which we briefly explain below.

8.1.2.2.1   Hello Protocol   The hello protocol is used to verify the identity of the neighbor node and the status of the horizontal links to that node, and to determine the status of the RCC. PNNI hello packets specify the ATM end system address, node ID, and port ID. Hello packets are periodically exchanged between neighbor nodes over all physical links and VPCs, and are also sent over all SVCCs used as RCCs.

As described before, a PG is identified by its *peer group identifier*. If the exchanged PG IDs are different, then the nodes belong to different PGs. A *border node* has at least one link that crosses the PG boundary.

8.1.2.2.2   Database Synchronization   Database synchronization is the exchange of information between neighbor nodes resulting in the nodes' having identical topology databases. When neighboring nodes conclude that they are in the same PG, they proceed to synchronize their *topology database*. The topology database includes detailed topology information about the PG in which the local node resides plus more abstract topology information representing the remainder of the PNNI routing domain.

8.1.2.2.3   Flooding   Flooding, used as the advertising mechanism in PNNI, is the reliable hop-by-hop propagation of PTSEs throughout a PG. It ensures that each node in a PG maintains an identical topology database.

The flooding procedure is as follows. First, PTSEs are encapsulated within *PNNI topology state packets* (PTSPs) for transmission. When a PTSP is received, its PTSEs are examined. Each PTSE is acknowledged by encapsulating information from its PTSE header within an *acknowledge packet*. If the PTSE is new or of more recent origin than the node's current copy, it is installed in the topology database and flooded to all neighbor nodes. A PTSE sent to a neighbor is periodically retransmitted until acknowledged.

8.1.2.2.4 Topology Information A topology database consists of a collection of all PTSEs received, which represent that node's present view of the PNNI routing domain. The topology database provides all the information required to compute a route from the given node to any address reachable in or through that routing domain, including nodal information, topology state information, and reachability information.

Each node generates a PTSE that describes its own identity and capabilities, information used to elect the PGL, and information used in establishing the PNNI hierarchy. For example,

- ATM end system address of the node,
- leadership priority,
- nodal information flag,
- node ID of preferred PGL, and,
- next higher-level binding information (if it is PGL).

Topology state parameters are classified as either attributes or metrics and include link-state parameters and nodal-state parameters, as shown in Figure 8.13.

Reachability information includes addresses and address prefixes that describe the destinations to which calls may be routed. This information is advertised in PTSEs by nodes in the PNNI routing domain and has

- internal reachable addresses (local reachability within the PNNI routing domain), and

Topology State Parameters		
	Topology Attributes	
Topology Metrics	Performance/Resource-Related	Policy-Related
Cell delay variation (CDV) Maximum cell transfer delay (maxCTD) Administrative weight (AW)	Cell loss ratio for CLP = 0 (CLP_0) Cell loss ratio for CLP = 0 + 1 (CLP_{0+1}) Maximum cell rate (maxCR) Available cell rate (AvCR) Cell rate margin (CRM) Variance factor (VF) Restricted branching flag	Restricted transit flag

CRM = allocated bandwidth − Σ sustainable cell rate (SCR);
VF = CRM/(variance of aggregate rate)

Fig. 8.13 Topology state information.

- exterior reachable addresses (reachability derived from other protocol exchanges outside this PNNI routing domain).

Border nodes advertise their uplinks in PTSEs flooded in their respective PGs.

8.1.2.2.5 Address Summarization and Reachability

Address summarization reduces the amount of addressing information that needs to be distributed in a PNNI network and thus contributes to scaling in large networks. It consists of using a single *reachable address prefix* to represent a collection of end system and/or node addresses that begin with the given prefix. The reachable address prefix associated with a node can be (1) an address prefix that either is explicitly configured at that node or takes on some default value (called *the summary address*), or (2) an address that doesn't match any of the node's summary addresses (called a *foreign address*).

Figure 8.14 illustrates these concepts, where the attachments to nodes A.2.1, A.2.2, and A.2.3 represent end systems. The alphanumeric associated

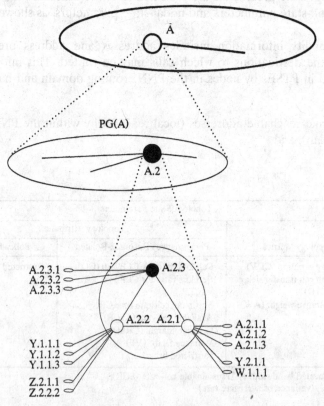

Fig. 8.14 Illustration of address summarization.

TABLE 8.3   Address Summarization and Reachability

Summary Addrs at A.2.1	Summary Addrs at A.2.2	Summary Addrs at A.2.3
P⟨A.2.1⟩ (configured) P⟨Y.2⟩ (configured)	P⟨Y.1⟩ (configured) P⟨Z.2⟩ (configured)	P⟨A.2.3⟩ (configured)

Reachable Addrs prefixes	Reachable Addrs prefixes	Reachable Addrs prefixes
flooded by node A.2.1 P⟨A.2.1⟩ P⟨Y.2⟩ P⟨W.1.1⟩	flooded by node A.2.2 P⟨A.2.2⟩ P⟨Y.1⟩ P⟨Z.2⟩	flooded by node A.2.3 P⟨A.2.3⟩

Reachability information advertised by LGN[a] A.2
P⟨A.2⟩ P⟨Y⟩ P⟨Z.2⟩ P⟨W.1.1⟩

[a]Logical group node.

with each end system represents that end system's ATM address. For example, ⟨A.2.1.3⟩ represents an ATM address, while P⟨A.2.1⟩, P⟨A.2⟩, and P⟨A⟩ represent successively shorter prefixes of that same ATM address. Table 8.3 explains the summarization with the reachability information, where for the chosen summary address list at A.2.1, P⟨W.1.1.1⟩ is a foreign address, since it doesn't match any of the summary addresses of node A.2.1.

8.1.2.2.6 Path Computation   PNNI signaling is based on source routing, and therefore the ingress ATM switch computes a path across the ATM network to an egress switch, which has advertised connectivity to the called party. This source route, which is a list of nodes on the path called a designated transit list (DTL), is included in the signaling messages to establish connections. Nodes internal to the network do not make routing decisions, but rather forward setup requests according to the DTL.

A consequence of using source routing is that the path computation algorithm is not specified as part of the PNNI specification. The implementation offers different strategies that the network administrator can select according to the needs of a particular configuration. The use of source routing thus makes it possible to build networks in which switches from different vendors coexist [3].

8.1.2.3 PNNI Evolution   The PNNI protocol was developed to address the needs of applications with real-time requirements such as guaranteed bandwidth and bounded delay. These requirements place special demands on routing and signaling that are addressed by PNNI, and that are not found in protocols in use in existing data networks [2, 3, 4].

The PNNI mechanisms include dynamic topology information, hierarchy, DTL (i.e., source routing), and crankback. It is important to note that these mechanisms are not needed to solve problems specific to ATM, but rather to solve problems introduced by the requirements of the applications. When applications requiring bandwidth or delay guarantees become widespread, it will be necessary to solve these problems in all networking environments, including the Internet [4].

The idea that PNNI routing would be a good choice for use in IP networks in addition to ATM has motivated the ATM Forum to develop *integrated PNNI* (I-PNNI) for routing in both environments [4]. The Internet Engineering Task Force (IETF) has also started to investigate the QoS routing problem for IP, as described in the following section.

8.2 QoS ROUTING FOR INTEGRATED SERVICES NETWORKS

With increasingly diverse QoS requirements, it is impractical to continue to rely on conventional routing paradigms that are primarily concerned with connectivity [5, 6, 7]. A typical resource reservation process has two steps: finding resources and making reservations. Resource reservations can only be made when routing has found paths with sufficient resources to meet user requirements. Therefore, modern routing strategies must take into consideration a wide range of QoS requirements.

Path selection within routing is typically formulated as a shortest-path optimization problem, that is, to determine a series of network links connecting the source and destination such that a particular objective function (e.g., hop count, delay, cost) is minimized [5]. Because the problem of calculating a path subject to multiple constraints has been proven NP-complete (i.e., intractable because there are no polynomial-time algorithms that can solve the problem) for many common parameter combinations [5, 6, 7], usually a compromise is made by choosing a subset of QoS parameters.

This section focuses on the selection of an appropriate path based on link metrics information and flow requirements in the Internet [5–18]. Specifically, it describes the metrics required to support QoS. It then presents and compares a set of path selection algorithms, which represent different trade-offs between accuracy and computational complexity. The overall goal is to allow deployment of QoS routing capabilities with the minimum possible disturbance to the existing routing infrastructure.

8.2.1 Selection of Metrics

In QoS-based routing, the path selection algorithms select paths that optimize one or multiple metrics. Let $d(i, j)$ be a metric of link (i, j). For any path $P = (i, j, k, \ldots, l, m)$, a metric $d(P)$ is

- *additive* if $d(P) = d(i, j) + d(j, k) + \cdots + d(l, m)$,
- *multiplicative* if $d(P) = d(i, j)d(j, k) \cdots d(l, m)$, and
- *concave* if $d(P) = \min\{d(i, j), d(j, k), \ldots, d(l, m)\}$.

For example, metrics such as delay, delay jitter, and hop count are additive, reliability (and indirectly packet loss probability) is multiplicative, and bandwidth is concave.

As mentioned above, finding an optimal path subject to constraints on two or more additive and/or multiplicative metrics in any possible combination is NP-complete. As a result, any algorithm that selects any two or more of delay, delay jitter, hop count, and loss probability as metrics and tries to optimize them simultaneously is NP-complete. The only computationally feasible combinations of metrics are bandwidth and any of the above.

However, the proof of NP-completeness is based on the assumptions that (1) all the metrics are independent and (2) the delay and delay jitter of a link are known *a priori*. Although such assumptions may be true in circuit-switched networks, metrics such as bandwidth, delay, and delay jitter are not independent in packet-switched networks. As a result, polynomial algorithms for optimizing delay and delay jitter exist [14]. Nevertheless, the work in [6] still serves as an indication of the complexity of a path selection algorithm: a path selection algorithm is still complex in data networks, even if it is no longer NP-complete.

Fortunately, simple algorithms for optimizing bandwidth and hop count metrics exist. These are the most useful metrics. With them, QoS routing is computationally feasible. [14] has proposed the following three QoS-based routing schemes:

1. A scheme that chooses the *widest–shortest* path, i.e., a path with the minimum number of hops. If there are several such paths, the one with the maximum residual (reservable) bandwidth is chosen.
2. A scheme that chooses the *shortest–widest* path, i.e., a path with the maximum residual (reservable) bandwidth. If there are several such paths, the one with the minimum number of hops is chosen.
3. A scheme that chooses the *shortest–distance* path. The distance of a k-hop path P is defined as

$$\text{dist}(P) = \sum_{i=1}^{k} \frac{1}{r_i},$$

where r_i is the bandwidth of link i.

The first scheme is basically the same as today's dynamic routing. It tries to preserve network resources by choosing the minimum-hop paths. The hop count of a path is important because the more hops a flow traverses, the more resources it consumes. For example, a 1-Mbit/s flow that traverses two hops consumes twice as much resources as one that traverses a single hop. The second scheme tries to distribute the load evenly by choosing the least loaded paths. The third scheme makes a tradeoff between preserving network resources and balancing network load. It emphasizes preserving network resources by choosing the minimum-hop-count path when the network load is heavy and emphasizes balancing network load by choosing the least loaded path when the network load is light.

For illustration, the widest–shortest-path scheme is considered below as the path selection algorithm to achieve the goal of minimum disturbance to the existing routing infrastructure. Specifically, the algorithms focus on selecting a path that is capable of satisfying the bandwidth requirement of the flow, while at the same time trying to minimize the number of hops used. This focus is useful in most instances but does not fully capture the complete range of potential QoS requirements. For example, a delay-sensitive flow of an interactive application could be put on a path using a satellite link, if that link provided a direct path and had plenty of unused bandwidth. This clearly would not be a desirable choice. To prevent such poor choices, delay-sensitive flows are assigned to a policy that eliminates from the network all links with high delay (e.g., satellite links), before invoking the path selection algorithm. Such a policy-based approach can be extended to criteria other than delay (e.g., security). In general, each existing policy would then present to the path selection algorithm a corresponding pruned network topology. While the use of policies to handle specific requirements allows considerable simplification of the optimization task to be performed by the path selection algorithm, it is not always the most effective. [9, 13] have shown a simple yet effective solution that can be constructed when the delay is assumed to be of a particular form.

8.2.2 Weighted Graph Model

A network can be modeled as a graph $G = (V, E)$. Nodes (V) of the graph represent switches, routers, and hosts. Edges (E) represent communication links. The edges are undirected only if the communication links are always symmetric. A symmetric link has the same properties (capacity, propagation delay, etc.) and the same traffic volume in both directions. For most real networks the communication links are asymmetric, and hence every link is represented by two directed edges in opposite directions.

Figure 8.15 shows a weighted graph model. Every link has a state measured by the QoS metrics of concern. For example, the link state could be bandwidth, delay, cost, or any possible combination of these. Every node also has a state. The node state can be either measured independently or combined into the state of the adjacent links. In the latter case, the residual bandwidth is the minimum of the link bandwidth and the CPU bandwidth. The CPU bandwidth is defined as the maximum rate at which the node can

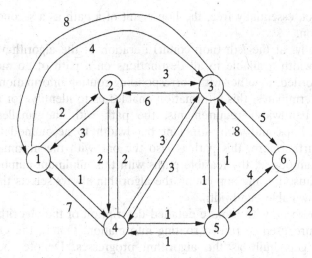

Fig. 8.15 A weighted graph model.

pump data into the link. The delay of a link consists of the link propagation delay and the queuing delay at the node. The cost of a link is determined by the total resource consumption at the link and node.

8.2.3 Path Selection Algorithm

There are several path selection algorithms [9, 13]. In the following we present the Bellman–Ford algorithm, which precomputes the QoS paths (static versions) for each flow requirement, and the Dijkstra-based algorithm, which computes the QoS paths on demand. (See [19, 20] for some background on these two algorithms.) On-demand computations are triggered by the given QoS requirements of flows.

8.2.3.1 Algorithm for Exact Precomputed QoS Paths (Static Bellman–Ford Algorithm) Consider a path selection algorithm that is based on a Bellman–Ford (BF) shortest-path algorithm, adapted to compute paths of maximum available bandwidth for all hop counts. For a given network topology and link metrics, the algorithm allows us to precompute all possible QoS paths, and also has a reasonably low computational complexity. Specifically, the algorithm allows us to precompute for any destination a minimum hop count path with maximum bandwidth, and has a computational complexity comparable to that of a standard shortest-path algorithm.

It is a property of the BF algorithm that, at its *h*th iteration, it identifies the optimal (i.e., maximum available bandwidth) path between the source and each destination, among paths of at most *h* hops. Therefore, we can also take advantage of the fact that the BF algorithm progresses by increasing hop count, to get, essentially free, the hop count of a path as a second optimization criterion.

Specifically, at the kth (hop count) iteration of the algorithm, the maximum bandwidth available to all destinations on a path of no more than k hops is recorded, together with corresponding routing information. After the algorithm terminates, this information enables us to identify, for all destinations and bandwidth requirements, the path with the smallest possible number of hops and with sufficient bandwidth to accommodate the new request. Furthermore, this path is also the one with the maximal available bandwidth among all the feasible paths with this minimum number of hops. This is because for any hop count, the algorithm always selects the one with maximum available bandwidth.

Let us proceed with a more detailed description of the algorithm and the data structure used to record routing information, that is, the QoS routing table that gets built as the algorithm progresses. Denote the available bandwidth on the edge between vertices n and m by $b(n, m)$. The vertex corresponding to the router where the algorithm is being run, i.e., the computing router, is denoted as the source node. The algorithm proceeds to precompute paths from this source node to all possible destination networks and for all possible bandwidth values. At each (hop count) iteration, intermediate results are recorded in a QoS routing table, which has the following structure:

- An $H \times K$ matrix, where H is the maximal allowed (or possible) number of hops for a path and K is the number of destination nodes.
- The (h, n) entry is built during the hth iteration (hop count value) of the algorithm, and consists of two fields:
  - bw (Bandwidth): The maximum available bandwidth on a path of at most h hops between the source node and destination node n.
  - nb (Neighbor): The routing information associated with the h-hops (or shorter) path to destination node n, whose available bandwidth is bw. In the context of hop-by-hop path selection, the neighbor information is simply the identity of the node adjacent to the source node on that path.

When the algorithm is invoked, the routing table is first initialized with all bw fields set to 0 and neighbor fields cleared. For each iteration h and each destination n, the algorithm first copies the bw and neighbor fields from row $h - 1$ into row h. Then the algorithm looks at each possible link (m, n) and checks the maximal available bandwidth on an (at most) h-hop path to node n, whose final hop is that link. This amounts to taking the lesser of the available bandwidth of the link (m, n) and the bw field in entry $(h - 1, m)$. If the resulting value is higher than the bw field in entry (h, n), then a better (larger bw value) path has been found for destination n and with at most h

TABLE 8.4   QoS Routing Table of the Weighted Graph Model in Figure 8.15

	Destination Nodes									
	2		3		4		5		6	
h	bw	nb	bw	nb	bw	nb	bw	nb	bw	nb
0	0	—	0	—	0	—	0	—	0	—
1	2	2	4	3	1	4	0	—	0	—
2	4	3	4	3	3	3	1	3	4	3
3	4	3	4	3	3	3	1	3	4	3

hops. The bw field of entry (h, n) is then updated to reflect the new value. In the case of hop-by-hop routing, the neighbor field records the identity of the first hop (next hop from the source) on the best path identified thus far for destination n and with h (or less) hops. The algorithm stops when there are no changes in all values between entries in row $h - 1$ and the corresponding ones in row h or the maximum number of iterations can be set explicitly to control the worst-case complexity. Table 8.4 illustrates all values in the QoS routing table for each iteration h of the weighted graph model in Figure 8.15.

8.2.3.2 Algorithm for On-Demand Computation of QoS Paths (Dynamic Dijkstra Algorithm)   The algorithm described above allows precomputation of QoS paths. However, it may be feasible in some instances (e.g., limited number of requests for QoS paths) to instead perform such computations *on demand*, that is, on receipt of a request for a QoS path. Below a standard Dijkstra algorithm is described to illustrate how the algorithm can, for a given destination and bandwidth requirement, generate a minimum-hop-count path that can accommodate the required bandwidth and also has a maximum bandwidth.

The algorithm essentially performs a minimum-hop-count path computation on a graph from which all edges whose available bandwidth is less than that requested by the flow triggering the computation have been removed. This can be performed either through a preprocessing step, or while running the algorithm by checking the available bandwidth value for any edge that is being considered.

The algorithm and the data structure used to record routing information are detailed as follows. Let us denote the available bandwidth on the edge between vertices n and m by $b(n, m)$. The algorithm maintains a vector t with dimension K equal to the number of destination nodes. Each entry n of the vector t consists of three fields:

- bw (Bandwidth): The maximum available bandwidth on a path between the source node s and destination node n.
- hc (Hop Count): The minimal number of hops on a path between the source node s and destination node n.

- nb (Neighbor): This is the associated routing information. With a hop-by-hop routing approach, the neighbor information is simply the identity of the node adjacent to the source node s on that path.

Additionally, the algorithm maintains

- the special values hc[s] and bw[s] for the source node s, where hc[s] is set to zero and bw[s] is set to infinity before running the algorithm;
- the set S of vertices whose the minimal number of hops on a path from the source node s has not yet been determined. Initially, the set S contains all vertices (V) in the graph.

```
Dijkstra(G, t, b, f, s)
/* Initialization */
for (each destination n in the vector t) do
begin
    hc[n] = Infinity;
    bw[n] = undefined;
    nb[n] = undefined;
end
hc[s] = 0;
bw[s] = infinity;
/* Compute QoS paths */
S = the set that contains all vertices in the graph G;
while (S is not empty) do
begin
    u = the vertex in S whose value in the field hc is minimum;
    S = S - {u};
    for (each vertex v adjacent to u) do
    begin
        if ((b(u, v) ≥ bandwidth requirement f) and (hc[v] > [u] + 1) then
        begin
            hc[v] = hc[v] + 1;
            bw[v] = min{bw[u], "b(u, v)};
            if (the vertex u is the source node s) then
                nb[v] = v;
            else
                nb[v] = nb[u];
        end
    end
end
```

Fig. 8.16 Pseudo-code for the Dijkstra algorithm.

TABLE 8.5   An Example to Illustrate the Entries of the Vector t at Each Iteration of the On-Demand Computation of the Dijkstra Algorithm by Using the Weighted Graph Model in Figure 8.15

Iteration	u	S	Node 2			Node 3			Node 4		
			hc	bw	nb	hc	bw	nb	hc	bw	nb
0	—	1, 2, 3, 4, 5, 6	∞	—	—	∞	—	—	∞	—	—
1	1	2, 3, 4, 5, 6	1	2	2	1	4	3	∞	—	
2	2	3, 4, 5, 6	1	2	2	1	4	3	2	2	2
3	3	4, 5, 6	1	2	2	1	4	3	2	3	3
4	4	5, 6	1	2	2	1	4	3	2	3	3
5	6	5	1	2	2	1	4	3	2	3	3
6	5		1	2	2	1	4	3	2	3	3

Iteration	u	S	Node 5			Node 6		
			hc	bw	nb	hc	bw	nb
0	—	1, 2, 3, 4, 5, 6	∞	—	—	∞	—	—
1	1	2, 3, 4, 5, 6	∞	—	—	∞	—	—
2	2	3, 4, 5, 6	∞	—	—	∞	—	—
3	3	4, 5, 6	∞	—	—	2	4	3
4	4	5, 6	∞	—	—	2	4	3
5	6	5	3	4	3	2	4	3
6	5		3	4	3	2	4	3

Source node s: node 1, hc[s] = 0, bw[s] = ∞. Given the bandwidth requirement $f = 2$ of a flow.

The pseudo-code for the algorithm is shown in Figure 8.16. Table 8.5 shows the values of the vector t at each iteration to illustrate the way the algorithm works, using the graph in Figure 8.15. For a given bandwidth requirement $f = 2$, the QoS paths from the source node $s = 1$ to all other vertices can be computed. The algorithm stops when the set S is empty.

8.2.3.3 *Algorithm for Approximate Precomputed QoS Paths (Static Dijkstra Algorithm)*   Next outlined is a Dijkstra algorithm that allows pre-computation of QoS routes for all destinations and bandwidth values. Allowing precomputation of paths helps lower computational overhead. The cost, however, is a loss in the accuracy of the precomputed paths. That is, the paths being generated may have a larger hop count than needed. This loss in accuracy comes from the need to rely on quantized bandwidth values that are used when computing a minimum-hop-count path. In other words, the range of possible bandwidth values that can be requested by a new flow is mapped into a fixed number of quantized values, and minimum-hop-count paths are

generated for each quantized value. A new flow is then assigned to the minimum-hop-count path that can carry the smallest quantized value larger than or equal to what is requested. For example, one could assume that bandwidth values of the graph in Figure 8.15 after being quantized are 2, 4, 6, and 8. The minimum-hop-count paths to all destinations are computed for each of these four values. A new flow with the bandwidth requirement $f = 3$ is assigned to the quantized value of 4, whose minimum-hop-count paths are used. The structure of the routing table is as follows:

- A $K \times Q$ matrix, where K is the number of destinations and Q is the number of quantized bandwidth values.
- The (n, q) entry contains information that identifies the minimum-hop-count path to destination n that is capable of accommodating a bandwidth request of at least bw[q] (the qth quantized bandwidth value). It consists of two fields:
  - hc (Hop Count): The minimal number of hops on a path between the source node and destination n that can accommodate a request of at least bw[q] units of bandwidth.
  - nb (Neighbor): This is the routing information associated with the minimum hop count path to destination node n, whose available bandwidth is at least bw[q].

The algorithm operates, again, on a directed graph, and the metric associated with each edge (n, m) in the graph is as before the available bandwidth $b(n, m)$. For each index q, the algorithm deletes from the original network topology all links (n, m) for which $b(n, m) < $ bw[q], and then runs on the remaining topology a Dijkstra minimum-hop-count algorithm between the source node and all other nodes in the graph. Note that, as with the Dijkstra algorithm used for on-demand path computation, the elimination of links such that $b(n, m) < $ bw[q] could also be performed while running the algorithm.

After the algorithm terminates, the qth column in the routing table is updated. In order to ensure that the path with the maximal available bandwidth is always chosen among all minimum-hop-count paths that can accommodate a given quantized bandwidth, a slightly different update mechanism for the neighbor field needs to be used in some instances. Specifically, when for a given row (i.e., destination node n) the value of the hop count field in column q is found to be the value in column $q + 1$ (here we assume $q < Q$)—that is, paths that can accommodate bw[q] and bw[$q + 1$] have the same hop count—then the algorithm copies the value of the neighbor field from entry $(n, q + 1)$ into entry (n, q).

Table 8.6 provides an example to illustrate the entries of the QoS routing table at each iteration for a specific quantized bandwidth value $q = 2$ by

TABLE 8.6   An Example to Illustrate the Entries of the QoS Routing Table at Each Iteration of the Precomputed Paths of Dijkstra's Algorithm for a Specific Quantized Bandwidth Value = 2 by Using the Weighted Graph Model in Figure 8.15

Iteration	u	S	Node 2 hc	Node 2 nb	Node 3 hc	Node 3 nb	Node 4 hc	Node 4 nb	Node 5 hc	Node 5 nb	Node 6 hc	Node 6 nb
0	—	1, 2, 3, 4, 5, 6	∞	—	∞	—	∞	—	∞	—	∞	—
1	1	2, 3, 4, 5, 6	1	2	1	3	∞	—	∞	—	∞	—
2	2	3, 4, 5, 6	1	2	1	3	2	2	∞	—	∞	—
3	3	4, 5, 6	1	2	1	3	2	2	∞	—	∞	—
4	4	5, 6	1	2	1	3	2	2	∞	—	2	3
5	6	5	1	2	1	3	2	2	3	3	2	3
6	5		1	2	1	3	2	2	3	3	2	3

Source node s: node 1, hc[s] = 0, bw[s] = ∞, Quantized bandwidth value = 2.

using the weighted graph model in Figure 8.15. Note that the 2-hop path from the source node $s = 1$ to destination node 4 computed from the algorithm provides available bandwidth equal to 2, which is not optimal (the path should go through node 3, which provides bandwidth 3). This indicates a loss in accuracy of this algorithm.

8.2.4   Computational Complexity

The approaches described above are based on either Bellman–Ford or Dijkstra shortest-path algorithms. The Dijkstra algorithm has been traditionally considered more efficient for standard shortest-path computations because of its lower worst-case complexity. The benefit of using a Dijkstra algorithm in QoS path selection is a greater synergy with the existing open shortest path first (OSPF) implementation. On-demand path computation of Dijkstra algorithm provides advantages in yielding better routes and minimizing the need for storage of data structures if there are few requests for QoS paths. The precomputed-path computation of the Dijkstra algorithm helps lower computational overhead but causes a loss in the accuracy of the precomputed paths. This is essentially the standard tradeoff between memory space, accuracy, and processing cycle. In the following, a number of considerations are reviewed in the context of multicriteria QoS paths, which indicate that a BF approach may often provide a lower-complexity solution.

The asymptotic worst-case complexity of a binary-heap implementation of the Dijkstra algorithm is $O(E \log V)$, where V is the number of vertices in the graph, and E the number of edges. Detailed discussion can be found in [20]. The asymptotic worst-case bound for the BF algorithm is $O(H\ E)$, where H is the maximum number of iterations of the algorithm and thus is an upper bound on the number of hops in a shortest path. Although theoretically H can be as large as $V - 1$, in practice it is usually considerably smaller than V. In some network scenarios, an upper bound h of small size (i.e., $h \ll V$) is imposed on the allowed number of hops, in which case the complexity reduces to $O(h\,E)$, or effectively to $O(E)$. As a consequence, as noted in [19] and experienced in [12], in practical networking scenarios, the BF algorithm offers an efficient solution to the shortest-path problem, one that often outperforms the Dijkstra algorithm.

In the context of QoS path selection, the potential benefits of the BF algorithm are even more apparent. As mentioned before, efficient selection of a suitable path for flows with QoS requirements cannot usually be handled using a single-objective optimization criterion. While multiobjective path selection is known to be an intractable problem, the BF algorithm allows us to handle a second objective, namely the hop count, which is reflective of network resources, at no additional cost in complexity. In contrast, the Dijkstra algorithm requires some modifications (or approximations, e.g., bandwidth quantization) in order to be able to deal with hop counts as a second objective. When on-demand computations of QoS paths are practical, Dijkstra's algorithm provides an (exact) solution of complexity $O(E \log V)$. If QoS paths are precomputed with Q quantized bandwidth values, the corresponding asymptotic worst-case complexity is $O(QE \log V)$, which is comparable to the complexity of the BF algorithm.

8.2.5 Further Reading

Chen and Nahrstedt [10] provide a good review of the QoS routing literature. The main part of this section comes from the work in [13] and [9]. These two papers provide a thorough discussion on the path selection algorithms discussed here. Details of pseudo-code of the algorithms can be found in [13], which also describes other aspects of how QoS routes are established and managed: the associated link advertisement mechanisms, options in balancing the requirements for accurate and timely information with the associated control overhead, and route establishment. Analyses of the complexity are in [20].

REFERENCES

1. ATM Forum, *User-Network Interface (UNI) Specification Version* 3.1, Sep. 1994.
2. ATM Forum, *Private Network–Network Interface Specification Version* 1.0 (*PNNI* 1.0), Mar. 1996.
3. "IBM PNNI control point (switched network services)," White paper, IBM Corp., Mar. 1997.
4. P. Dumortier, "Toward a new IP over ATM routing paradigm," *IEEE Commun. Mag.*, pp. 82–86, Jan. 1998.

5. W. C. Lee, M. G. Hluchyj, and P. A. Humblet, "Routing subject to quality of service constraints in integrated communication networks," *IEEE Netw.*, pp. 46–55, Jul./Aug. 1995.

6. Z. Wang and J. Crowcroft, "Quality-of-service routing for supporting multimedia applications," *IEEE J. Select. Areas Commun.*, vol. 14, no. 7, pp. 1228–1234, Sept. 1996.

7. R. Vogel, R. G. Herrtwich, W. Kalfa, H. Wittig, and L. C. Wolf, "QoS-based routing of multimedia streams in computer networks," *IEEE J. Select. Areas Commun.*, vol 14, no. 7, pp. 1235–1244, Sep. 1996.

8. G. Apostolopoulos, R. Guerin, S. Kamat, and S. K. Tripathi, "Quality of service based routing: a performance perspective," *Proc. ACM SIGCOMM*, 1998.

9. R. Guerin, A. Orda, and D. Williams, "QoS routing mechanisms and OSPF extensions," *Proc. 2nd Global Internet Miniconf. (Joint with GLOBECOM'97)*, Nov. 1997.

10. S. Chen and K. Nahrstedt, "An overview of quality of service routing for next-generation high-speed networks: problems and solutions," *IEEE Netw.*, pp. 64–79, Nov./Dec. 1998.

11. D. H. Lorenz and A. Orda, "QoS routing in networks with uncertain parameters," *IEEE/ACM Trans. Netw.*, vol. 6, no. 6, pp. 768–778, Dec. 1998.

12. B. V. Cherkassky, A. V. Goldberg, and T. Radzik, "Shortest paths algorithms: theory and experimental evaluation," *Proc. 5th Annual ACM SIAM Symp. on Discrete Algorithms*, Arlington, VA, pp. 516–525, Jan. 1994..

13. R. Guerin, S. Kamat, A. Orda, T. Przygienda, and D. Williams, "QoS routing mechanisms and OSPF extensions," Internet draft, Jan. 1998, draft-guerin-QoS-routing-ospf-03.txt.

14. Q. Ma, "QoS routing in the integrated services networks," Ph.D. thesis, Carnegie Mellon University, CMU-CS98-138, Jan. 1998.

15. G. Apostolopoulos, R. Guerin, S. Kamat, S. K. Tripathi, "Improving QoS routing performance under inaccurate link state information," *Proc. 16th Int. Teletraffic Congr.*, Jun. 1999.

16. R. Guerin, L. Li, S. Nadas, P. Pan, and V. Peris, "The cost of QoS support in edge devices: an experimental study," *Proc. IEEE INFOCOM*, Mar. 1999.

17. G. Apostolopoulos, R. Guerin, and S. Kamat, "Implementation and performance measurements of QoS routing extensions to OSPF," *Proc. IEEE INFOCOM*, Mar. 1999.

18. R. Guerin and A. Orda, "QoS-based routing in networks with inaccurate information: theory and algorithms," *IEEE/ACM Trans. Netw.*, vol. 7, no. 3, pp. 365–374, Jun. 1999. Previous version, *Proc. IEEE INFOCOM*, 1997.

19. D. Bertsekas and R. G. Gallager, *Data Networks*, Prentice-Hall, Englewood Cliffs, NJ, 2nd Edition, 1992.

20. T. H. Cormen, C. E. Leiserson, and R. L. Rivest, *Introduction to Algorithms*, MIT Press, Cambridge, MA, 1990.

CHAPTER 9

DIFFERENTIATED SERVICES

The rapid transformation of the Internet into a ubiquitous commercial infrastructure has not only created rapidly rising bandwidth demand but also significantly changed consumer demands for QoS. Consequently, service providers need not only to evolve their networks to higher speed, but also need to provide increasingly sophisticated services to serve different requirements of QoS for different customers. At the same time, Internet service providers (ISPs) would like to maximize the utilization of the costly network resources in a manner that enables them to control the usage of network resources in accordance with service pricing and revenue potential. *Differentiated services* (Diffserv or DS) is a set of technologies that allows network service providers to offer services with different kinds of network QoS objectives to different customers and their traffic streams.

According to this model, network traffic is classified and conditioned at the entry to a network and assigned to different behavior aggregates [1]. Each such aggregate is assigned a single DS codepoint (i.e., one of the markups possible with the DS bits). Different DS codepoints signify that the packet should be handled differently by the interior router. Each type of processing that can be provided to the packet is called a *per-hop behavior* (PHB). In the core of the network, packets are forwarded according to the PHBs associated with the codepoints. The PHB to be applied is indicated by a Diffserv codepoint (DSCP) in the IP header of each packet [7]. The DSCP markings are applied either by a trusted customer or by the boundary routers on entry to the Diffserv network.

The advantage of such a scheme is that many traffic streams can be aggregated to one of a small number of behavior aggregates (BAs), which are each forwarded using the same PHB at the router, thereby simplifying the processing and associated storage. In addition, there is no signaling, other than what is carried out in the DSCP of each packet, and no other related processing is required in the core of the Diffserv network, since QoS is invoked on a packet-by-packet basis.

Section 9.1 explains the concepts of service level agreement and traffic conditioning agreement. The basic architecture of Diffserv is presented in Section 9.2. Section 9.3 and Section 9.4 describe the network boundary traffic conditioning and the PHBs, respectively. A conceptual model of a Diffserv router is discussed in Section 9.5.

9.1 SERVICE LEVEL AGREEMENT AND TRAFFIC CONDITIONING AGREEMENT

9.1.1 Service Level Agreement

A *service level agreement* (SLA) is a formal definition of the relationship that exists between two organizations, usually a supplier of services and its customer. The SLA is a contract that specifies the forwarding service a customer should receive. A customer may be a user organization (source domain) or another DS domain (upstream domain). Here, a DS domain is a contiguous set of nodes that operate with a common set of service provision policies and PHB definitions.

A SLA can be dynamic or static. Static SLAs are negotiated on a regular basis (e.g., monthly or yearly). Dynamic SLAs use a signaling protocol to negotiate the service on demand.

The SLA typically contains:

- The type and nature of service to be provided, which includes the description of the service to be provided, such as facilities management, network services, or help desk support.
- The expected performance level of the service, which includes two major aspects: reliability and responsiveness. Reliability includes availability requirements—when the service is available, and what are the bounds on service outages that may be expected. Responsiveness includes how soon the service is performed in the normal course of operations.
- The process for reporting problems with the service, which forms a big part of a typical SLA. It includes information about the person to be contacted for problem resolution, the format in which complaints have to be filed, and the steps to be undertaken to solve the problem quickly.
- The time frame for response and problem resolution, which specifies a time limit by which someone would start investigating a problem that was reported, etc.

- The process for monitoring and reporting the service level, which outlines how performance levels are monitored and reported—that is, who will do the monitoring, what types of statistics will be collected, how often they will be collected, and how past or current statistics may be accessed.
- The credits, charges, or other consequences for the service provider on not meeting its obligation (that is, failing to provide the agreed-upon service level).
- Escape clauses and constraints, including the consequences if the customer does not meet his or her obligation, which qualifies access to the service level. *Escape clauses* are conditions under which the service level does not apply, or under which it is considered unreasonable to meet the requisite SLAs—for example, when the service provider's equipment has been damaged in flood, fire, or war. They often also impose some constraints on the customer's behavior. For example, a network operator may void the SLA if the customer is attempting to breach the security of the network.

9.1.2 Traffic Conditioning Agreement

A SLA also specifies the *traffic conditioning agreement* (TCA), which contains the rules used to realize the service—what the client must do to achieve desired service, and what the service provider will do to enforce the limits. A TCA specifies classifier rules and any corresponding traffic profiles and metering, marking, discarding, and/or shaping rules that are to apply to the traffic streams selected by the classifier. A TCA encompasses all of the traffic conditioning rules explicitly specified within a SLA along with all of the rules implicit in the relevant service requirements and/or from a DS domain's service provisioning policy.

A *traffic profile* specifies the temporal properties of a traffic stream selected by a classifier. It provides rules for determining whether a particular packet is in profile or out of profile. For example, a profile based on a token bucket may look like this:

$$codepoint = X, \quad \text{use token-bucket } r, b$$

The above profile indicates that all packets marked with DS codepoint X should be measured against a token bucket meter with rate r and burst size b. In this example out-of-profile packets are those packets in the traffic stream that arrive when insufficient tokens are available in the bucket. The concept of being in- or out-of-profile can be extended to more than two levels; for example, multiple levels of conformance with a profile may be defined and enforced.

As packets enter the domain, they will be classified into a traffic aggregate by the specified filter at the domain ingress interface of the border router.

The filter must be associated with a traffic profile that specifies the committed information rate (CIR) and a description on how it is to be measured. For example, the measurement may be based on a committed burst size (CBS) or an averaging time interval (T1).

The traffic profile may also include other traffic parameters. These parameters may place additional constraints on packets to which the assurance applies or may further differentiate traffic that exceeds the CIR. Such parameters could include the peak information rate (PIR), the peak burst size (PBS), the excess burst size (EBS), or even a second averaging time interval (T2).

9.2   BASIC ARCHITECTURE OF DIFFERENTIATED SERVICES

Diffserv divides a network into several domains. A *DS domain* [1] is a continuous set of nodes that operate with a common set of resource provision policies and PHB definitions. It has a well-defined boundary, and so there are two types of nodes associated with a DS domain: boundary nodes and interior nodes. *Boundary nodes* connect the DS cloud to other domains. *Interior nodes* are connected to other interior nodes or boundary nodes, but they must be within the same DS domain. The boundary nodes are assigned the duty of classifying ingress traffic so that incoming packets are marked appropriately to choose one of the PHB groups supported inside the domain. They also enforce the TCAs between their own DS domain and the other domains it connects to. The TCA specifies the rules used to realize the service, such as metering, marking, and discarding. Interior nodes map the DS codepoints of each packet into the set of PHBs and perform appropriate forwarding behavior. Any non-DS-compliant node inside a DS domain results in unpredictable performance and a loss of end-to-end QoS. A DS domain is generally made up of an organization's intranet or an ISP, that is, a network controlled by a single entity. Diffserv is extended across domains by SLAs between them. An SLA specifies rules such as traffic remarking, actions to be taken for out-of-profile traffic, etc. The TCAs between domains are decided from this SLA.

Figure 9.1 shows the basic elements of a Diffserv network. Depending on direction of traffic flow, DS boundary nodes can be both ingress nodes and egress nodes. Traffic enters the DS cloud through an ingress node and exits through an egress node. An ingress node is responsible for enforcing the TCA between the DS domain and the domain of the sender node. An egress node shapes the outgoing traffic to make it compliant with the TCA between its own DS domain and the domain of the receiver node.

Flows are classified by predetermined rules so that they can fit into a limited set of class flows. The boundary routers use the 8-bit type of service (ToS) field of the IP header, called the DS field in Diffserv terminology, to mark the packet for preferential treatment by the interior routers. Six bits

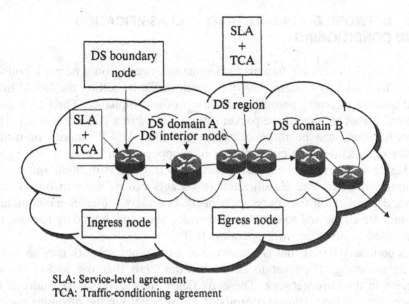

SLA: Service-level agreement
TCA: Traffic-conditioning agreement

Fig. 9.1 Basic elements of a differentiated services network.

are used for DSCP, and the other two bits are reserved for future use as shown in Figure 9.2. Only the boundary routers need to maintain per-flow states and perform the shaping and the policing. This is usually advantageous, since the links between the customer and service provider are usually slow, so additional computational delay is not that much of a problem for routers interfacing to these links. Therefore, it is affordable to do the computationally intensive traffic shaping and policing strategies at the boundary routers. But once inside the core of the service providers, packets need to be routed (or forwarded) very quickly, and so the routing must incur minimum computational delay at any router or switch. Since the number of flows at the boundary router is much smaller than that in the core network, it is also advantageous to do flow control at the boundary routers.

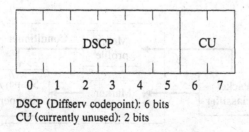

DSCP (Diffserv codepoint): 6 bits
CU (currently unused): 2 bits

Fig. 9.2 The Diffserv field in the 8-bit IP ToS field.

9.3 NETWORK BOUNDARY TRAFFIC CLASSIFICATION AND CONDITIONING

Traffic conditioners are various QoS functions needed on a network boundary. The boundary routers classify or mark traffic by setting the DSCP field and monitor incoming network traffic for profile compliance. The DSCP field indicates what treatment the packet should receive in a Diffserv domain. The QoS functions can be those of packet classifier, DSCP marker, or traffic metering function, with either shaper or dropper action.

Figure 9.3 presents the logical structure of traffic classification and conditioning functions. The classification of packets can be done in one of two ways, depending on the connectivity of the boundary router. Some boundary routers are connected to customer networks, and some boundary routers are connected to other network operators (ISPs).

A boundary router that is connected to a customer network uses six fields in an incoming IP packet to determine the PHB that the packet should receive in the core network. These six fields are the IP source address, IP destination address, transport protocol, Diffserv field in the incoming packet header, source port in the transport headers, and destination port in the transport header. A rule that maps a packet to a PHB does not need to specify all six fields. We refer to such a rule as a *classification rule*. When a classification rule does not specify any value for a field, that field's value is not used for the purpose of classification.

Boundary routers could use just one field in the incoming IP packet to determine the PHB for their network. This field could be the Diffserv field contained in the incoming packet. A boundary router would simply change the Diffserv field to some other value that corresponds to a specific PHB at the core routers. This type of classification would be the one expected at the exchange points of other ISPs. The neighboring ISP domain might have been using a different set of PHBs or might use different Diffserv field values to represent the same PHB.

In addition to the classification process, the access routers should change the Diffserv field value so that it corresponds to the correct PHB in the core network. It can also exert rate control or shaping on all the packets that have

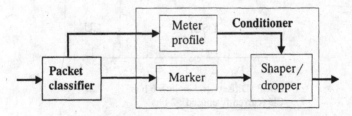

Fig. 9.3 Packet classifier and traffic conditioner.

been determined to be in a specific profile. Thus, a boundary router may limit the total number of packets that can be sent into each class. If it is found that a packet cannot be mapped into any of the PHBs because a limit would be exceeded, it may either be mapped into a different PHB or discarded.

9.4 PER-HOP BEHAVIORS AND SOME IMPLEMENTATION EXAMPLES

A PHB is a description of the externally observable forwarding behavior of a DS node applied to a particular DS behavior aggregate. According to RFC 2475 [1], a PHB is the means by which a node allocates resources to aggregate streams; the given example is that a PHB specifies a percentage of the capacity of a link. The interior routers in the Diffserv model only need to forward packets according to the specified PHBs.

If only one behavior aggregate occupies a link, the observable forwarding behavior will generally only depend on the congestion of the link. Distinct behavioral patterns are only observed when multiple behavioral aggregates compete for buffer and bandwidth resources on a node as shown in Figure 9.4. There are two flows: source 1 → destination 1 and source 2 → destination 1. They have different classifications and will be treated differently by the interior router (e.g., using different scheduling and/or discarding preference). A network node allocates resources to the behavior aggregates with the help of the PHBs. PHBs can be defined either in terms of their resources (e.g., buffer and bandwidth), in terms of their priority relative to other PHBs, or in terms of their relative traffic properties (e.g., delay and loss). Multiple PHBs are lumped together to form a *PHB group* [1] to ensure consistency. PHBs are implemented at nodes through some buffer management or packet scheduling mechanisms. A particular PHB group can be implemented in a

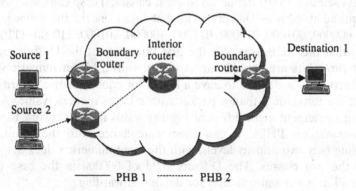

Fig. 9.4   The per-hop behavior in Diffserv.

variety of ways, because PHBs are defined in terms of behavior characteristics and are not implementation-dependent.

The standard for Diffserv describes PHBs as the building blocks for services. The focus is on enforcing an SLA between the user and the service provider. Customers can mark the DS byte of their packets to indicate the desired service, or have them marked by the boundary router based on a multifield (MF) classification, such as IP destination and source addresses, transport numbers, protocol ID, and so on. Inside the core, packets are forwarded according to their behavior aggregates. These rules are derived from the SLA. When a packet goes from one domain to another, the DS byte may be rewritten by the new network boundary routers. A PHB for a packet is selected at a node on the basis of its DS codepoint. For example, DS codepoint (101110) indicates an expedited forwarding (EF) PHB as described in detail below. The mapping from DS codepoint to PHB may be 1-to-1 or N-to-1. All codepoints must have some PHBs associated with them. If not, they are mapped to a default PHB. Examples of the parameters of the forwarding behavior each traffic class should receive are the bandwidth partition and the drop priority. Examples of implementations of these are weighted fair queuing (WFQ) for bandwidth partition and random early detection (RED) for drop priority. The four most popularly used PHBs are default behavior, class selector, assured forwarding (AF), and EF.

9.4.1 Default Behavior

The default or best-effort PHB corresponds to the default best-effort packet forwarding in traditional IP network. Packets belonging to this PHB can be forwarded in any manner without any restrictions. The codepoint recommended by IETF for best-effort PHB is 0x000000.

9.4.2 Class Selector

The class selector PHBs define up to eight classes. These class selector PHBs are required to have a Diffserv field that takes one of the following eight values: 000000, 001000, 010000, 011000, 100000, 101000, 110000, 111000.

Note that this PHB uses only the first three bits of the Diffserv field and the last three bits are zeroed out. A packet with a higher numeric value in the Diffserv field is defined to have a better (or equal) relative priority in the network for forwarding than a packet with a lower numeric value. A router need not implement eight different priority levels in the network to support the class selector PHBs. It can claim compliance with the standards by supporting only two priority levels, with the eight numeric values mapping to one of the two classes. The Diffserv field of 000000 in the case of class selector PHB is the same as that for default forwarding.

9.4.3 Assured Forwarding

The AF standard extended the basic in or out marking in RED with in and out (RIO) into a structure of four forwarding classes and, with each forwarding class, three drop precedences. Each forwarding class is allocated a minimum amount of buffers and bandwidth. Customers can subscribe to the services built with AF forwarding classes, and their packets will be marked with the appropriate AF DSCPs. Table 9.1 summarizes the recommended AF codepoint values for four AF forwarding classes and three levels of drop precedence in each class.

The three drop priorities within each forwarding class are used to choose which packets to drop during congestion. When backlogged packets from an AF forwarding class exceed a specified threshold, packets with the higher drop priority are dropped first, and then packets with the lower drop priority. Drop priorities in AF are specific to the forwarding class; comparing drop priorities in two different AF classes may not always be meaningful. For example, when a DS node starts to drop the packets with the highest drop priority in one forwarding class, the packets in other forwarding classes may not experience any packet dropping at all. Each forwarding class has its bandwidth allocation. Dropping takes place only in the forwarding class in which traffic exceeds its own resources.

A DS node must implement all four AF classes. For each AF class, the minimum amount of forwarding resources (buffers and bandwidth) that are allocated to an AF class must be configurable. The bandwidth allocation must be guaranteed over both short and long time scales.

In general, a DS node may reorder packet of different AF classes but should not reorder packets with different drop priorities but in the same class. The boundary nodes should avoid splitting traffic from the same application flow into different classes, since it will lead to packet reordering with a microflow in the network.

The AF PHB is used to provide *assured services* to the customers, so that the customers will get reliable services even in times of network congestion. IP packets are forwarded with high probability as long as the aggregate traffic from each site does not exceed the subscribed information rate.

TABLE 9.1   Recommended AF Codepoint Value

Drop Precedence	Codepoint Value			
	Class 1	Class 2	Class 3	Class 4
Low	001010	010010	011010	100010
Medium	001100	010100	011100	100100
High	001110	010110	011110	100110

9.4.4 Expedited Forwarding

EF PHB is defined as a forwarding treatment for a Diffserv aggregate where the departure rate of the aggregate's packets from any Diffserv node must equal or exceed a configurable rate [3, 4]. Queues in the network contribute latency and jitter to the traffic. In order to reduce loss, latency, and jitter, one should reduce queueing in the system. A queue-free service will guarantee bounded traffic rates. The EF traffic *should* receive its minimum rate of departure irrespective of how much other traffic at the node might be. The traffic should be conditioned (via policing and shaping the aggregated traffic) so that the maximum arrival rate at any node is less than the predetermined departure rate. Formally, the average departure rate measured over a time period should be greater than or equal to the configured rate. DS ingress routers must negotiate a rate less than this configured rate with adjacent upstream routers. The measured time period is equal to the time required to send a maximum transmission unit (MTU) at the configured rate. To enforce this condition, the ingress router must strictly police all incoming traffic. Packets in excess of the configurated rate are dropped. The default EF configured rate is 0, which means that all packets marked with EF are dropped.

The EF PHB is used to provide *premium service* to the customer. It is a low-delay, low-jitter service providing nearly constant bit rate. The SLA specifies a peak bit rate that customer applications will receive, and it is the customers' responsibility not to exceed the rate, in violation of which packets are dropped.

EF PHB is implemented in a variety of ways. For example, if priority queuing is used, then there must be an upper bound (configured by the network administrator) on the rate of EF traffic to be allowed. EF traffic exceeding the bound is dropped.

9.4.5 PHB Implementation with Packet Schedulers

The section describes some typical packet schedulers and how they can be used to support the PHBs. Here we only show two implementation examples. Other possible implementations of packet scheduling, such as WFQ and its variants, can be found in Chapters 4 and 5. Implementations of buffer management, such as RED can be found in Chapter 6.

9.4.5.1 *Static Priority Queues* Consider a static priority queue scheduler with two levels of priorities. Such a scheduler serves packets in the higher-priority queue if they are available and only serves the lower-priority queue if the higher-priority queue is empty.

This scheduler can support the class selector PHBs by mapping packets with the Diffserv fields 1000000, 101000, 110000, and 111000 into the higher-priority queue and those with the Diffserv fields 000000, 001000, 010000, and

011000 into the lower-priority queue. Other mappings are also possible. The Diffserv standards require that packets with the Diffserv fields 110000 and 111000 should map to the higher-priority queue, and packets with the Diffserv field 000000 should map to the lower-priority queue. Packets with other Diffserv fields may be mapped to either of the two queues. A network administrator should ensure that the mapping is consistent across all the routers within a single administrative domain.

The latter example cannot support the AF PHB, because there are only two levels instead of four. It can support some of the function of the EF PHB by mapping it to the higher-priority level. However, it cannot enforce an upper limit on the rate at which the packets in the higher-priority level can be supported. As a result, a simple two-level priority queue cannot support the EF or AF PHB.

If the static priority queue is augmented with rate controls, it can be used to implement the EF upper rate limit. The rate-controlled, static-priority queuing system can be used to implement EF. If at least four rate-controlled priority queues exist, they can be used to implement the four AF classes with limits on their maximum rate utilization as well.

9.4.5.2 *Weighted Round-Robin* A round-robin scheduler with weights assigned to each of multiple queues would be the closest match to implementation of the EF and AF PHBs. The round-robin queues could be assigned weights so that they would be able to serve each of the queues in the normal order. Class selector PHBs could be supported: packets with the Diffserv fields 110000 and 111000 would be mapped to a queue with a larger weight than the queue to which packets with the Diffserv field 000000 (best effort) were mapped.

If a router were to implement only the EF and the default PHBs, it would need to implement only two queues, which would be served round-robin with specific weights assigned to them. If the router were to implement EF, AF, and the default, it would need six queues overall, each with a different assigned weight.

In a typical use of the weighted round-robin (WRR) scheme, called *rate-proportional allocation*, the weight is proportional to the allocated bandwidth. For example, suppose flows A and B in a WRR system should have a ratio of $\frac{1}{3}$. For a link with a capacity of 400 Mbit/s or more, flows A and B will get minimum bandwidths of 100 and 300 Mbit/s. However, the WRR algorithm itself does not require rate-proportional allocation. In fact, non-rate-proportional allocation may also be used to provide controlled priority at fine granularity. Suppose we assign a weight of 1 to both flows A and B (the ratio of weights is now 1). If all other conditions remain the same, flows A and B will each be allocated 200 Mbit/s. If flows A and B continue to send 100 and 300 Mbit/s of traffic over the 400-Mbit/s link, flows A and B will still get 100- and 300-Mbit/s bandwidth respectively.

9.5 CONCEPTUAL MODEL

This section introduces a conceptual model of a Diffserv router and describes its various components. Note that a Diffserv interior router is likely to require only a subset of these components: the model presented here is intended to cover the case of both Diffserv boundary and interior routers. The conceptual model includes abstract definitions of the following components:

- Traffic classification elements
- Inbound traffic metering functions
- Actions of marking, absolute dropping, counting, and multiplexing
- Queueing elements, including capabilities of algorithmic dropping and scheduling
- Certain combinations of the above functional datapath elements into higher-level blocks known as *traffic conditioning blocks* (TCBs)

The components and combinations of components described in this section form building blocks that need to be manageable by Diffserv configuration and management tools. Figure 9.5 shows the major functional blocks of a Diffserv router. Service providers can build different network services from the main building blocks: setting of bits, conditioning of packets, and forwarding of packets.

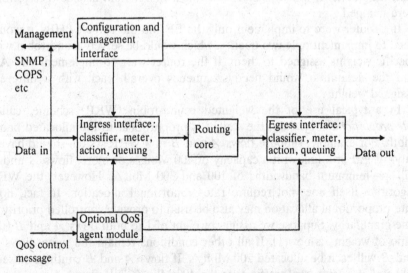

Fig. 9.5 Major functional blocks of a Diffserv router.

9.5.1 Configuration and Management Interface

Diffserv operating parameters are monitored and provisioned through this interface. Monitored parameters include statistics regarding traffic carried at various Diffserv service levels. These statistics may be important for accounting purposes and/or for tracking compliance to traffic conditioning agreements (TCAs) negotiated with customers. The parameters accommodated are primarily the TCA parameters for *classifiers* and *meters* and the associated PHB configuration parameters for *actions* and *queueing elements*. The network administrator interacts with the Diffserv configuration and management interface via one or more management protocols, such as SNMP (simple network management protocol) or COPS (common open policy service) [9], or through other router configuration tools such as serial terminal or telnet consoles.

9.5.2 Optional QoS Agent Module

Diffserv routers may snoop or participate in either per-microflow or per-flow-aggregate signaling of QoS requirements, using, for example, the resource reservation protocol (RSVP). A QoS agent component of a Diffserv router, if present, might be active only in the control plane and not in the data plane.

9.5.3 Diffserv Functions at Ingress and Egress Interfaces

Each router may perform four QoS control functions in the datapath of traffic in each direction:

- Classify each message according to some set of rules.
- If necessary, determine whether the data stream is within or outside its rate by metering the stream.
- Perform a set of resulting actions, including applying a drop policy appropriate to the classification and queue in question, and perhaps additionally marking the traffic with a DSCP [7].
- Enqueue the traffic for output in the appropriate queue. The scheduling of output from this queue may lead to shaping of the traffic or may simply cause it to be forwarded with some minimum rate or maximum latency assurance.

9.5.4 Shaping and Policing

Diffserv nodes may apply shaping, policing, and/or marking to traffic streams that exceed the bounds of their TCA in order to prevent one traffic stream from seizing more than its share of resources from a Diffserv network. In this model, *shaping*, sometimes considered as a traffic conditioning action, delays the traffic by buffering them in a queue. *Policing* is modeled as either a concatenation of a meter with an absolute dropper or as a concatenation of an algorithmic dropper with a scheduler. These elements will discard packets that exceed the TCA.

9.5.5 Traffic Classification

Traffic classification policy identifies the subset of network traffic that may receive differentiated service by being conditioned and/or mapped to one or more behavior aggregates (by DS codepoint remarking) within the DS domain [1]. There are two types of classifiers:

- Behavioral Aggregate Classifiers: These select packets on the basis of their DS codepoints.
- Multifield Classifiers: These select packets based on values of multiple header fields. A common type of MF classifier is a 6-tuple classifier that classifies according to six fields from the IP and TCP or UDP headers (destination address, source address, IP protocol, source port, destination port, and DSCP). MF classifiers may classify on other fields, such as media access control (MAC) addresses, virtual local area network (VLAN) tags, link-layer traffic class fields, or other higher-layer protocol fields.

Classifiers are 1-to-N (fanout) devices: they take a single traffic stream as input and generate N logically separate traffic streams as output. Classifiers are parameterized by filters and output streams. Packets from the input stream are sorted into various output streams by filters that match the contents of the packet or possibly match other attributes associated with the packet. Classifiers are configured by some management procedure on the basis of the relevant TCA. It is also the classifier's job to authenticate the basis on which it classifies packets.

9.5.6 Meters

Diffserv network providers may choose to offer services to customers based on a temporal (i.e., rate) profile within which the customer submits traffic for the service. Meters measure the *temporal properties* of the stream against the appropriate traffic profile from the TCA. Further processing is done by action elements based on whether the packet is in or out of profile. The meter passes this information to the other components along with the packet.

9.5.7 Action Elements

The classifiers and meters described up to this point are fanout elements that are generally used to determine the appropriate action to apply to a packet.

The set of possible actions that can then be applied include *marking, absolute dropping, multiplexing, counting*, and *null action*.

1. DSCP Marker: DSCP markers set a codepoint (e.g., the DSCP in an IP header). DSCP markers may also act on unmarked packets (e.g., those submitted with DSCP of zero) or may re-mark previously marked packets. In particular, the model supports the application of marking based on a preceding classifier match. The mark set in a packet will determine its subsequent PHB treatment in downstream nodes of a network and possibly also in subsequent processing stages within this router.

2. Absolute Dropper: Absolute droppers simply discard packets. There are no parameters for these droppers. An absolute dropper is a terminating point of the datapath and has no outputs. It is probably desirable to forward the packet through a counter action first for instrumentation purposes.

3. Multiplexor: A multiplexor is a simple logical device for merging traffic streams. It is parameterized by its number of incoming ports.

4. Counter: One basic action is to count the data packets that are being processed. The statistics result might be used later for customer billing, service verification, or network engineering purposes. Counters can be used to count packets about to be dropped by an absolute dropper or to count packets arriving at or departing from some other functional element.

5. Null Action: A null action performs no action on the packet. Such an element is useful in the event that the configuration or management interface does not have the flexibility to omit an action element in a datapath segment.

9.5.8 Queueing Elements

Queueing elements modulate the transmission of packets belonging to the different traffic streams and determine their ordering, possibly storing them temporarily or discarding them. Packets are usually stored either because there is a resource constraint (e.g., available bandwidth) that prevents immediate forwarding, or because the queueing block is being used to alter the temporal properties of a traffic stream (i.e., for shaping). Packets are discarded either because of buffering limitations, because a buffer threshold is exceeded (including when shaping is performed), as a feedback control signal to reactive control protocols such as TCP, or because a meter exceeds a configured profile (i.e., for policing). Queueing systems perform three distinct, but related, functions: they store packets, they modulate the departure of packets belonging to various traffic streams, and they selectively discard packets.

9.5.9 Traffic Conditioning Blocks

The classifier, meter, action, algorithmic dropper, queue, and scheduler functional datapath elements described above can be combined into TCBs. A TCB is an abstraction of a set of functional datapath elements that may be used to facilitate the definition of specific traffic conditioning functionality. It can be considered as a template that can be replicated many times for different traffic streams or different customers. When the Diffserv treatment for a given packet needs to have such building blocks repeated, this is performed by cascading multiple TCBs: an output of one TCB may drive the input of a succeeding one.

The function of traffic conditioning is to ensure that the traffic entering a DS domain complies with the TCA between the sender's domain and the receiver's domain, and with the domain's service provision policy. The conditioner of a DS boundary node marks a packet with its appropriate codepoint.

REFERENCES

1. S. Blake, D. Black, M. Carlson, E. Davies, Z. Wang, and W. Weiss, "An architecture for differentiated services," RFC 2475, Internet Engineering Task Force (IETF), Dec. 1998.

2. J. Heinanen, F. Baker, W. Weiss, and J. Wroclawski, "Assured forwarding PHB group," RFC 2597, Internet Engineering Task Force (IETF), Jun. 1999.

3. V. Jacobson, K. Nichols, and K. Poduri, "An expedited forwarding PHB," RFC 2598, Internet Engineering Task Force (IETF), Jun. 1999.

4. B. Davie, A. Charny, F. Baker, J. Bennet, K. Benson, J. L. Boudec, A. Chin, W. Courtney, S. Davari, V. Firoin, C. Kalmanek, K. K. Ramakrishnam, and D. Stiliadis, "An expedited forwarding PHB," Internet draft, Aug. 2001. draft-ietf-diffserv-rfc2598bis-00.txt.

5. B. Braden, D. Clark, J. Crowcroft, B. Davie, S. Deering, D. Estrin, S. Floyd, V. Jacobson, G. Minshall, C. Partridge, L. Peterson, K. Ramakrishnan, S. Shenker, J. Wroclawski, and L. Zhang, "Recommendations on queue management and congestion avoidance in the Internet," RFC 2309, Internet Engineering Task Force (IETF), Apr. 1998.

6. D. Grossman, A. Smith, S. Blake, and Y. Bernet, "An informal management model for Diffserv router," Internet draft, Nov. 2000. draft-ietf-diffserv-model-05.txt.

7. K. Nichols, S. Blake, F. Baker, and D. Black, "Definition of the differentiated services field (DS field) in the IPv4 and IPv6 header," RFC 2474, Internet Engineering Task Force (IETF), Dec. 1998.

8. F. Baker, A. Smith and K. Chan, "Management information base for the differentiated services architecture," Internet draft, Nov. 2000. draft-ietf-diffserv-mib-06.txt.

9. D. Durham, J. Boyle, R. Cohen, S. Herzog, R. Rajan, and A. Sastry, "The COPS (common open policy service) protocol," RFC 2748, Internet Engineering Task Force (IETF), Jan. 2000.

10. X. Xiao and L. Ni, "Internet QoS: a big picture," *IEEE Network*, vol. 13, no. 2, pp. 8–18, Mar./Apr. 1999.

11. V. Kumar, T. Lakshman, and D. Stiliadis, "Beyond best effort: router architectures for the differentiated services of tomorrow's Internet," *IEEE Commun. Mag.*, vol. 36, no. 5, pp. 152–164, May 1998.

12. S. Keshav and R. Sharma, "Issues and trends in router design," *IEEE Commun. Mag.*, vol. 36, no. 5, pp. 144–151, May 1998.

13. E. Rosen, A. Viswanathan, and R. Callon, "Multiprotocol label switching architecture," RFC 3031, Internet Engineering Task Force (IETF), Jan. 2001.

14. S. Vegesna, *IP Quality of Service*, Cisco Press, Indianapolis, Dec. 2000.

15. B. Davie, S. Davari, P. Vaananen, R. Krishnan, P. Cheval, and J. Heinanen, "MPLS support of differentiated services," Internet draft, Feb. 2001. draft-ietf-mpls-diff-ext-08.txt.

CHAPTER 10

MULTIPROTOCOL LABEL SWITCHING

Multiprotocol label switching (MPLS) techniques are developed to support, with a single network, different kinds of network protocols, such as Internet protocol (IP), ATM, frame relay, and so on. A router that supports MPLS is called a *label-switching router* (LSR). A group of LSRs with the same MPLS level forms an *MPLS domain*. An MPLS domain can be further subdivided into several levels of subdomains in a nested manner. The edge router or an MPLS domain can be either an ingress router or an egress router.

When a packet enters an MPLS domain, the ingress router of the domain will analyze the packet's network layer header and assign the packet to a particular forwarding equivalent class (FEC). The FEC is used to describe an association of discrete packets with a destination address and a class of traffic. It allows the grouping of packets into classes. The matching of the FEC with a packet is achieved by using a label to identify each FEC. For different classes of service, different FECs and their associated labels are used. After that, there is no further analysis of the packet's network layer header at subsequent hops within the same MPLS domain. Rather, the label is used in each hop as an index to a table, which specifies the next hop and a new label. In other words, all forwarding in the MPLS domain is driven by the labels. The path through one or more LSRs followed by packets in a particular FEC is called a *label-switched path* (LSP). It should be noted that the label associated with the FEC is changeable over the MPLS domain as long as each router maintains a label mapping table so that the router can recognize which FEC the incoming label is to be mapped. Only then can the router do the new label assignment and forward the packet to the next hop. Before a packet leaves a MPLS domain, its MPLS header is removed. This

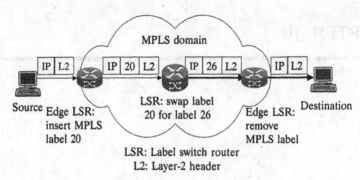

Fig. 10.1 Illustration of MPLS.

whole process is shown in Figure 10.1, where labels (20 and 26) are inserted between the IP packet and layer 2 header (L2).

As a result, there must be a certain agreement between neighbor routers before packet arrival in order to set up a valid table for packet forwarding. If a router in the MPLS domain receives a packet with an unrecognizable or empty label, it will either be forced to analyze the packet network layer header for the next hop, or simply drop the packet.

A labeled packet traversing an MPLS network does not necessarily carry only a single label. In case the MPLS domain is further subdivided into several MPLS subdomains, it is useful to have a more general model in which a labeled packet carries a number of labels, organized as a last-in, first-out stack. This is called the *label stack*. The basic utility of the label stack is for the MPLS network to support a hierarchy architecture, such as the concept of a *tunnel*. This will be described in details in Section 10.1.5. Nevertheless, the processing of a labeled packet is completely independent of the level of hierarchy. The processing is always based on the top label, without regard for the possibility that some other labels may have been above it in the past, or that some other labels may be below it at present [1].

An unlabeled packet can be thought of as a packet whose label stack is empty—in which the label stack has depth 0. If a packet's label stack is of depth m, we refer to the label at the bottom of the stack as the level 1 label, to the label above it (if such exists) as the level 2 label, and to the label at the top of the stack as the level m label.

The MPLS forwarding has a number of advantages over the conventional connectionless forwarding.

1. It is not necessary for MPLS routers to inspect the packet's network layer headers. Rather, they are only required to do a table lookup and label replacement.
2. Packets that enter the network from different routers are distinguishable, so that forwarding decisions that depend on the ingress router can

be easily made. This is not true of conventional forwarding, since packets are memoryless about their ingress routers.

3. The label table of an MPLS switch may contain more information than just the next hop and the new label for a particular FEC. For example, an entry may include the precedence or class of service predetermined for the corresponding FEC. On the contrary, conventional forwarding can only consider information encoded in the packet header.

4. To improve the performance it is sometimes desirable, rather than letting the route be chosen by dynamic routing algorithms from hop to hop in the network, to force a packet to follow a route that is explicitly chosen before the packet enters the network. This capability is necessary to support traffic engineering. In conventional forwarding, this requires the packet to carry an encoding of its route along with it. In MPLS, labels associated with the corresponding tables can be used to represent the route, so that the indication of the explicit route need not be carried with the packet. This can reduce the size of the packet overhead.

In general, MPLS allows a label to represent not only the FEC, but also (fully or partially) the precedence or class of service of the packet.

Section 10.1 describes the basic architecture of MPLS. Section 10.2 discusses the issue of label distribution. Section 10.3 describes the MPLS mechanisms to support differentiated services. Section 10.4 presents the label forwarding model for Diffserv LSRs. Section 10.5 discusses two MPLS application: traffic engineering and virtual private networks.

10.1 BASIC ARCHITECTURE

10.1.1 Label and Label Binding

Suppose that R_u and R_d are upstream and downstream LSRs, respectively, in an MPLS domain. Then R_u and R_d may agree that when R_u transmits a packet with label L to R_d, the value L is mapped a particular FEC F in which the packet is one of the members. That is, they should agree to a *binding* between label L and FEC F for packets moving from R_u to R_d. With such an agreement, L becomes R_u's *outgoing label*, and R_d's *incoming label*, representing FEC F.

Note that label L represents FEC F only between R_u and R_d. That is, L does not necessarily represent F for any packets other than those sent from R_u to R_d. On the other hand, F is not necessarily represented by L for any paths other than that between R_u and R_d. In other words, the representation of F by L is *local* from R_u to R_d.

As a result, the decision to bind a particular label L to a particular FEC F is made by the LSR that is downstream with respect to that binding, in order not to doubly assign any label value. The downstream LSR then informs the upstream LSR of the binding. Thus, labels are assigned downstream, and label bindings are distributed from downstream routers to upstream routers.

RSVP-TE [11] can be used as a signaling protocol to establish label-switched paths (LSPs). It has two basic message types: Path and Resv messages. A Path message travels from a sender to a receiver and includes a request to bind labels to a specific LSP. Labels are allocated downstream and distributed by means of the RSVP Resv message.

10.1.2 Label Stack

10.1.2.1 Encoding the Label Stack The label stack is represented as a sequence of *label stack entries*. Each label stack entry is represented by four octets. This is shown in Figure 10.2. In a non-ATM environment, each label stack entry contains a 20-bit label, a 3-bit *experimental* field (formerly known as the class of service, or CoS, field), a 1-bit label stack indicator, and an 8-bit time-to-live (TTL) field. In an ATM environment, the entry contains only a label encoded in the VCI/VPI field as shown in Figure 10.3.

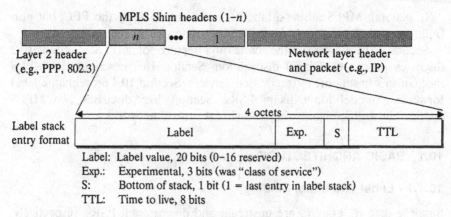

Label: Label value, 20 bits (0–16 reserved)
Exp.: Experimental, 3 bits (was "class of service")
S: Bottom of stack, 1 bit (1 = last entry in label stack)
TTL: Time to live, 8 bits

Fig. 10.2 MPLS encapsulation.

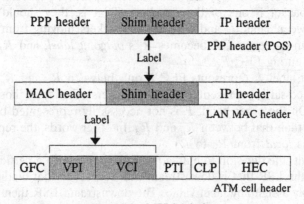

Fig. 10.3 MPLS labels.

The label stack entries appear after the data link layer headers, but before any network layer headers. The top of the label stack appears earliest in the packet, and the bottom appears latest. The network layer packet immediately follows the label stack entry, which has the S-bit set.

Each label stack entry is broken down into the following fields:

1. Bottom of stack (S): This bit is set to one for the last entry in the label stack (i.e., for the bottom of the stack), and zero for all other label stack entries.

2. Time to live (TTL): This eight-bit field is used to encode a time-to-live value. TTL must be set to the value of the IP TTL field when the packet is first labeled, and is decremented at each MPLS network hop. When last label is popped off the stack, MPLS TTL is copied to the IP TTL field.

3. Experimental Use: This three-bit field is reserved for experimental use.

4. Label Value: This 20-bit field carries the actual value of the label. When a labeled packet is received, the label value at the top of the stack is looked up. As a result of a successful lookup one learns:

   (a) the next hop to which the packet is to be forwarded;

   (b) the operation to be performed on the label stack before forwarding (this operation may be to replace the top label stack entry with another, or to pop an entry off the label stack, or to replace the top label stack entry and then push one or more additional entries on the label stack).

In addition to learning the next hop and the label stack operation, one may also learn the outgoing data link encapsulation, and possibly other information that is needed in order to forward the packet properly. The network layer must be inferable from value of the bottom label of the stack. Because pushing multiple labels may cause the length of the frame to exceed the layer-2 MTU, the LSR must support the maximum IP datagram size for labeling as a parameter, and any unlabeled datagrams greater in size than this parameter must be fragmented.

10.1.2.2 *Basic Operation of Label Stack* As we mentioned in the previous section, a labeled packet can carry a label stack rather than a single label in order to support a hierarchical structure of MPLS networks.

Suppose that a label switched path of level m for a particular packet P is an ordered sequence of routers, $\langle R_1, \ldots, R_n \rangle$. Then these routers have the following properties:

1. R_1 is the LSP ingress router, so it pushes a label onto P's label stack, resulting in a label stack of depth m.

2. For all i, $1 < i < n$, P has a label stack of depth m when received by LSR R_i, and has a label stack of depth no less than m during the

transit from R_1 to R_{n-1}. Note that, there could be an $(m + 1)$th or deeper level of subnetworks between any two level m LSPs. An example is given in Section 10.1.6.

3. For all i, $1 < i < n$, R_i transmits P to R_{i+1} by using the label at the top of the label stack (the level m label) as an index into its label mapping table for R_{i+1} and a new level-m-label value.

4. For all i, $1 < i < n$, if a system S receives P after it is transmitted by R_i but before it is received by R_{i+1} (e.g., R_i and R_{i+1} might be connected via an MPLS subnetwork, and S might be one of the LSRs), then S's forwarding decision is based on neither the level m label nor the network layer header, but a label stack on which additional labels have been pushed (i.e., on a level $m + k$ label, where $k > 0$).

The reason to define such LSP properties is that whenever an LSR pushes a label onto an already labeled packet, it needs to make sure that the new label corresponds to a FEC whose LSP egress is the LSR that assigned the label that is now second in the stack.

10.1.3   Route Selection

Route selection refers to the method used for selecting the LSP for a particular FEC. The proposed MPLS protocol architecture supports two options for route selection: (1) hop-by-hop routing, and (2) explicit routing.

Hop-by-hop routing allows each node to independently choose the next hop for each FEC. This is the usual mode today in existing IP networks.

In an explicitly routed LSP, each LSR does not choose the next hop independently; rather, a single LSR, generally the LSP ingress or the LSP egress, specifies several (or all) of the LSRs in the LSP. If a single LSR specifies the entire LSP, the LSP is *strictly* explicitly routed. If a single LSR specifies only a part of the LSP, the LSP is *loosely* explicitly routed.

The sequence of LSRs followed by an explicitly routed LSP may be chosen by configuration, or may be selected dynamically by a single node. Explicit routing is useful for a number of purposes, such as policy routing or traffic engineering. In MPLS, the explicit route needs to be specified at the time that labels are assigned, but the explicit route does not have to be specified with each IP packet. This makes MPLS explicit routing much more efficient than the alternative of IP source routing.

10.1.4   Penultimate Hop Popping

We note again that if $\langle R_1, \ldots, R_n \rangle$ is a level m LSP for packet P, then the label stack of level m can be popped at R_{n-1} rather than R_n. That is, P can be transmitted from the penultimate LSR, with a label stack of depth $m - 1$ to the egress LSR.

This is appropriate because the purpose of the level m label is to get the packet from R_1 to R_n. That is, once R_{n-1} has decided to send the packet to R_n, the label has no longer any function, and thus need no longer be carried.

Moreover, if we do not apply penultimate hop popping, it may cause redundant work for R_n, because when R_n receives a packet, it first looks up the top label and determines as a result of that lookup that the "next hop" is indeed R_n itself. Only then it will pop the stack and examine the remaining part of the label stack. In other words, this would require R_n to do double lookups. If there is another label (label of level $m - 1$) on the stack, R_n will look this up and forward the packet according to that lookup. In this case, R_n is also an intermediate node forthe packet's level $m - 1$ LSP. However, if there is no other label on the stack, then the packet is forwarded according to its network layer destination address, or it may start a new travel over another MPLS domain.

The penultimate-hop-popping scheme requires that when an LSR looks up the label, knowing that itself is the penultimate hop as well as what the egress hop is, the LSR pops the stack and forwards the packet to the egress. In this case, the LSP egress node receives the packet, on which the top label is now the one it needs to lookup in order to make its own forwarding decision. With this scheme, both the penultimate and egress nodes need to do only a single table lookup.

Not all the MPLS switches, however, are able to pop the label stack, and this scheme cannot be universally required. Therefore the penultimate-hop-popping scheme can only be applied if this is specifically requested by the egress node and the penultimate node is capable of doing so.

10.1.5 LSP Tunnels

For the purpose of traffic engineering or network control, packets of a particular FEC are sometimes required to follow a specified route from an upstream router R_u to a downstream router R_d, despite the probability that R_d may be adjacent to neither R_u nor the packet destination. This concept is known as creating a *tunnel* from R_u to R_d, where R_u is called the *transmit end point*, and R_d the *receive end point*, of the tunnel. The packet so handled is then called a *tunneled packet*.

If a tunneled packet follows the hop-by-hop route from R_u to R_d, we say that it is in an *hop-by-hop routed tunnel*. If, on the other hand, it travels from R_u to R_d over a route other than the hop-by-hop route, we say that it is in an *explicitly routed tunnel*.

It is very simple to implement a tunnel as an LSP, and use label switching rather than network layer encapsulation to cause the packet to travel through the tunnel. For example, a tunnel could be an LSP $\langle R_1, \ldots, R_n \rangle$, where R_1 and R_n are the transmit end point and receive end point, respectively, of the tunnel. This is called an *LSP tunnel*.

The set of packets that are to be sent through the LSP tunnel constitutes an FEC, and each LSR in the tunnel must assign a label to that FEC (i.e., must assign a label to the tunnel). The criterion for assigning a particular packet to an LSP tunnel is a local matter at the tunnel's transmit end point. To put a packet into an LSP tunnel, the transmit end point pushes a label for the tunnel onto the label stack and sends the labeled packet to the next hop in the tunnel.

If it is not necessary for the tunnel's receive end point to be able to determine which packets it receives through the tunnel, then, as discussed earlier, the label stack may be popped at the penultimate LSR in the tunnel.

10.1.6 An Example: Hierarchy of LSP Tunnels

Consider an LSP $\langle R_1, R_2, R_3, R_4 \rangle$ (see Fig. 10.4). Let us suppose that R_1 receives the unlabeled packet P, and pushes on its label stack the label to cause it to follow this path, and that this is in fact the hop-by-hop path. However, let us further suppose that R_2 and R_3 are not directly connected, but are *neighbors* by virtue of being the end points of an LSP tunnel. Assume also the actual sequence of LSRs traversed by P is $\langle R_1, R_2, R_2^1, R_2^2, R_3, R_4 \rangle$.

When P travels from R_1 to R_2, it will have a label stack of depth 1. R_2, switching on the label, determines that P must enter the tunnel. Thus, R_2 first replaces the incoming label with a label that is meaningful to R_3. Then it pushes on a new label. This level 2 label has a value that is meaningful to R_2^1. Switching is done on the level 2 label by R_2^1 and R_2^2. R_2^2, which is the penultimate hop in the R_2–R_3 tunnel, pops the label stack before forwarding the packet to R_3. When R_3 sees packet P, P has only a level 1 label, having

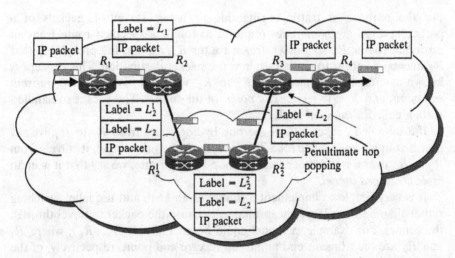

Fig. 10.4 Hierachy of LSP tunnels.

now exited the tunnel. Since R_3 is the penultimate hop in P's level 1 LSP, it pops the label stack, and R_4 receives P unlabeled.

The label stack mechanism allows LSP tunneling to nest to any depth.

10.1.7 Next-Hop Label Forwarding Entry

The *next-hop label forwarding entry* (NHLFE) [1] is used when forwarding a labeled packet. It contains the following information:

1. the packet's next hop;
2. one of the following operations to perform on the label stack:
   (a) replace the label at the top of the label stack with a specified new label (e.g., when the LSR is one of the internal nodes of the MPLS domain);
   (b) pop the label stack (e.g., when the LSR is the penultimate node);
   (c) replace the label at the top of the label stack with a specified new label, and then push one or more specified new labels onto the label stack (e.g., when the LSR is a transmit end point of an LSP tunnel);
3. the precedence or class of service of the packet.

Note that at a given LSR, the packet's "next hop" might be that LSR itself. In this case, it is necessary for the LSR to pop the top level label and then make the forwarding decision according to the remaining part of the label stack. The packet may still be a labeled packet. Or it may be the native IP packet, on whose header the LSR may need to operate in order to forward the packet. This can only occur if the LSR is an MPLS egress node.

The *incoming label map* (ILM) maps each incoming label to a set of NHLFEs when forwarding labeled incoming packets. On the other hand, the *FEC-to-NHLFE* (FTN) maps each FEC to a set of NHLFEs when forwarding unlabeled incoming packets that are to be labeled before being forwarded.

If the mapping to a set of NHLFEs (either ILM or FTN) contains more than one element, then exactly one element of the set must be chosen before the packet is forwarded. One can do so for load balancing over multiple equal-cost paths.

In order to forward a labeled packet, a LSR examines the label at the top of the label stack. It uses the ILM to map this label to an NHLFE. Using the information in the NHLFE, it determines where to forward the packet, and performs an operation on the packet's label stack. It then encodes the new label stack into the packet, and forwards the result.

In order to forward an unlabeled packet, a LSR analyzes the network layer header to determine the packet's FEC. It then uses the FTN to map this to an NHLFE. Using the information in the NHLFE, it determines where to forward the packet, and performs an operation on the packet's label stack. It then encodes the new label stack into the packet, and forwards the result.

10.2   LABEL DISTRIBUTION

The request for a label–FEC binding can be initiated by either the upstream LSR or the downstream LSR. To make sure that, however, each LSR can uniquely interpret each incoming label, the label assignment and distribution must be in a downstream-to-upstream direction.

A *label distribution protocol* is then a set of procedures by which a downstream LSR informs its upstream peer of the label–FEC bindings it has assigned. Two LSRs using a label distribution protocol to exchange label–FEC binding information are known as *label distribution peers* with respect to that binding [6].

It should be clear that two LSRs may be label distribution peers with respect to some set of bindings, but not with respect to some other set of bindings. Moreover, the label distribution protocol may also encompass any negotiations in which two label distribution peers need to engage. Suppose that a particular binding of label L to FEC F, distributed by R_d to R_u, has associated with it some attributes. If R_u is the downstream node of another LSR R_i with respect to F, then R_u must also distribute a binding of a label (not necessarily equal to L) to FEC F, associated with the corresponding attributes that it received from R_d, to R_i.

It is not assumed that there is only a single label distribution protocol in an MPLS network. Rather, a number of different label distribution protocols are being standardized. Existing protocols, such as BGP [5] and RSVP [11], have been extended so that the label distribution can be adapted to them. New protocols have also been defined in order to do label distribution [6, 10]. In this section, we discuss only the general concepts related to label distribution protocols.

10.2.1   Unsolicited Downstream vs Downstream-on-Demand

An upstream LSR is allowed to explicitly request a label binding for a particular FEC from its next hop with respect to that FEC. This is known as *downstream-on-demand* label distribution. On the other hand, the MPLS architecture also allows a downstream LSR to distribute bindings to its upstream LSRs that have not explicitly requested them. This is known as *unsolicited downstream* label distribution.

Note that, with an MPLS implementation, which type of label distribution technique is used may depend on which characteristics of the interfaces the implementation can support. It is nevertheless acceptable for both techniques to be used in the same network at the same time, as long as that the label distribution peers have agreed on which is to be used.

10.2.2   Label Retention Mode: Liberal vs Conservative

An LSR R_d may have distributed a label binding for a particular FEC to another LSR R_u even though R_d is not currently R_u's next hop with respect to that FEC.

In this case, R_u can have two alternative actions. The first one, which is known as *liberal label retention mode*, is to maintain and keep track of such a binding; the second, known as *conservative label retention mode*, is to discard the binding.

When the liberal label retention mode is applied, R_u may immediately begin using the binding again once R_d becomes its next hop for that FEC eventually. When the conservative label retention mode is applied, on the contrary, R_u must reacquire the binding once R_d becomes its next hop.

Liberal label retention mode allows a quicker adaptation to routing changes, while conservative label retention mode allows an LSR to maintain fewer labels.

10.2.3   LSP Control: Ordered vs Independent

Some FECs correspond to address prefixes that are distributed via a dynamic routing algorithm. The setup of the LSPs for these FECs can be done in one of two ways, *independent LSP control* or *ordered LSP control*.

In independent LSP control, each LSR, upon noting that it recognizes a particular FEC, makes an independent decision to bind a label to that FEC and distribute that binding to its label distribution peers. This corresponds to the way that conventional IP datagram routing works: each node makes an independent decision as to how to treat each packet, and relies on the routing algorithm to converge rapidly so as to ensure that each datagram is correctly delivered.

In ordered LSP control, an LSR only binds a label to a particular FEC if it is the egress LSR for that FEC, or if it has already received a label binding for that FEC from its next hop for that FEC. If we want to ensure that traffic in a particular FEC follows a path with some specified set of properties for the purpose of traffic engineering, then ordered control must be used.

With independent control, some LSRs may begin label switching of traffic in the FEC before the LSP is completely set up, and thus some traffic in the FEC may follow a path that does not have the specified set of properties. Ordered control also needs to be used if the recognition of the FEC is a consequence of the setting up of the corresponding LSP.

Ordered control and independent control are fully interoperable. However, unless all LSRs in an LSP are using ordered control, the overall effect on network behavior is largely that of independent control, since one cannot be sure that an LSP is not used until it is fully set up.

10.2.4   Label Distribution Peering and Hierarchy

With reference to the example given in Section 10.1.6, again, we consider that packet P travels along a level 1 LSP $\langle R_1, R_2, R_3, R_4 \rangle$, and when going from R_2 to R_3 travels along a level 2 LSP $\langle R_2, R_2^1, R_2^2, R_3 \rangle$. From the perspective of the level 2 LSP, R_2's label distribution peer is R_2^1. From the perspective of the level 1 LSP, R_2's label distribution peers are R_1 and R_3. That is, an LSR can have label distribution peers at each layer of hierarchy. Note that in this example, R_2 and R_2^1 must be interior gateway neighbors, but R_2 and R_3 need not be.

When two label distribution peers are interior gateway neighbors, they are referred to as *local label distribution peers*. Otherwise, they are called *remote label distribution peers*. In the above example, R_2 and R_2^1 are local label distribution peers, while R_2 and R_3 are remote label distribution peers.

To distribute label–FEC bindings, an LSR performs label distribution with its local label distribution peer by sending label distribution protocol messages, which are addressed to the peer directly.

On the other hand, the LSR can perform label distribution with its remote label distribution peers in one of the following ways:

1. Explicit Peering: In explicit peering, the LSR distributes labels to a peer by sending label distribution protocol messages that are addressed to the peer directly. This is exactly as the LSR would do for local label distribution peers. This technique is most useful when the number of remote label distribution peers is small, or the number of higher-level label bindings is large, or the remote label distribution peers are in distinct routing areas or domains. Of course, the router needs to know which labels to distribute to which peers.

2. Implicit Peering: In implicit peering, the LSR does not send label distribution protocol messages that are addressed to its peer directly. Rather, it distributes higher level labels to its remote label distribution peers, encodes a higher level label as an attribute of a lower-level label, and then distributes the lower-level label, along with this attribute, to its local label distribution peers. The local label distribution peers then propagate the information to their local label distribution peers. This process continues till the information reaches the remote peer. This technique is most useful when the number of remote label distribution peers is large. Implicit peering does not require an N-square peering mesh to distribute labels to the remote label distribution peers, because the information is propagated through the local label distribution peering. However, implicit peering requires the intermediate nodes to store information that they might not be directly interested in.

10.2.5 Selection of Label Distribution Protocol

As mentioned above, MPLS can be implemented with different kinds of label distributon protocols. It does not establish standard rules for choosing which label distribution protocol to use in which circumstances. Nevertheless, it is still worthwhile to point out the following considerations.

10.2.5.1 *Border Gateway Protocol (BGP)* In many scenarios, it is desirable to bind labels to FECs that can be identified with routes to address prefixes. If there is a standard, widely deployed routing algorithm that distributes those routes, it can be argued that label distribution is best achieved by piggybacking the label distribution on the distribution of the routes themselves.

BGP distributes such routes. If a BGP speaker needs to also distribute labels to its BGP peers, using BGP to do the label distribution [5] has a number of advantages. In particular, it permits BGP route reflectors to distribute labels, thus providing a significant scalability advantage over using label distributions protocol between BGP peers.

10.2.5.2 *Labels for RSVP Flowspecs* When RSVP is used to set up resource reservations for particular flows, it is desirable to label the packets in those flows, so that the RSVP filter spec does not need to be applied at each hop. Some suggested that having RSVP distribute the labels as part of its path-reservation setup process is the most efficient method of distributing labels for this purpose [11]. Figure 10.5 shows an example of how labels are carried in one RESV message in order from R_5 to R_1.

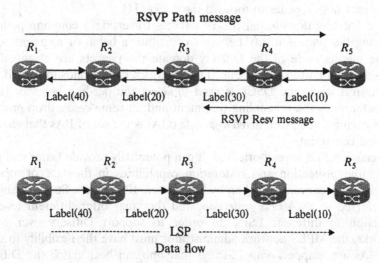

Fig. 10.5 Label distribution using RSVP.

10.2.5.3 *Labels for Explicitly Routed LSPs* In some applications of MPLS, particularly those related to traffic engineering, it is desirable to set up an explicitly routed path, from ingress to egress. It is also desirable to apply resource reservations along that path.

There are two approaches to this:

1. Start with an existing protocol that is used for setting up resource reservations, and extend it to support explicit routing and label distribution.
2. Start with an existing protocol that is used for label distribution, and extend it to support explicit routing and resource reservations.

Finally, it should be always kept in mind that a label distribution protocol is used between nodes in an MPLS network to establish and maintain the label bindings. In order for MPLS to operate correctly, label distribution information needs to be transmitted reliably, and the label distribution protocol messages pertaining to a particular FEC need to be transmitted in sequence. Flow control is also desirable, as is the capability to carry multiple label messages in a single datagram.

10.3 MPLS SUPPORT OF DIFFERENTIATED SERVICES

In an MPLS domain, a label-switched path (LSP) can be established for a data stream using MPLS signaling protocols. At the ingress label switched router (LSR), each packet is assigned a label and is transmitted downstream. At each transit LSR along the LSP, the basic function of the label is to map to the next hop in order to forward the packet [4].

In a Diffserv domain, all the IP packets traversing a common path and requiring the same Diffserv behavior constitute a behavior aggregate (BA). At the ingress node of the Diffserv domain the packets are classified and marked with a Diffserv codepoint (DSCP), which corresponds to their BA. At each transit node, the DSCP is used to select the per-hop behavior (PHB) that determines the scheduling treatment and, in some cases, drop probability for each packet. An *ordered aggregate* (OA) is the set of BAs that share an ordering constraint.

Because MPLS is path-oriented, it can potentially provide faster and more predictable protection and restoration capabilities in the face of topology changes than conventional hop-by-hop-routed IP systems. Such capabilities are referred to as *MPLS protection*, and they may offer different levels of protection to different LSPs. In order to support Diffserv over MPLS networks, the MPLS network administrator must have the flexibility to select how BAs are mapped onto LSPs so that one can best match the Diffserv, traffic engineering, and protection objectives within one's particular network.

CHAPTER 10

MULTIPROTOCOL LABEL SWITCHING

Multiprotocol label switching (MPLS) techniques are developed to support, with a single network, different kinds of network protocols, such as Internet protocol (IP), ATM, frame relay, and so on. A router that supports MPLS is called a *label-switching router* (LSR). A group of LSRs with the same MPLS level forms an *MPLS domain*. An MPLS domain can be further subdivided into several levels of subdomains in a nested manner. The edge router or an MPLS domain can be either an ingress router or an egress router.

When a packet enters an MPLS domain, the ingress router of the domain will analyze the packet's network layer header and assign the packet to a particular forwarding equivalent class (FEC). The FEC is used to describe an association of discrete packets with a destination address and a class of traffic. It allows the grouping of packets into classes. The matching of the FEC with a packet is achieved by using a label to identify each FEC. For different classes of service, different FECs and their associated labels are used. After that, there is no further analysis of the packet's network layer header at subsequent hops within the same MPLS domain. Rather, the label is used in each hop as an index to a table, which specifies the next hop and a new label. In other words, all forwarding in the MPLS domain is driven by the labels. The path through one or more LSRs followed by packets in a particular FEC is called a *label-switched path* (LSP). It should be noted that the label associated with the FEC is changeable over the MPLS domain as long as each router maintains a label mapping table so that the router can recognize which FEC the incoming label is to be mapped. Only then can the router do the new label assignment and forward the packet to the next hop. Before a packet leaves a MPLS domain, its MPLS header is removed. This

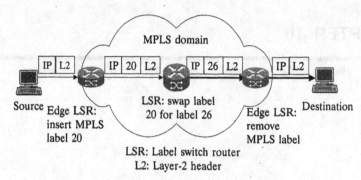

Fig. 10.1 Illustration of MPLS.

whole process is shown in Figure 10.1, where labels (20 and 26) are inserted between the IP packet and layer 2 header (L2).

As a result, there must be a certain agreement between neighbor routers before packet arrival in order to set up a valid table for packet forwarding. If a router in the MPLS domain receives a packet with an unrecognizable or empty label, it will either be forced to analyze the packet network layer header for the next hop, or simply drop the packet.

A labeled packet traversing an MPLS network does not necessarily carry only a single label. In case the MPLS domain is further subdivided into several MPLS subdomains, it is useful to have a more general model in which a labeled packet carries a number of labels, organized as a last-in, first-out stack. This is called the *label stack*. The basic utility of the label stack is for the MPLS network to support a hierarchy architecture, such as the concept of a *tunnel*. This will be described in details in Section 10.1.5. Nevertheless, the processing of a labeled packet is completely independent of the level of hierarchy. The processing is always based on the top label, without regard for the possibility that some other labels may have been above it in the past, or that some other labels may be below it at present [1].

An unlabeled packet can be thought of as a packet whose label stack is empty—in which the label stack has depth 0. If a packet's label stack is of depth m, we refer to the label at the bottom of the stack as the level 1 label, to the label above it (if such exists) as the level 2 label, and to the label at the top of the stack as the level m label.

The MPLS forwarding has a number of advantages over the conventional connectionless forwarding.

1. It is not necessary for MPLS routers to inspect the packet's network layer headers. Rather, they are only required to do a table lookup and label replacement.
2. Packets that enter the network from different routers are distinguishable, so that forwarding decisions that depend on the ingress router can

For example, the network administrator should be allowed to decide whether different sets of BAs are to be mapped onto the same LSP or mapped onto separate LSPs.

As a result, in addition to the packet forwarding (according to the FEC specified), when a packet arrives at an LSR, two additional pieces of information, the service class and the drop precedence of that packet, should be conveyed to the LSR so that it is able to handle the packet and meet the above objectives.

The service class of a packet is usually predetermined and either explicitly carried by the packet with its header, or inferred by the packet label with a label-to-service-class mapping table preestablished at each transit LSR. On the contrary, the drop precedence of a packet is more packet-dependent and may be determined dynamically. Thus, it is usually carried with the packet header.

A 3-bit experimental (EXP) field of the MPLS shim header can be used for carrying the above information when necessary. In respect of the EXP field, two types of LSPs can be combined in general to support Diffserv over MPLS networks:

1. LSPs that can transport multiple OAs, so that the EXP field of the MPLS shim header conveys to the LSR the PHB to be applied to the packet. That PHB contains both information about the packet's scheduling treatment and its drop precedence. This kind of PHB scheduling classes (PSCs) is also referred to as EXP-inferred PSCs, and the corresponding LSPs are called EXP-inferred LSPs (E-LSPs). With E-LSPs, the label represents the combination of a FEC and the set of BAs transported over the E-LSPs. If all the supported BAs are transported over an E-LSP, the label then represents the complete FEC. In E-LSP scenario, a single LSP can be used to support up to eight BAs of a given FEC, regardless of how many OAs these BAs span. With such LSPs, the EXP field of the MPLS shim header is used by the LSR to determine the PHB to be applied to the packet. This includes both the PSC and the drop preference.

2. LSPs that only transport a single OA, so that the packet's scheduling treatment is inferred by the LSR exclusively from the packet's label value, while the packet's drop precedence is conveyed in the EXP field of the MPLS shim header or in the encapsulating link layer's specific selective drop mechanism. This kind of PSCs is also referred to as label-only-inferred PSCs and the corresponding LSPs are called label-only-inferred LSPs (L-LSPs). With L-LSPs, the label represents the combination of a FEC and an OA. In the L-LSP scenario, a separate LSP can be established for every single FEC–OA pair. With such LSPs, the PSC is explicitly signaled at label establishment time so that, after label establishment, the LSR can infer exclusively from the label value the PSC to be applied to a labeled packet. When the shim header is

used, the drop precedence to be applied by the LSR to the labeled packet is conveyed inside the labeled packet MPLS shim header using the EXP field. When the shim header is not used (e.g., in MPLS over ATM), the drop precedence to be applied by the LSR to the labeled packet is conveyed inside the link layer header encapsulation using link-layer specific drop precedence fields (e.g., in ATM CLP).

For a given FEC, unless medium-specific restrictions apply, any combinations of these two types of LSPs is allowed within an MPLS–Diffserv domain. The network administrator selects the actual combination of LSPs from the set of allowed combinations and selects how the BAs are actually transported over this combination of LSPs, in order to best match the environment and objectives in terms of Diffserv support, traffic engineering, and MPLS protection as introduced above.

Moreover, there may be more than one LSP carrying the same OA with a particular FEC for purposes of, for example, load balancing of the OA. Then, in order to respect ordering constraints, all packets of a given microflow, possibly spanning multiple BAs of a given OA, must be transported over the same LSP. Conversely, each LSP must be capable of supporting all the (active) BAs of a given OA.

Regardless of which label-binding protocol is used, E-LSPs and L-LSPs may be established without bandwidth reservation or with bandwidth reservation.

Establishing an E-LSP or L-LSP with bandwidth reservation means that bandwidth requirements for the LSP are signaled at LSP establishment time. Such signaled bandwidth requirements may be used by LSRs at establishment time to perform admission control of the signaled LSP over the Diffserv resources provided (e.g., via configuration, SNMP, or policy protocols) for the relevant PSC(s). Such signaled bandwidth requirements may also be used by LSRs at establishment time to perform adjustment to the Diffserv resources associated with the relevant PSC(s) (e.g., adjust the PSC scheduling weights).

Note that establishing an E-LSP or L-LSP with bandwidth reservation does not mean that per-LSP scheduling is necessarily required. Since E-LSPs and L-LSPs are specified here for support of Diffserv, the required forwarding treatment (scheduling and drop policy) is defined by the appropriate Diffserv PHB. This forwarding treatment must be applied by the LSR at the granularity of the BA and must be compliant with the relevant PHB specification.

When bandwidth requirements are signaled at establishment of an L-LSP, the signaled bandwidth is obviously associated with the L-LSP's PSC. Thus, LSRs that use the signaled bandwidth to perform admission control may perform admission control over Diffserv resources that are dedicated to the PSC (e.g., over the bandwidth guaranteed to the PSC through its scheduling weight).

When bandwidth requirements are signaled at establishment of an E-LSP, the signaled bandwidth is associated collectively to the whole LSP and therefore to the set of transported PSCs. Thus, LSRs that use the signaled bandwidth to perform admission control may perform admission control over global resources that are shared by the set of PSCs (e.g., over the total bandwidth of the link).

10.4 LABEL-FORWARDING MODEL FOR DIFFSERV LSRs

Since different OAs of a given FEC may be transported over different LSPs, the label-swapping decision of a Diffserv LSR clearly depends on the forwarded packet's BA. Also, since the IP DS field of a forwarded packet may not be directly visible to an LSR, the way to determine the PHB to be applied to a received packet and to encode the PHB into a transmitted packet is different than in a non-MPLS Diffserv router.

Thus, in order to describe label forwarding by Diffserv LSRs, the LSR Diffserv label-switching behavior can be modeled as comprising four stages:

1. Incoming PHB determination
2. Outgoing PHB determination with Optional Traffic Conditioning
3. Label forwarding
4. Encoding of Diffserv information into the encapsulation layer

Obviously, to enforce the Diffserv service differentiation, the LSR must also apply the forwarding treatment corresponding to the outgoing PHB.

This model is presented here to describe the functional operations of Diffserv LSRs and does not constrain actual implementation. Each stage is described in more detail as follows.

10.4.1 Incoming PHB Determination

This stage determines to which BA the received packet belongs. Incoming PHB can be determined by considering a label stack entry or IP header.

10.4.2 Outgoing PHB Determination with Optional Traffic Conditioning

The traffic conditioning stage is optional and may be used on an LSR to perform traffic conditioning, including BA demotion or promotion. For the purpose of specifying Diffserv-over-MPLS forwarding, we simply note that the PHB to be actually enforced, and conveyed to downstream LSRs, by an LSR (referred to as the *outgoing PHB*) may be different from the PHB that had been associated with the packet by the previous LSR (referred to as the *incoming PHB*).

When the traffic conditioning stage is not present, the outgoing PHB is simply identical to the incoming PHB.

10.4.3 Label Forwarding

Label swapping is performed by LSRs on incoming labeled packets using an incoming label map (ILM), where each incoming label is mapped to one or more NHLFEs. Moreover, label imposition is performed by LSRs on incoming unlabelled packets using a FEC-to-NHLFEs (FTN) map, where each incoming FEC is mapped to one or more NHLFEs.

A *Diffserv context* for a label is defined as comprising:

1. LSP type (E-LSP or L-LSP)
2. Supported PHBs
3. Encaps-to-PHB mapping for an incoming label
4. Set of PHB-to-encaps mappings for an outgoing label

A Diffserv context can be stored in the ILM for each incoming label. Also, the NHLFE may also contain any other information needed in order to properly dispose of the packet. In accordance with this, a Diffserv context can be stored in the NHLFE for each outgoing label that is swapped or pushed. This Diffserv context information is populated into the ILM and the FTN at label establishment time.

If the label corresponds to an E-LSP for which no EXP ↔ PHB mapping has been explicitly signaled at LSP setup, the set of supported PHBs is populated with the set of PHBs of the preconfigured EXP ↔ PHB mapping If the label corresponds to an E-LSP for which an EXP ↔ PHB mapping has been explicitly signaled at LSP setup, the set of supported PHBs is populated with the set of PHBs of the signaled EXP ↔ PHB mapping. If the label corresponds to an L-LSP, the set of supported PHBs is populated with the set of PHBs forming the PSC that is signaled at LSP setup.

When the ILM (or the FTN) maps a particular label to a set of NHLFEs that contains more than one element, exactly one element of the set must be chosen before the packet is forwarded. In accordance with this, an incoming label (FEC) may be mapped, for Diffserv purposes, to multiple NHLFEs (for instance, where different NHLFEs correspond to egress labels supporting different sets of PHBs). When a label (FEC) maps to multiple NHLFEs, the Diffserv LSR *must* choose one of the NHLFEs whose Diffserv context indicates that it supports the outgoing PHB of the forwarded packet.

When a label (FEC) maps to multiple NHLFEs that support the outgoing PHB, the procedure for choosing one among those is outside the scope of this discussion. This situation may be encountered where it is desired to do load balancing of a BA over multiple LSPs. In such situations, in order to respect ordering constraints, all packets of a given microflow must be transported over the same LSP.

10.4.4   Encoding Diffserv Information into the Encapsulation Layer

This stage determines how to encode the fields that convey Diffserv information in the transmitted packet (e.g., MPLS shim EXP, ATM CLP, frame relay DE).

10.5   APPLICATIONS OF MULTIPROTOCOL LABEL SWITCHING

The key idea behind MPLS is the use of a forwarding paradigm based on label swapping that can be combined with a range of different control modules. Each control module is responsible for assigning and distributing a set of labels, as well as for maintaining other relevant control information. For example, an MPLS router might include:

- A *unicast routing* module, which builds up the routing table using the conventional IP routing protocol, assigns labels to the routes, distributes labels using the label distribution protocol (LDP), etc.
- A *traffic engineering* module, which enables explicitly specified label-switched paths to be set up through a network for traffic engineering purposes.
- A *virtual private network* (VPN) module, which builds VPN-specific routing tables using the border gateway protocol (BGP) and distributes labels corresponding to VPN routes.

Because MPLS allows different modules to assign labels to packets using a variety of criteria, it decouples the forwarding of a packet from the contents of the packet's IP header. This property is essential for such features as traffic engineering and VPN support.

10.5.1   Traffic Engineering

The term *traffic engineering* refers to the ability to control where traffic flows in a network, with the goal of reducing congestion and getting the most use out of the available facilities. Traditionally IP traffic is routed on a hop-by-hop basis, and the current *interior gateway protocol* (IGP) always uses the shortest path to forward traffic. The path the IP traffic takes may not be optimal, because it depends on static link metric information without any knowledge of the available network resources or the requirements of the traffic that needs to be carried on that path. Using shortest paths may cause the following problems:

1. The shortest paths from different sources overlap at some links, causing congestion on those links.

·2. The traffic from a source to a destination may exceed the capacity of the shortest paths, while a longer path between these two routers is underutilized.

10.5.1.1 *An Example of Traffic Engineering* In Figure 10.6, there are two paths from router C to router E, indicated as paths 1 and 2. If the router selects one of these paths as the shortest path from C to E ($C-D-E$), then it will carry all traffic that is destined for E through that path. The resulting traffic volume on that path may cause congestion, while another path ($C-F-G-H-E$) is underloaded. To maximize the performance of the overall network, it may be desirable to shift some fraction of the traffic from one link to another. While one could set the cost of path $C-D-E$ equal to the cost of path $C-F-G-H-E$, such an approach to load balancing becomes cumbersome if not impossible in networks of complex topology. Explicitly routed paths, implemented using MPLS, can be used as a more straightforward and flexible way of addressing this problem, allowing some fraction of the traffic on a congested path to be moved to a less congested path.

The solution to the traffic engineering problem relies on the fact that labels and label-switched paths can be established by a variety of different control modules, as discussed earlier. For example, the traffic engineering control module can establish a label-switched path from from B to C to F to G to H to E (path 1) and another from A to C to D to E (path 2) as shown in Figure 10.7. By setting policies that select certain packets to follow these paths, traffic flow across the network can be managed. In order to do traffic engineering effectively, the Internet Engineering Task Force (IETF) has introduced *constraint-based routing* [7] and an enhanced link-state IGP [8]. In order to control the path of LSPs effectively, each LSP can be assigned one

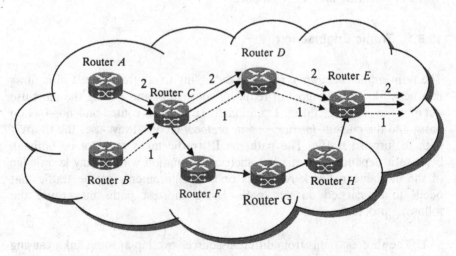

Fig. 10.6 Congestion problem caused by selecting the shortest path.

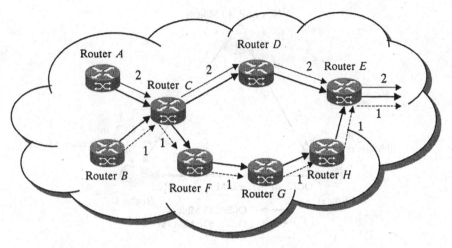

Fig. 10.7   Solution to the congestion problem using traffic engineering.

or more attributes. These attributes will be considered in computing the path for the LSP. The attributes and their meanings are summarized below:

- *Bandwidth:*   The minimum reservable bandwidth of a path for the LSP to be set up along that path.
- *Path Attribute:*   An attribute that decides whether the path of the LSP should be manually specified or dynamically computed by constraint-based routing.
- *Setup Priority:*   The attribute that decides which LSP will get the resource when multiple LSPs compete for it.
- *Holding Priority:*   The attribute that decides whether a resource held by an established LSP should be preempted by a new LSP.
- *Affinity (Color):*   An administratively specified property of an LSP.
- *Adaptability:*   Whether to switch the LSP to a more optimal path when one becomes available.
- *Resilience:*   The attribute that decides whether to reroute the LSP when the current path is affected by failure.

10.5.1.2 Constraint-Based Routing   Constraint-based routing has two basic elements: route optimization and route placement. Route optimization is responsible for selecting routes for traffic demands subject to a given set of constraints. Once the routes are decided, route placement implements these routes in the network so that the traffic flows will follow them. Constraint-based routing computes routes that are subject to constraints such as band-width and administrative policy. Because constraint-based routing considers more factors than network topology in computing routes, it may find a longer

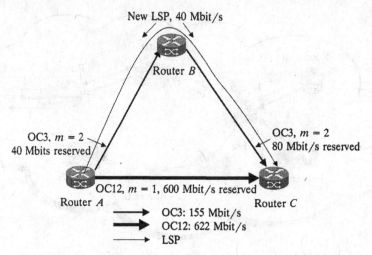

Fig. 10.8 Constraint-based routing.

but lightly loaded path rather than the heavily loaded shortest path. Thereby, network traffic is distributed more evenly and network resources are utilized more efficiently.

For example, in Figure 10.8, the shortest path between router A and router C is through link A–C with IGP metric $m = 1$. But because the reservable bandwidth on the shortest path is only $622 - 600 = 22$ Mbit/s, it cannot satisfy the requested bandwidth (40 Mbit/s) of the new LSP. Constraint-based routing will select longer path A–B–C instead, because the shortest path does not meet the bandwidth constraint.

It should be noted that the reservable bandwidth of a link is equal to the maximum reservable bandwidth set by network administrators minus the total bandwidth reserved by LSPs traversing the link. It does not depend on the actual amount of available bandwidth on that link. For example, if the maximum reservable bandwidth of a link is 155 Mbit/s, and the total bandwidth reserved by LSPs is 50 Mbit/s, then the reservable bandwidth of the link is 105 Mbit/s, regardless of whether the link is actually carrying 50 Mbit/s of traffic or more or less. In other words, in order to avoid routing instability [7], constraint-based routing does not compute LSP paths based on the instantaneous residual bandwidth of links.

Constraint-based routing can be off line or on line. With offline constraint-based routing, an offline server computes paths for LSPs periodically (e.g., hourly/daily). LSPs are then configured to take the computed paths. With on line constraint-based routing, routers may compute paths for LSPs at any time dynamically.

10.5.1.3 Enhanced Link-State IGPs
In order for constraint-based routing to compute LSP paths subject to constraints, an enhanced link-state IGP

must be used to propagate link attributes in addition to normal link-state information. Common link attributes may include:

1. Reservable bandwidth
2. Link affinity (color), that is, an administratively specified property of the link

Enhanced link-state IGPs will flood information more frequently than normal IGPs because the changes in reservable bandwidth or link affinity can cause the enhanced IGP to flood information. Therefore, a tradeoff must be made between the needs for accurate information and for avoiding excessive flooding. When the enhanced IGP builds an LSP's forwarding table, it will consider LSPs originated by the LSR, so that the LSPs can actually be used to carry traffic

10.5.1.4 Traffic Engineering Solution With the help of MPLS, constraint-based routing, and an enhanced IGP, traffic engineering can be done much more effectively. The two problems discussed at the beginning of Section 10.5.1 can be solved.

First, by setting the maximum reservable bandwidth of each link, and by setting the bandwidth requirement for each LSP, constraint-based routing will automatically avoid placing too many LSPs on any link. This solves the first problem. For example, in Figure 10.9, constraint-based routing will automatically choose LSP $B \to E$ on a longer path to avoid the congestion in link $C \to E$ due to overlap of the LSPs using the shortest path.

Second, if the traffic from router C_1 to router B_1 exceeds the capacity of any single path from C_1 to B_1, while a longer path is underloaded as shown

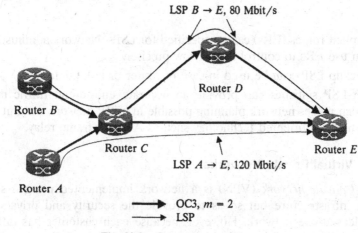

LSP $B \to E$, 80 Mbit/s

Router D

Router B

Router C

Router E

LSP $A \to E$, 120 Mbit/s

Router A

→ OC3, $m = 2$
→ LSP

Fig. 10.9 Congestion avoidance.

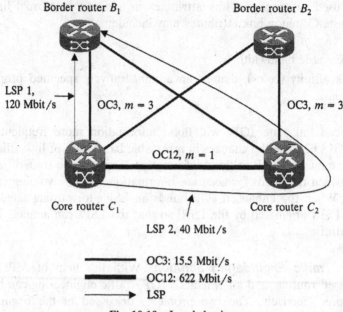

Fig. 10.10   Load sharing.

in Figure 10.10, then multiple LSPs can be configured from C_1 to B_1 to use the resource of the underloaded LSP link, and load ratio of these two LSPs can be specified as desired, so that the load can be distributed optimally. This is called *load sharing*. It solves the second problem. For example, if the total traffic from router C_1 to router B_1 is 160 Mbit/s, two LSPs can be configured from C_1 to B_1 if the routed LSP can provide enough bandwidth. So load sharing can be done among multiple paths of different cost, and the load ratio can be specified as described.

Besides these, MPLS provides the following advantages in traffic engineering:

1. Explicit routes (ERs) can be specified for LSPs. Network administrators can use ERs to control traffic flow precisely.
2. Backup LSPs can be used in case of router or link failure.
3. Per-LSP statistics can provide an accurate end-to-end traffic matrix, which makes network planning possible in an IP network without using connection-oriented technology such as ATM or frame relay.

10.5.2   Virtual Private Networks

A *virtual private network* (VPN) is a network implemented using a shared network infrastructure but so as to provide the security and privacy of a private leased-line network. However, because each customer has different security concerns, number of sites, number of users, routing complexity, mission-critical applications, traffic patterns, traffic volumes, staff networking expertise, and willingness to outsource network services, service providers must offer subscribers a portfolio that contains a number of different VPN service delivery models to satisfy a broad range of customer requirements. Over the years, a number of diverse VPN models have been proposed [12, 13]:

1. Traditional VPNs
   - Frame relay (Layer 2)
   - ATM (Layer 2)
2. VPNs based on customer premises equipment (CPE)
   - IP security protocol (IPSec) tunnels over the Internet (layer 3)
   - Point-to-point transport protocol (PPTP) or layer 2 transport protocol (L2TP) (layer 2)
3. Provider-provided VPNs:
   - BGP/MPLS VPNs (layer 3)

From the implementation point of view, they can be further classified into two VPN implementation models: the overlay model and the peer-to-peer model. The overlay model is that where the service provider provides emulated leased lines to the customer. The design and provision of virtual circuits across the backbone must be complete prior to any traffic flow. In the case of an IP network, this means that even though the underlying technology is connectionless, it requires a connection-oriented approach to provision the service. This model has scaling problems when the number of circuits or tunnels between customer devices is large and the IGP design is extremely complex and difficult to manage. On the other hand, the peer-to-peer model is that service provider and customer exchange layer 3 routing information and the provider relays the data between the customer sites and without the customer's involvement. This model suffers from lack of isolation between the customers and the need for coordinated IP address space between them.

MPLS combines the benefits of layer 2 switching with layer 3 routing and switching, which make it possible to construct a technology that combines the benefits of an overlay VPN (such as security and isolation among customers) with the benefits of simplified routing in a peer-to-peer VPN. MPLS-based VPN architecture provides the capability to deliver private network services over a public IP network. This section shows that MPLS-based VPNs are powerful enough to provide secure connectivity and relatively simple configuration for both intranets and extranets.

10.5.2.1 Terminology

- *Site*: A contiguous part of the customer network.
- *Intranet*: A VPN interconnecting corporate sites.
- *Extranet*: A VPN connecting a corporate site or sites to external business partners or suppliers. The Internet is the ultimate insecure extranet VPN.

- *Customer Edge (CE) Router*: A router at a customer site that connects to the service provider (via one or more provider edge routers).
- *Provider Edge (PE) Router*: A router in the service provider network to which CE routers connect.
- *Provider Core Router*: A router in the service provider network inter-connecting PE routers but, generally, not itself a PE router.
- *Entry and Exit PE routers*: The PE routers by which a packet enters and exits the service provider network.

10.5.2.2 Basic Implementation of MPLS-based VPN A VPN is a collection of policies, and these policies control connectivity among a set of sites. A customer site is connected to the service provider network by one or more *ports* (or *interfaces*), and the service provider associates each port with a VPN routing table, which is also called a VPN routing and forwarding (VRF) instance. Figure 10.11 describes a topology of a MPLS-based VPN. There are five customer sites and two component VPNs: VPN A and VPN B, which are supported by shared service provider networks using MPLS technology. We will use this as an example to show how to build a connection of VPN A from a host at site 1 to a server at site 5.

The PE router exchanges routing information with CE routers using a static or dynamic routing protocol. While a PE router maintains VPN routing information, it is only required to maintain VPN routes for those VPNs to which it is directly connected. And each PE router maintains a VRF for each of its directly connected sites. Each customer connection [such as frame relay *permanent virtual circuit* (PVC) or ATM PVC] is mapped to a specific VRF. Therefore, it is a port on the PE router, and not a site, that is associated with a single VRF. PE routers have the ability to maintain multiple forwarding tables that support the per-VPN segregation of routing information.

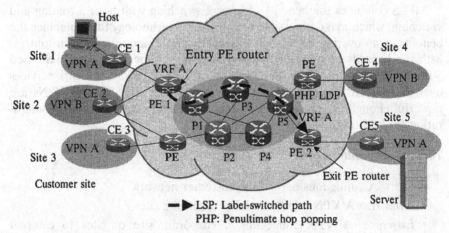

Fig. 10.11 Architecture of MPLS-based virtual private network.

After learning local VPN routes from CE routers, a PE router exchanges VPN routing information with other PE routers using the internal border gateway protocol (IBGP). Finally, when using MPLS to forward VPN data traffic across the provider's backbone, the ingress PE router functions as the ingress LSR, and the egress PE router functions as the egress LSR.

10.5.2.3 *Operation Example*

Before traffic is transferred across the VPN, the VPN route distribution and label switched path (LSP) need to be established.

10.5.2.3.1 *Distribution of VPN Routing Information*

VPN route distribution consists of two parts. The first part is responsible for the exchange of routing information between the CE and PE routers at the edges of the provider's backbone. The second part is responsible for the exchange of routing information between the PE routers across the provider's backbone.

In this example, PE 1 is configured to associate a VRF (e.g., VRF A) with the interface or subinterface over which it learns routes from CE 1. When CE 1 advertises the route for PE 1, PE 1 installs a local route to VRF A.

Then PE 1 advertises the route to PE 2 using IBGP. Before advertising the route, PE 1 selects an MPLS label to advertise with the route.

MPLS-based VPN also supports overlapping address spaces. Each site belongs to a VPN. A route distinguisher (RD) number is used to prefix the IP addresses for the site. It is configured on the interface (or subinterface) connecting to the site. This gives a way to tell duplicate private addresses apart. For example, subnet 10.1.1.0 for VPN 20 is different than subnet 10.1.1.0 for VPN 100. From the MPLS VPN provider's point of view, they are really 20:10.1.1.0 and 100:10.1.1.0, which are quite different. Putting the 8-byte route distinguisher in front of a 4-byte IP address gives us a 12-byte routing prefix. These are regarded as the VPN-IPv4 family of addresses.

When PE 2 receives PE 1's route advertisement, it determines if it should install the route to VRF A by performing route filtering based on the BGP extended community attributes carried with the route. If PE 2 decides to install the route in VRF A, it then advertises the route to CE 5.

10.5.2.3.2 *LSP Establishment*

In order to use MPLS to forward VPN traffic across the provider's backbone, LSPs must be established between the PE router that advertises the route and the PE router that learns the route. LSPs can be established and maintained across the service provider's network using either the label distribution protocol (LDP) or the resource reservation protocol (RSVP) as described in Section 10.2. The provider uses the LDP if it wants to establish a best-effort LSP between two PE routers. The provider uses the RSVP if it wants to either assign bandwidth to the LSP or use traffic engineering to select an explicit path for the LSP. RSVP-based LSPs support specific QoS guarantees and/or specific traffic engineering objectives.

10.5.2.3.3 Traffic Forwarding After route distribution and LSP establishment, a host at site 1 wants to communicate with a server at site 5. When a packet arrives at CE1, it performs a longest-match route lookup and forwards the IPv4 packet to PE 1. PE 1 receives the packet, performs a route lookup in the corresponding VRF, and obtains the following information:

- The MPLS label
- The BGP next hop for the route
- The outgoing subinterface for the LSP from PE 1 to PE 2

User traffic is forwarded from PE 1 to PE 2 using MPLS with a label stack containing two labels. For this data flow, PE 1 is the ingress LSR for the LSP, and PE 2 is the egress LSR for the LSP. Before transmitting a packet, PE 1 pushes the label to the bottom of the label stack. This label is orginally installed in VRF A when PE 1 receives PE 2's IBGP advertisement. Next, PE 1 pushes the label associated with the LDP or RSVP-based LSP to PE 2 to the top of the stack. Then PE 1 forwards the MPLS packet on the outgoing interface to the first provider core router along the LSP from PE 1 to PE 2. Core routers switch packets across the core of the provider's backbone network, according to the top label. The penultimate router to PE 2 pops the top label and forwards the packet to PE 2.

When PE 2 receives the packet, it pops the label, creating a native a IPv4 packet. PE 2 uses the bottom label to identify the directly attached CE. Finally, PE 2 forwards the native IP packet to CE 5, which forwards the packet to the server at site 5.

10.5.2.4 Advantages of MPLS-based VPNs The key objective of MPLS-based VPNs is to simplify network operations for customers while allowing the service provider to offer scalable, revenue-generating, value-added services. It has many advantages, including the following.

- Customers can choose their own addressing plans, which may or may not overlap with those of other customers or the service provider.
- The MPLS-based VPN model is highly scalable with increasing numbers of sites and customers. It also supports the *any-to-any* model of communication among sites within a VPN without requiring the installation of a full mesh of permanent virtual circuits (PVCs) or backhauling of traffic across the service provider network.
- From the customer's point of view, a significant advantage of the MPLS VPN model is that in many cases, routing can be dramatically simplified relative to the PVC model. Rather than managing routing over a topologically complex virtual backbone composed of many PVCs, an MPLS-based VPN customer can generally use the service provider backbone as the default route to all of that company's sites.

- The CE router at each customer site does not directly exchange routing information with other CE routers. Customers do not have to deal with intersite routing issues, because those are the responsibility of the service provider.
- Providers of VPN services often need to provide a range of QoS to their customers. MPLS-based VPNs support QoS using the emerging Diffserv techniques. These techniques allow customer traffic to be sorted into categories based on a wide range of policies such as site of origin, application type, and so forth, as it enters the provider network. Within the network, the classes of traffic are identified by header bits or by different labels, which the routers use to determine queuing treatment and thus QoS parameters such as delay and loss.

REFERENCES

1. E. Rosen, A. Viswanathan, and R. Callon, "Multiprotocol label switching architecture," RFC 3031, Internet Engineering Task Force (IETF), Jan. 2001.
2. D. Awduche, J. Malcolm, J. Agogbua, M. O'Dell, and J. McManus, "Requirements for traffic engineering over MPLS," RFC 2702, Internet Engineering Task Force (IETF), Sep., 1999.
3. X. Xiao, A. Hannan, B. Bailey, and L. Ni, "Traffic engineering with MPLS," *IEEE Netw.*, vol. 14, no. 2, pp. 28–33, Mar./Apr. 2000.
4. B. Davie, S. Davari, P. Vaananen, R. Krishnan, P. Cheval, and J. Heinanen, "MPLS support of differentiated services," Internet draft, Feb. 2001, draft-ietf-mpls-diff-ext-08.txt.
5. Y. Rekhter and E. Rosen, "Carrying label information in BGP-4," Internet draft, Jan. 2001, draft-ietf-mpls-bgp4-mpls-05.txt1.
6. L. Andersson, P. Doolan, N. Feldman, A. Fredette, and B. Thomas, "LDP specification," RFC 3036, Internet Engineering Task Force (IETF), Jan. 2001.
7. E. Crawley, R. Nair, B. Jajagopalan, and H. Sandick, "A framework for QoS-based routing in the Internet," RFC 2386, Internet Engineering Task Force (IETF), Aug., 1998.
8. T. Li, G. Swallow, and D. Awduche, "IGP requirement for traffic engineering with MPLS," Internet draft, Feb. 1999, draft-li-mpls-igp-te-00.txt.
9. T. Bates, R. Chandra, D. Katz, and Y. Rekhter, "Multiprotocol extension for BGP-4," RFC 2283, Internet Engineering Task Force (IETF), Feb. 1998.
10. B. Jamoussi, O. Aboul-Magd, L. Andersson, P. Ashwood-Smith, F. Hellstrand, K. Sundell, R. Callon R. Dantu, L. Wu, P. Doolan, T. Worster, N. Feldman, A. Fredette, M. Girish, E. Gray, J. Halpern, J. Heinanen, T. Kilty, A. Malis, and P. Vaananen, "Constraint-based LSP setup using LDP," Internet draft, Jul. 2000, draft-ietf-mpls-cr-ldp-04.txt.
11. D. O. Awduche, L. Berger, D.-H. Gan, T. Li, V. Srinivasan, and G. Swallow, "RSVP-TE: extensions to RSVP for LSP Tunnels," Internet draft, Feb. 2001, draft-ietf-mpls-rsvp-lsp-tunnel-08.txt.

12. C. Semeria, "RFC 2547bis: BGP/MPLS VPN fundamentals," *Juniper Networks White Paper*, Mar. 2000.

13. I. Pepelnjak and J. Guichard, *MPLS and VPN Architecture*, Cisco Press, Indianapolis, Oct. 2000.

14. S. Blake, D. Black, M. Carlson, E. Davies, Z. Wang, and W. Weiss, "An architecture for differentiated services," RFC 2475, Internet Engineering Task Force (IETF), Dec. 1998.

15. B. Braden, et al., "Recommendations on queue management and congestion avoidance in the Internet" RFC 2309, Internet Engineering Task Force (IETF), Apr. 1998.

16. D. Grossman, A. Smith, S. Blake, and Y. Bernet, "An informal management model for Diffserv router," Internet draft, Nov. 2000, draft-ietf-diffserv-model-05.txt.

17. K. Nichols, S. Blake, F. Baker, and D. Black, "Definition of the differentiated services field (DS field) in the IPv4 and IPv6 header," RFC 2474, Internet Engineering Task Force (IETF), Dec. 1998.

18. F. Baker, A. Smith, and K. Chan, "Management information base for the differentiated services architecture," Internet draft, Nov. 2000, draft-ietf-diffserv-mib-06.txt.

19. X. Xiao and L. Ni, "Internet QoS: a big picture," *IEEE Netw.*, vol. 13, no. 2, pp. 8–18, Mar./Apr. 1999.

20. S. Vegesna, *IP Quality of Service*, Cisco Press, Indianapolis, Dec. 2000.

APPENDIX

SONET AND ATM PROTOCOLS

The asynchronous transfer mode (ATM) was standardized by CCITT (the International Consultative Committee for Telephone and Telegraphy), currently called ITU-T (International Telecommunications Union–Telecommunication), as the multiplexing and switching principle for the broadband integrated services digital network (B-ISDN). ATM is a connection-oriented transfer mode based on statistical multiplexing techniques. It is asynchronous in the sense that the recurrence of cells containing information from an individual user is not necessarily periodic. It is capable of providing high-speed transmission with low cell loss, low delay, and low delay variation. It also provides flexibility in bandwidth allocation for heterogeneous services ranging from narrowband to wideband services (e.g., video-on-demand, video conferencing, videophone, and video library).

ATM connections are either preestablished using management functions, or they are set up dynamically on demand using signaling, such as user-network interface (UNI) signaling and private network–network interface (PNNI) routing signaling. In the former case they are referred to as *permanent virtual connections* (PVCs); in the latter, as *switched virtual connections* (SVCs).

As shown in Figure A.1, two ATM end systems communicate through two ATM switches. The left end system sends messages generated at the application layer to the right one. The messages are first carried in IP packets,[1] which are then passed down to different layers below the IP layer. Control bytes are added at different layers to facilitate the communications through the ATM network. At the receiver, the control bytes are stripped off before

---

[1] In native ATM applications, messages can be directly carried by ATM cells without first being carried by IP packets.

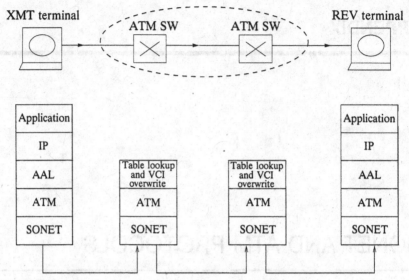

Fig. A.1   Protocol layers in the SONET (synchronous optical network) ATM network.

the final messages are recovered at the application layer (if no error occurs during the transmission). Inside the ATM network, there are only two layers, the physical layer (e.g., synchronous optical network, SONET) and the ATM layer, while at the end systems an additional layer (ATM adaptation layer, AAL) is added. ATM cells are routed through the ATM switches according to the routing information inside the cell header (5 bytes). The routing tables in every switch node on the path are updated either statically for PVCs or dynamically for SVCs. Once the routing tables are all updated, the connection between the two end systems is established and the left one can start to send traffic.

The detailed protocol conversion between the layers is shown in Figure A.2. A message generated at the application layer is passed down to the IP layer, where a 20-byte IP header is added. At the AAL, some AAL trailer overhead bytes are added and segmented into fixed-length data units (e.g., 48 bytes per unit). They are then passed down to the ATM layer, where a 5-byte cell header is added to every cell. A series of cells is then put into the payload of SONET frames, which repeats every 125 μs. SONET frames may be thought of as trains, and their overhead bytes as train engines[2] Cells are like cars in the trains, which may go to different destinations. As the trains arrive at a train station (i.e., an ATM switch), the engines are removed and cars are routed to different tracks (i.e., output links of the switch). These cars are packed together with others onto trains that go to the same station.

---

[2]The overhead bytes are actually evenly distributed in the SONET frame. For easy explanation, they are lumped together here.

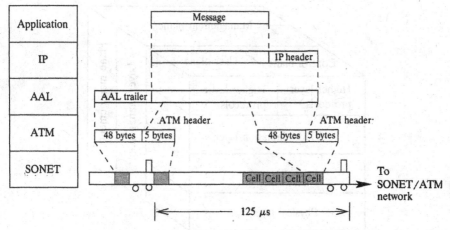

Fig. A.2 Conversion between protocol layers.

Section A.1 describes the ATM protocol reference model. Section A.2 describes SONET transmission frames and the functions related to the overhead bytes. Section A.3 describes the functions performed in different sublayers of the reference model. Section A.4 describes the ATM cell format and the related functions performed in the ATM layer. Section A.5 describes different AAL types (1, 2, 3/4, and 5).

A.1 ATM PROTOCOL REFERENCE MODEL

The ATM protocol reference model shown in Figure A.3 specifies the mapping of higher-layer protocols onto AAL, ATM, and its underlying physical layer. It is composed of four layers and three planes (user, control, and management planes).

- *U-plane*: The *user* plane provides for the transfer of user application information. It contains the physical layer, the ATM layer, and multiple ATM adaptation layers required for different services, such as constant-bit-rate (CBR) and variable-bit-rate (VBR) services.
- *C-plane*: The *control* plane protocols deal with call establishment and release and other connection control functions necessary for providing switched services. The C-plane structure shares the physical and ATM layers with the U-plane. It also includes AAL procedures and higher-layer signaling protocols, such as integrated local management interface (ILMI), UNI signaling, and PNNI routing protocols.
- *M-plane*: The *management* plane provides management functions and the capability to exchange information between the U-plane and C-plane.

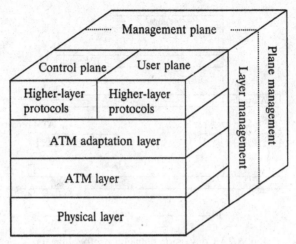

Fig. A.3   ATM protocol reference model.

The ILMI protocol uses the *simple network management protocol* (SNMP) to provide ATM network devices with status and configuration information concerning virtual path connections (VPCs), virtual channel connections (VCCs), registered ATM addresses, and the capabilities of ATM interfaces. UNI signaling specifies the procedures for dynamically establishing, maintaining, and clearing ATM connections at the UNI. The procedures are defined in terms of messages and the information elements used to characterize the ATM connection and ensure interoperability. The PNNI protocol provides the functions to establish and clear such connections, efficiently manage the resources of the network, and allow networks to be easily configured.

A.2   SYNCHRONOUS OPTICAL NETWORK (SONET)

ATM cells can be carried on different physical layers, such as digital signal level 1 (DS1), DS3, SONET, and others. Here, we only discuss how cells are carried in SONET transmission frames. SONET is a digital transmission standard based on a synchronous, rather than the plesiochronous, multiplexing scheme.

A.2.1   SONET Sublayers

Figure A.4 shows a SONET end-to-end connection with some typical SONET equipment, such as SONET multiplexor (mux) terminal, regenerator, add–drop mux (ADM), and digital cross-connect system (DCS). In practical applications, the communication is bidirectional. Let us consider an example

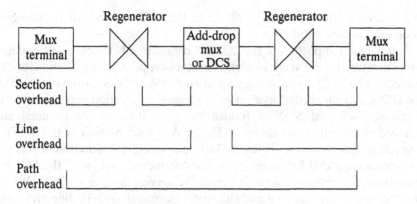

DCS: Digital cross-connect system

Fig. A.4   SONET end-to-end connection.

where non-SONET transmission signals from the left are multiplexed by a SONET mux terminal into SONET frames that are converted to optical signals. Due to attenuation and dispersion in optical fibers, the SONET bit stream needs to be *regenerated*: the optical signal is converted to an electrical signal, amplified, resampled, and then converted back to the optical signal. Because of the resampling, a phase-lock loop circuit is required to recover the clock from the incoming bit stream. The ADM is used to add and drop

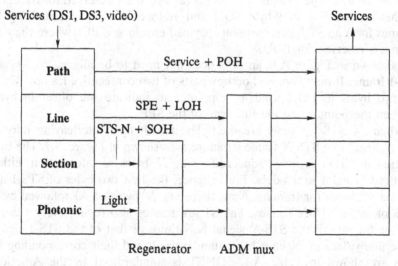

POH: Path Overhead   SPE: Synchronous payload envelope   DS1: Sigital signal 1
LOH: Line Overhead   STS: Synchronous transport signal   DS3: Sigital signal 3
SOH: Section Overhead   ADM mux: Add and drop multiplexor

Fig. A.5   SONET layers and associated overheads.

one or several low-rate tributaries, while the DCS is used for rerouting or grooming low-rate tributaries.

A *section* is the portion of the SONET connection between a terminal and a regenerator, or between two regenerators. A *line* is the portion of the connection between a mux terminal and an ADM or DCS, or between ADMs and DCSs. A *path* related to services is a logical connection between two end terminals, such that SONET frames are generated by one terminal and removed by the other. As shown in Figure A.5, each portion of the SONET connection has overhead bytes to facilitate operation administration and maintenance (OAM) functions in the corresponding portion of the SONET connection. For instance, messages from the service layer (e.g., DS1, DS3, or video streams) are augmented with path overhead (POH), line overhead (LOH), and section overhead (SOH) as they are passed down to different layers in SONET. At the receiver end, these overheads are stripped off as the messages are passed up to the service layer.

A.2.2 STS-*N* Signals

The basic SONET transmission signal is synchronous transfer signal level 1 (STS-1), and its transmission frame is shown in Figure A.6. Each frame has 810 bytes and is organized into 9 rows of 90 bytes (transmitted from left to right, top to bottom). Frames are repeated every 125 μs, corresponding to an 8-kHz voice sampling rate. The bit rate of the STS-1 is (810 bytes)/(125 μs), or 51.84 Mbit/s. The first three columns (27 bytes) are reserved for transport overhead, consisting of 9-byte SOH and 18-byte LOH. The remaining 87 columns form an STS-1 synchronous payload envelope (SPE), where the first column is reserved for POH.

As shown in Figure A.6, an SPE does not need to be aligned to a single STS-1 frame. It may *float* and occupy parts of two consecutive frames. There are two bytes in LOH used as a pointer to indicate the offset in bytes between the pointer and the first byte of the SPE.

When *N* STS-1 signals are multiplexed with byte interleaving into an STS-*N* signal, the STS-*N* frame structure is as shown in Figure A.7. The byte position in STS-1 is now "squeezed" with *N* bytes of information, either overhead bytes or user's data. For instance, the first two bytes of STS-1 are A1 and A2 bytes for framing. Now, there are *N* bytes of A1 followed by *N* bytes of A2 in STS-*N* frames. The STS-*N* frames also repeat every 125 μs, and the bit rate of the STS-*N* signal is N times of that of the STS-1 signal.

Certain values of *N* have been standardized, and their corresponding bit rates are shown in Table A.1. SONET is standardized by the American National Standards Institute (ANSI) and is deployed in North America, while Synchronous Digital Hierarchy (SDH) is standardized by ITU-T and is deployed in the other parts of the world. SONET and SDH are technically consistent. The major difference between them is in their terminologies.

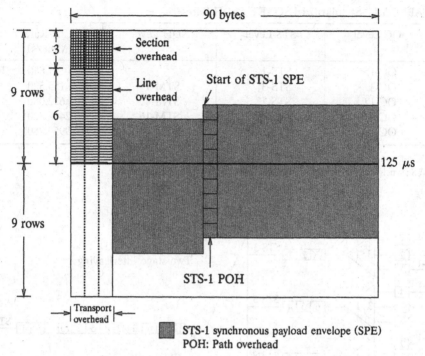

Fig. A.6 Two STS-1 (synchronous transfer signal level 1) frames.

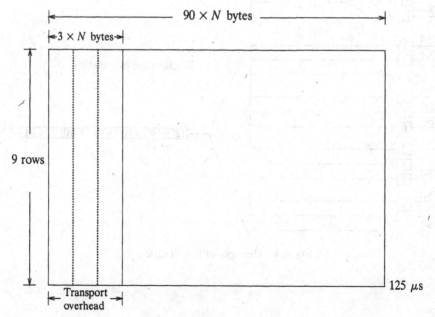

Fig. A.7 STS-N (synchronous transfer signal level N) frames.

TABLE A.1 Standardized SONET / SDH Rates

OC Level	STS Level	SDH Level	Line Rate (Mbit/s)
OC-1	STS-1		51.840
OC-3	STS-3	STM-1	155.520
OC-12	STS-12	STM-4	622.080
OC-48	STS-48	STM-16	2488.320
OC-192	STS-192	STM-64	9953.280

OC : optical carry;   SDH : synchronous digital hierarchy;
STS : synchronous transfer signal;   STM : synchronous transfer mode.

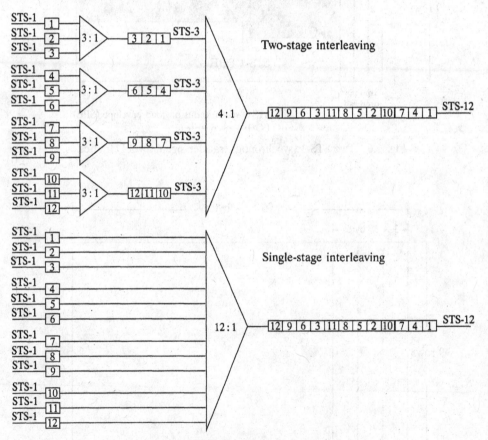

Fig. A.8 Example of byte interleaving.

Table A.1 shows the mapping between them. Synchronous transfer mode level 1 (STM-1) is the basic transmission signal in SDH, and its bit rate is 155.52 Mbit/s. Higher-capacity STMs are formed at rates equivalent to m times this basic rate. STM capacities for $m = 4$, $m = 16$, and $m = 64$ are defined and called STM-4, STM-16, and STM-64. Their corresponding rates are shown in Table A.1. The optical format of the STS-N signal is called optical carrier level N (OC-N).

Figure A.8 shows two examples of multiplexing 12 STS-1 signals into an STS-12 signal, one with two stages of multiplexing and one with a single stage. In order to have identical byte streams at the output of the mux, it has been standardized that the output byte stream has to be the same as the one from multiplexing multiple STS-3 signals.

A.2.3 SONET Overhead Bytes

Figure A.9 shows the section, line, and path overhead bytes. A1 (11110110) and A2 (00101000) are used for frame alignment. In order to reduce the probability of having A1 and A2 bytes in the payload frame and to avoid a long stream of 0s and 1s, the entire frame except the first 9 bytes is scrambled (see Section A.2.4). The J0 byte is allocated to a section trace. This byte is used to transmit repetitively a *section access point identifier* so that a section receiver can verify its continued connection to the intended transmitter. B1 is used to monitor the error in the regenerator section. B1 is computed using bit-interleaved even parity over all the bits of the previous STS-1 frame after scrambling and is placed in the B1 byte of the current frame before scrambling. The E1 byte is allocated in the first STS-1 of an STS-N signal for a 64-kbit/s orderwire channel used for voice communication between regenerators and terminal locations. The F1 byte is allocated in the first STS-1 of an STS-N signal and is reserved for user purposes (e.g., to provide temporary data or voice channel connections for special maintenance purposes). The D1–D3 bytes are located in the first STS-1 of an STS-N signal for section data communication. These three bytes form a 192-kbit/s message-based channel for alarms, maintenance, control, monitoring, administering, and other communication needed between section-terminating equipment.

The *pointer*, (H1 and H2 bytes), is used to find the first byte of the SPE. When their value is zero, the SPE starts immediately following the H3 byte. The H3 byte is used for frequency justification to compensate for clock deviations between the source and the mux terminals, which may occur in some circumstances (see Section A.2.4). The B2 bytes are used to monitor bit errors at the mux section. They are computed using bit-interleaved even parity over all bits of the previous STS-1 frame except for the first three rows of SOH and are placed in the B2 location of the current frame before scrambling. The K1, K2, and M0 bytes are related to automatic protection switching (APS) operations (see Section A.2.6). The D4-D12 bytes form a

Transport overhead

Path overhead

Section overhead	Framing A1	Framing A2	Section trace 10	Trace J1
	BIP-8 B1	Orderwire E1	User F1	BIP-8 B3
	Data com D1	Data com D2	Data com D3	Signal label C2
Line overhead	Pointer H1	Pointer H2	Pointer action H3	Path status G1
	BIP-8 B2	APS K1	APS K2	User F2
	Data com D4	Data com D5	Data com D6	Indicator H4
	Data com D7	Data com D8	Data com D9	User F3
	Data com D10	Data com D11	Data com D12	APS K3
	SSM S1	FEBE M0	Orderwire E2	Tandem connection N1

SSM: Synchronous status messzage
FEBE: Far end block error
BIP-8: Bit interleaving parity 8

Fig. A.9 Section, line, and path overhead bytes.

576-kbit/s channel available at the line-terminating equipment. The S1 byte is allocated for the *synchronous status message* (SSM) function. It indicates the type of clock generating the synchronization signal. The M0 byte is allocated for the *far end block error* (FEBE) function, which is assigned to the Nth STS-1 of an STS-N signal. The E2 byte provides an orderwire channel of 64 kbit/s for voice communication between line-terminating equipment.

The J1 byte is used to transmit repetitively a *path access point identifier* so that a path-receiving terminal can verify its continued connection to the intended transmitter. The B3 byte is allocated for a path error monitoring function. It is calculated using bit-interleaved even parity over all bits of the previous frame payload before scrambling and is placed at the B3 byte location of the current frame payload before scrambling. Bits 1–4 of the G1 byte convey the count of interleaved-bit blocks that have been detected in error (ranging from 0 to 8). This error count is obtained from comparing the calculated parity result and the B3 byte value of the incoming frame. Bytes F2 and F3 are allocated for user communication purposes between path elements and are payload-dependent. Byte H4 provides a generalized position indicator for payloads and can be payload-specific. Bits 1–4 of the K3 byte are allocated for APS signaling for protection at the path levels. Byte N1 is allocated to provide a *tandem connection monitoring* (TCM) function. The tandem connection sublayer is an optional sublayer that falls between the multiplex section and path layers and is application-specific.

A.2.4 Scrambling and Descrambling

Figure A.10 shows how scrambling and descrambling are done in SONET. A pseudo-random bit stream with $2^7 - 1$ bits is generated by a shift register (Seven D-type flip-flops) with some feedback lines as shown in Figure A.11.

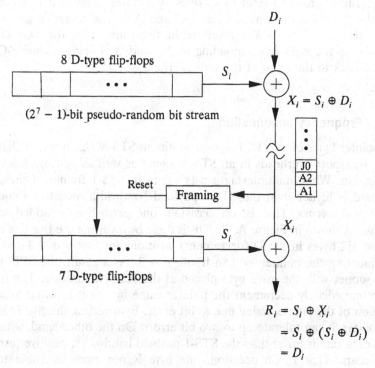

Fig. A.10 Frame synchronous scrambling.

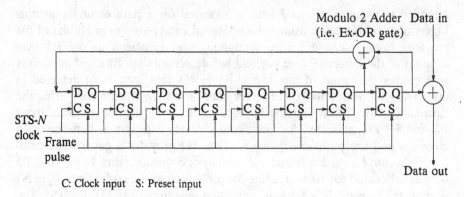

C: Clock input   S: Preset input

Fig. A.11   Circuit of the frame synchronous scrambler.

The pseudo-random bit stream (S_i) is then exclusive-ORed with the data bit stream (D_i). The scrambled bit stream (X_i) is sent to the network and is descrambled with the same pseudo-random bit stream. If no error occurs in the network, the received bit stream (R_i) will be obtained by exclusive-ORing the R_i with S_i. To synchronize the pseudo-random bit streams at the transmitter and the receiver, the pseudo-random bit stream generators (the shift registers) are both reset to all ones by a frame pulse that is generated from the detection of framing bytes (A1 and A2). The scramble generating polynomial is $1 + X^6 + X^7$, resulting in the outputs of the two D-type flip-flops on the right (corresponding to X^6 and X^7) being exclusive-ORed and fed back to the input of the shift register.

A.2.5   Frequency Justification

The pointer bytes of all STS-1 signals within an STS-N signal are to align the STS-1 transport overheads in an STS-N signal as well as perform frequency justification. When multiplexing a user's data to STS-1 frames, if the user's data rate is higher than the STS-1 payload bandwidth, negative frequency justification occurs. That is, on occasion, one extra byte is added to the payload, as shown in Figure A.12. This is done by inverting the five bits of the H1 and H2 bytes in the D (decrement) positions for frame $n + 2$ and then decrementing the pointer by 1 in frame $n + 3$. As a result, the SPE $n + 3$ starts sooner with the extra byte placed at the H3 byte location. The reason not to immediately decrement the pointer value by one is to avoid misinterpretation of the pointer value due to bit error. By inverting the five D bits of the pointer, it can tolerate up to two bit errors. On the other hand, when the user's data rate is lower than the STS-1 payload bandwidth, positive justification occurs. That is, on occasion, one byte is not made available to the

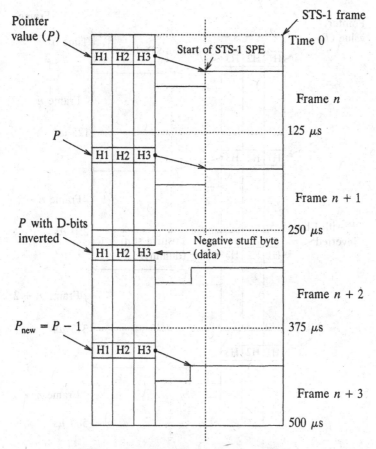

Fig. A.12 Negative STS-1 pointer justification.

payload. As shown in Figure A.13, this is done by inverting the five I (increment) bits in the H1 and H2 bytes in frame $n + 2$ and then incrementing the pointer by 1 in frame $n + 3$. As a result, SPE $n + 3$ starts later. The byte next to H3 is now left empty or stuffed and not allowed to be part of the SPE.

A.2.6 Automatic Protection Switching (APS)

Two types of maintenance signals are defined for the physical layer to indicate the detection and location of a transmission failure. These signals are *alarm indication signal* (AIS) and *remote defect indication* (RDI), and are applicable at the section, line, and path layers of the physical layer. AIS is used to alert associated termination points in the direction of transmission that a failure has been detected and alarmed. RDI is used to alert associated

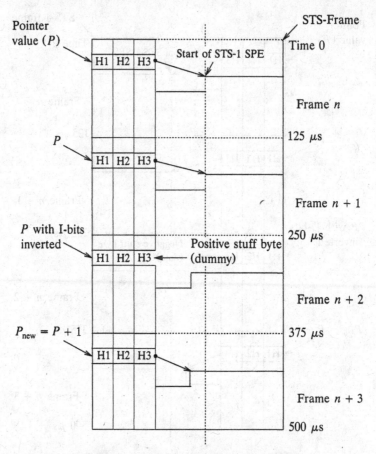

Fig. A.13 Positive STS-1 pointer justification.

termination points in the opposite direction of transmission that a defect has been detected. As shown in Figure A.14(a), when a fiber is cut or the transmission equipment fails, an AIS is generated and transmitted in the downstream direction with respect to terminal A, while an RDI signal is generated for the upstream direction. If failures occur at both directions as shown in Figure A.14(b), the AIS and RDI signals are generated and transmitted in both directions.

K1 and K2 bytes are allocated for APS signaling for the protection of the SONET equipment. For instance, line AIS is generated by inserting a 111 code in positions 6, 7, and 8 of the K2 byte in the downstream direction, while line RDI is generated by inserting a 110 code in positions 6, 7, and 8 of the K2 byte in the upstream direction. The M0 byte is used to convey the count of interleaved bit blocks that have been detected in error in the range of $[0, 24]$. This error count, obtained from comparing the calculated parity result and the B2 value of the incoming signal at the far end, is inserted in

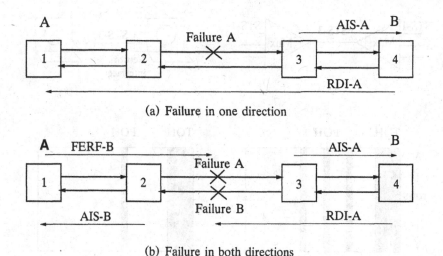

(a) Failure in one direction

(b) Failure in both directions

Fig. A.14   AIS (alarm indication signal) and RDI (remote defect indication) signals.

the M0 field and sent back. It reports to the near-end line-terminating point about the error performance of its outgoing signal.

A.2.7 STS-3 vs STS-3c

Superrate services, such as B-ISDN ATM service, require multiples of the STS-1 rate. They are mapped into an STS-Nc SPE and transported as a concatenated STS-Nc whose constituent STS-1s are lined together in fixed phase alignment. Figure A.15 shows the comparison of the STS-3 and STS-3c, where the former is multiplexed from three STS-1 signals, while the latter is used to carry a single ATM cell stream. For the STS-3, there are three POHs, one for each STS-1 signal, while for the STS-3c there is only one POH. Thus, the number of bytes in the payload of the STS-3c frame is more than that of the STS-3 frame by 18 bytes.

Figure A.16 shows the frame structure of an STS-3c frame and the corresponding overhead bytes. Since there is only one SPE for each STS-3c frame, only one set of pointers (H1 and H2 bytes) in the LOH is used; the other two sets (H1* and H2*) are not used. The entire STS-3c payload capacity (excluding SOH, LOH, and POH) can be filled with cells, yielding a transfer capacity for ATM cells of 149.760 Mbit/s (155.52 Mbit/s × 260/270). The remainder (5760 kbit/s) is available for the section, line, and path overhead. Because the STS-3c payload capacity (2340 bytes, or 260 × 9 bytes) is not an integer multiple of the cell length (53 bytes), a cell may cross a frame payload boundary.

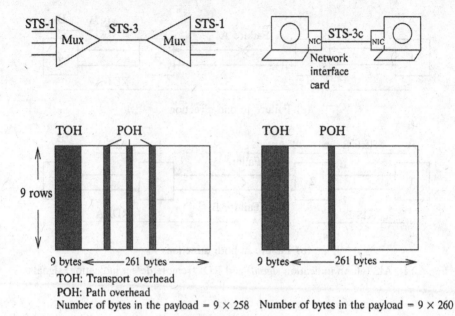

TOH: Transport overhead
POH: Path overhead
Number of bytes in the payload = 9×258 Number of bytes in the payload = 9×260

Fig. A.15   STS-3 vs STS-3c.

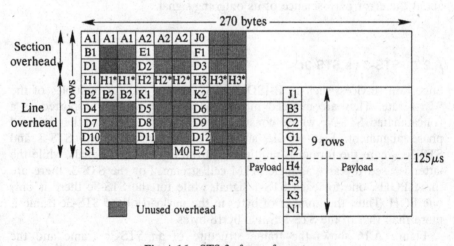

Fig. A.16   STS-3c frame format.

A.2.8 OC-*N* Multiplexor

Figure A.17 shows how an OC-*N* signal is composed and where the functions associated with path-terminating equipment, line-terminating equipment, and section-terminating equipment are performed as the user's payload passes through the path, line, and section layers. The user's payload is first augmented with path overhead to form an SPE, and the B3 byte is calculated at

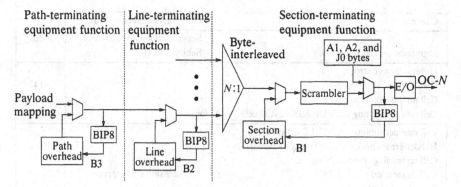

Fig. A.17 OC-N signal composition.

the path layer. The SPE is then augmented with line overhead, and the B2 byte is calculated at the line layer. Finally, the resulting partial frame is multiplexed in byte interleaving and augmented with section overhead, and the B1 byte is calculated. Note that the calculated B3, B2, and B1 bytes are placed at the corresponding positions in the next frame. Also note that the entire frame, except the framing bytes (A1 and A2) and J0 byte, is scrambled.

A.3 SUBLAYER FUNCTIONS IN THE REFERENCE MODEL

The functions of the physical layer (U-plane) are grouped into the *physical-medium-dependent* (PMD) sublayer and the *transmission convergence* (TC) sublayer (see Figure A.18). The PMD sublayer deals with aspects that depend on the transmission medium selected. The PMD sublayer specifies physical medium and transmission (e.g., bit timing, line coding) characteristics and does not include framing or overhead information.

The TC sublayer deals with physical layer aspects that are independent of the transmission medium characteristics. Most of the functions constituting the TC sublayer are involved with generating and processing some overhead bytes contained in the SONET frame. On transmission, the TC sublayer maps the cells to the frame format, generates the header error control (HEC, the last byte of the ATM cell header), and sends idle cells when the ATM layer has none to send. On reception, the TC sublayer delineates individual cells in the received bit stream, and uses the HEC to detect and correct received errors.

The HEC byte is capable of single-bit error correction and multiple-bit error detection. The transmitter calculates the HEC value across the entire ATM cell header and inserts the result in the HEC field. The value is the remainder of the division (modulo 2) by the generator polynomial $x^8 + x^2 + x + 1$ of the product of x^8 with the contents of the header excluding the

Convergence	Convergence sublayer (CS)	AAL
Segmentation and reassembly	Segmentation and reassembly Sublayer (SAR)	
Generic flow control Cell VPI/VCI translation Cell multiplex/demultiplex Cell rate decoupling (with UNASSIGNED cells: ATM forum		ATM
Cell rate decoupling (with IDLE cells: ITU-T) Header error check (HEC) generation/verification Cell scrambling/descrambling Cell delineation (using HEC) Path signal identification Frequency justification Frame scrambling/descrambling Frame generation/recovery	Transmission convergence (TC)	PHY
Bit timing Line coding Physical-medium-dependent scrambling/descrambling	Physical-medium-dependent (PMD)	

Fig. A.18 Protocol reference model sublayers and functions.

HEC field. The remainder is added (modulo 2) to an 8-bit pattern[3] (01010101) before being inserted in the HEC field. The receiver must subtract the same pattern from the 8-bit HEC before proceeding with the HEC calculation.

At the receiver two modes of operation are defined: correction mod and detection mode, as shown in Figure A.19. In correction mode only a single-bit error can be corrected, while detection mode provides for multiple-bit error detection. In detection mode all cells with detected errors in the header are discarded. When a header is examined and no error found, the receiver switches to correction mode.

Cell scrambling and descrambling permit the randomization of the cell payload to avoid continuous nonvariable bit patterns and improve the efficiency of the cell delineation algorithm.

The cell delineation function identifies cell boundaries based on the HEC field. Figure A.20 shows the state diagram of the HEC cell delineation method. In the HUNT state, the delineation process is performed by checking for the correct HEC in the incoming data stream. Once a correct HEC has been detected, it is assumed that one cell header has been found, and the receiver passes to the PRE-SYNC state. In the PRE-SYNC state, the HEC checking is done cell by cell. The transition occurs from the PRE-SYNC state to the SYNC state if m consecutive correct HECs are detected, or to the

---

[3] When the first four bytes of the cell header are all zeros and the remainder is also zero, modulo-2 adding the 8-bit pattern to the remainder can reduce the length of the stream of zeros.

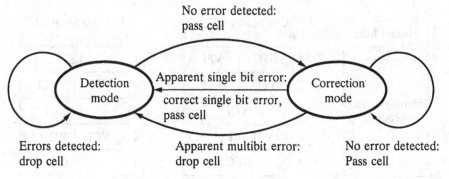

No error detected:
pass cell

Errors detected:
drop cell

Apparent multibit error:
drop cell

No error detected:
Pass cell

Fig. A.19   Receiver HEC bimodal operation.

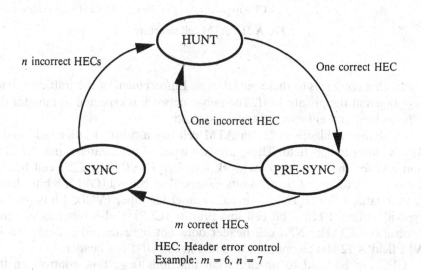

HEC: Header error control
Example: $m = 6, n = 7$

Fig. A.20   Cell delineation state diagram.

HUNT state if an incorrect HEC is detected. In the SYNC state, the receiver moves to the HUNT state if n consecutive cells have an incorrect HEC. For instance, m can be 6 and n can be 7.

A.4 ASYNCHRONOUS TRANSFER MODE

The ATM layer provides for the transparent transfer of fixed-size ATM cells between communicating upper layer entities (AAL entities). This transfer occurs on a preestablished ATM connection according to a traffic contract. A traffic contract is composed of a service class, a vector of traffic parameters (e.g., peak cell rate, sustainable rate, and maximum burst size), a confor-mance definition, and others. Each ATM end system is expected to generate

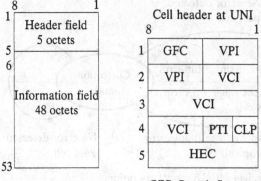

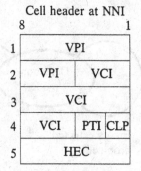

GFC: Generic flow control   PTI: Payload type identifier
VPI: Virtual path identifier   CLP: Cell loss priority
VCI: Virtual channel identifier   HEC: Header error control

Fig. A.21 ATM cell structure.

traffic that conforms to these parameters. Enforcement of the traffic contract is optional at the private UNI. The public network is expected to monitor the offered load and enforce the traffic contract.

As shown in Figure A.21, an ATM cell has a 5-byte header field and a 48-byte information field. There are two types of cell headers, one for UNI and one for the network-to-network interface (NNI). The UNI cell header contains the following fields: 4 bits generic flow control (GFC), 8 bits virtual path identifier (VPI), 16 bits virtual channel identifier (VCI), 3 bits payload type identifier (PTI), 1 bit cell loss priority (CLP), and 8 bits header error control (HEC). The NNI cell header does not have the GFC field, and its VPI field is 12 bits as opposed to 8 bits in the UNI cell header.

GFC can be used to provide local functions (e.g., flow control) on the customer site. It has local significance only; the value encoded in the field is not carried from end to end and will be overwritten by the ATM switches.

A.4.1 Virtual Path and Virtual Channel Identifier

ATM has a two-level hierarchy of ATM connection, whereby a group of virtual channel connections (VCCs) are usually bundled in a virtual path connection (VPC). Within a transmission link, there can be multiple VPCs, and within each VPC there can be multiple VCCs. VPCs are logical *pipes* between any two ATM switch nodes, which are not necessarily directly connected by a single physical link. Thus, it allows a distinction between physical and logical network structures and provides the flexibility to rear-range the logical structures according to the traffic requirements and end point locations. A VPC is often chosen to be a constant-rate pipe, but it is not restricted to be so, and it can be in any of the service categories. For

ATM switch

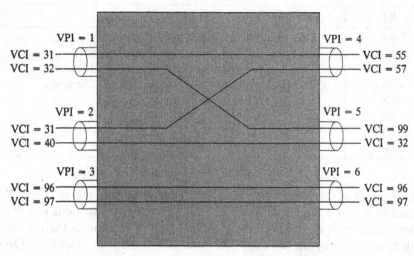

Fig. A.22   Virtual paths and virtual channels.

instance, a corporation might establish VPCs through a network service provider to interconnect various building locations of the corporation. The presence of VPCs allows for quicker establishment of switched VCCs. The VPCs also enable faster restoration.

As shown in Figure A.22, when a VPC (e.g., VPI = 3) is routed by a switch node, only the VPI will be used to look up the routing table and be replaced with a new value (VPI = 6), and its VCI values remain unchanged (VCI = 96 and 97). When a VCC (e.g., VPI = 2, VCI = 31) is routed through a switch, the combination of the VPI and VCI is used to look up the routing table and is replaced with a new value (VPI = 4, VCI = 57) at the output link.

The two identifiers, together with the physical link the cells arrive from, uniquely identify connections at each ATM switch. At UNI, there are 8 bits for VPI, while at NNI there are 12 bits. In general, VCI values are unique only in a particular VPI value, and VPI values are unique only in a particular physical link. The VPI or VCI has local significance only, and its value is replaced with a new value at a new physical link, which is known as *label swapping*. The reason for having the VPI and VCI change at every link is to support a large number of ATM end systems. For instance, the number of hosts supported by IPv4 is limited to 2^{32}, whereas in ATM the number is limited to 2^{24} at UNI or 2^{28} at NNI per transmission link.

A.4.2   Payload Type Identifier

The main purpose of the *payload type identifier* (PTI) is to discriminate between user cells (i.e., cells carrying user information) and nonuser cells. As shown in Table A.2, PTI values of 0 to 3 indicate user cells. PTI values of 2

TABLE A.2   Payload Type Identifier (PTI) Encoding

RTI	Interpretation
000	User data cell, congestion not experienced, SDU type = 0
001	User data cell, congestion not experienced, SDU type = 1
010	User data cell, congestion experienced, SDU type = 0
011	User data cell, congestion experienced, SDU type = 1
100	Segment OAM F5 flow-related cell
101	End-to-end OAM flow-related cell
110	Resource management cell
111	Reserved for future functions

and 3 indicate that congestion has been experienced in the network. PTI values of 4 and 5 are used for VCC level management functions (F5 flow), where OAM cells of F5 flow have the same VPI–VCI value as the user data cells transported by the VCC. A PTI value of 4 is used for identifying OAM cells communicated within the bounds of a VCC segment (i.e., a single link segment across the UNI), while a PTI value of 5 is used for identifying end-to-end OAM cells. A PTI value of 6 indicates the resource management (RM) cell.

A congestion flow control scheme used in the ATM network is *explicit forward congestion notification*. When a network node detects congestion on the link, it will mark the "congestion experienced" bit in the PTI of user data cells passing through the congested link (i.e., changing the pattern 000 to 010, or 001 to 011). Once a cell is marked "congestion experienced" at a node, it cannot be modified back to "congestion not experienced" by any downstream node along the path to the destination node. The receiver can set a congestion indication bit in the RM cells to inform the sender to slow down, increase, or maintain the rate. The RM cells are normally transmitted from the sender to the receiver every certain number of data cells it sends.

A.4.3   Cell Loss Priority

The CLP field may be used for loss priority indication by an ATM end system and for selective cell discarding in network equipment. This bit in the ATM cell header indicates two levels of priority for ATM cells. Cells with CLP = 0 are higher priority than those with CLP = 1. Cells with CLP = 1 may be discarded during periods of congestion to preserve the loss rate of the cells with CLP = 0.

A.4.4   Predefined Header Field Values

Table A.3 shows some predefined header field values. The cell rate decoupling function at the sending entity adds unassigned cells to the assigned cell stream (cells with valid payload) to be transmitted. In other words, it

TABLE A.3 Predefined Header Field Values

Use	Value[a]			
	Octet 1	Octet 2	Octet 3	Octet 4
Unassigned cell indication	00000000	00000000	00000000	0000xxx0
Metasignaling (default)[b, c]	00000000	00000000	00000000	00010a0c
Metasignaling[c, d]	0000yyyy	yyyy0000	00000000	00010a0c
General broadcast signaling (default)[b]	0000000	00000000	00000000	00100aac
General broadcast signaling[d]	0000yyyy	yyyy0000	00000000	00100aac
Point-to-point signaling (default)[b]	00000000	00000000	00000000	00100aac
Point-to-point signaling[d]	0000yyyy	yyyy0000	00000000	00100aac
Invalid pattern	xxxx0000	00000000	00000000	0000xxx1
Segment OAM F4 flow cell[d]	0000aaaa	aaaa0000	00000000	00110a0a
End-to-end OAM F4 flow cell[d]	0000aaaa	aaaa0000	00000000	01000a0a

[a]"a" indicates that the bit is available for use by the appropriate ATM layer function; "x" indicates "don't care" bits; "y" indicates VPI value other than 00000000; "c" indicates that the originating signaling entity shall set the CLP bit to 0 (the network may change the value of the CLP bit).
[b]Reserved for signaling with the local exchange.
[c]The transmitting ATM entity shall set bit 2 of octet 4 to 0. The receiving ATM entity shall ignore bit 2 of octet 4.

transforms a noncontinuous stream of assigned cells into a continuous stream of assigned and unassigned cells. At the receiving entity, the opposite operation is performed for both unassigned and invalid cells. The rate at which the unassigned cells are inserted or extracted depends on the bit rate (rate variation) of the assigned cell and/or the physical layer transmission rate.

Metasignaling cells are used by the metasignaling protocol for establishing and releasing SVCs. For PVCs, metasignaling is not used.

The VPC operation flow (F4 flow) is carried via specially designated OAM cells. The OAM cells of the F4 flow have the same VPI value as the user-data cells transported by the VPC, but are identified by two unique preassigned virtual channels within this VPC. At the UNI, the virtual channel identified by a VCI value 3 is used for VP level management functions between ATM nodes on both sides of the UNI (i.e., single VP link segment) while the virtual channel identified by a VCI value 4 can be used for VP level end-to-end management functions.

A.5 ATM ADAPTATION LAYER

The AAL performs the functions necessary to adapt the capabilities provided by the ATM layer to the needs of higher-layer applications. Since the ATM layer provides an indistinguishable service, the AAL is capable of providing

	Service class			
Attribute	Class A	Class B	Class C	Class D
Timing relation between source and destination	Required		Not required	
Bit rate	Constant		Variable	
Connection mode	Connection-oriented			Connectionless
AAL(s)	AAL1	AAL2	AAL3/4 or AAL5	AAL3/4 or AAL5
Example(s)	DS1, E1, $n \times$ 64-kbit/s emulation	Packet video, audio	Frame relay X.25	IP, SMDS

Class A—constant-bit-rate (CBR) service with end-to-end timing, connection-oriented
Class B—variable-bit-rate (VBR) service with end-to-end timing, connection-oriented
Class C—variable-bit-rate (VBR) service with no timing requested, connection-oriented
Class D—variable-bit-rate (VBR) service with no timing requested, connectionless

Fig. A.23 ITU ATM and B-ISDN service classes.

different functions for different service classes. For instance, VBR users may require such functions as protocol data unit (PDU) delineation, bit error detection and correction, and cell loss detection. CBR users typically require source clock frequency recovery, detection, and possible replacement of lost cells. These services are classified by ITU-T based on three parameters: timing relationship between source and destination, bit rate, and connection mode. Four classes have been defined and shown in Figure A.23.

- *Class A*:   This class corresponds to CBR, connection-oriented services with a timing relationship between source and destination. Typical services of this class are voice, circuit emulation, and CBR video.
- *Class B*:   This class corresponds to VBR connection-oriented services with a time relationship between source and destination. Packet video is a typical example of this service class.
- *Class C*:   This class corresponds to VBR connection-oriented services with no timing relationship between source and destination. X.25 and frame relay are typical examples of this service class.
- *Class D*:   This class corresponds to VBR connectionless services with no timing relationship between source and destination. IP data transfer is a typical example of this type of service.

As shown in Figure A.24, the AAL is subdivided into two sublayers: the segmentation and reassembly (SAR) sublayer and the convergence (CS) sub-layer. The SAR sublayer performs the segmentation of user information into the ATM cell payloads. The CS sublayer maps the specific user requirements onto the ATM transport network. The functions performed at the CS

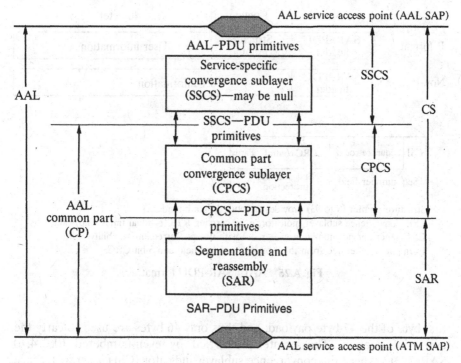

Fig. A.24   Generic AAL protocol sublayer model

sublayer differ for each of the services, whereas the SAR sublayer provides the same functions to all the services.

The CS sublayer is further divided into two parts: the common part convergence sublayer (CPCS), which is common to all users of AAL services, and the service specific convergence sublayer (SSCS), which depends on the characteristics of the user's traffic. Figure A.24 shows how these sublayers are related to each other. The SSCS sublayer can be null, meaning that it need not be implemented, whereas the CPCS sublayer is always present. It is apparent that the protocol functionality performed at the SAR and CPCS is common to all AAL users.

A.5.1   AAL Type 1 (AAL1)

AAL1 is used for CBR services that require tight delay and jitter control between the two end systems, e.g., emulated $n \times 64$-kbit channels or T1 or E1 facilities. As shown in Figure A.25, the 47-byte SAR-PDU payload used by CS has two formats, called P and non-P. In the non-P format, the entire SAR–PDU is filled with user information.

The P-format is used in the structured data transfer mode, where a structured data set, a byte stream, is transferred between source and destination. The structured data set is pointed at by a pointer that is placed at the

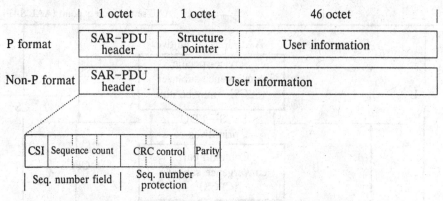

Structure pointer (0 to 93) provides structured data boundaries.
CSI: Convergence sublayer indicator, e. g., carrying RTS (residual timestamp) values.
CRC generator polynomial $= x^3 + x + 1$ with one-bit correction capability.
Even parity generated from the 4-bit seq. number field and 3-bit CRC.

Fig. A.25   AAL1 SAR-PDU format.

first byte of the 47-byte payload, and thus only 46 bytes are used to carry the user's information. The pointer is carried by even-numbered $(0, 2, 4, 6)$ SAR–PDU, where the convergence sublayer indicator (CSI) is set to 1. Since the pointer is transferred in every two PDUs, it needs to point to any byte in the structured data set, that is, up to 93 bytes $(46 + 47)$. Thus, 7 bits are used in the one-byte pointer to address the first byte of an $n \times 64$-kbit/s structure.

The first byte of the SAR–PDR consists of a 1-bit CSI, a 3-bit sequence count, a 3-bit CRC (cyclic redundant check), and a parity bit. The CSI bit carries the CS indication. The sequence count carries the sequence number of the SAR–PDUs (0 to 7). The CSI bit and the sequence count are protected by a 3-bit CRC. The resulting 7-bit field is protected by an even-parity check bit. The CSI and the sequence count values are provided by the CS sublayer. Such a 4-bit sequence number protection field is able to correct single-bit errors and to detect multiple-bit errors. By using the sequence count, the receive will be able to detect any cell loss if the number of lost cells is not an integer multiple of 16.

The AAL1 CS provides two methods to support asynchronous CBR services where the clocks are not locked to a network clock. The two methods are *adaptive clock* and *synchronous residual timestamp* (SRTS). With the SRTS technique, an accurate reference network is supposed to be available at both the transmitter and receiver. The timing information about the difference between the source clock and the network rate is conveyed by the CSI bit. The difference is bounded to 4-bit precision, and thus the source clock's stability tolerance is limited to 2 parts per million (ppm). The four bits of timing information are carried in the CSI field of four SAR–PDUs with an odd sequence count $(1, 3, 5, 7)$. The receiver can thus regenerate the source clock rate with a given accuracy by using the CSI field.

A.5.2 AAL Type 2 (AAL2)

AL2 is recently standardized and is designed for low-bit-rate, delay-sensitive applications that generate short, variable-length packets. A motivating application for AAL2 was low-bit-rate voice, where the delay in filling the payload of an ATM cell with the encoded speech from a single voice source would have degraded performance because of a large assembly delay. Thus, a key attribute of AAL2 is the ability to multiplex higher-layer packets from different native connections, called *logical link connections* (LLCs) (up to 255), onto a single ATM virtual channel connection (VCC) without regard to the cell boundaries. In other words, AAL2 does not require that each encapsulated packet fit within a single ATM payload, but rather allows it to span across payloads. A connection identification (CID) field is used in the packet header to identify the LLC to which a packet belongs. A length indicator (LI) field is used to identify the boundaries of variable-length LLC packets.

Figure A.26 shows the AAL2 SAR−PDU format, where the STF is located at the beginning of each ATM cell and the packet header precedes each

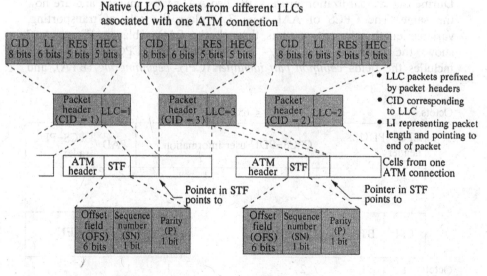

CID: Connection identification ATM: Asynchronous transfer mode
HEC: Header error check STF: Start field
LI: Length indicator LLC: Logical link connection
RES: Reserved

Fig. A.26 AAL2 SAR−PDU format.

native packet. The packet header is 3 bytes long. The CID field is 8 bits long and identifies the LLC for the packet. The LI field comprises 6 bits and indicates the length of the LLC packet. When the LI points beyond the end of the current ATM cell, the packet is split between cells. The HEC field comprises 5 bits and provides error detection over the packet header. Five bits are reserved (RES) for future use.

In the STF, the offset field (OSF) is 6 bits in length. It indicates the remaining length of the packet that (possibly) started in the preceding cell from this ATM connection and is continuing in the current cell. Thus, the OSF points to the start of the first new packet and provides immediate recovery of the packet boundary after an event causing loss of packet delineation. The 1-bit sequence number (SN) field provides a modulo-2 sequence numbering of cells. The one parity (P) bit provides odd parity and covers the STF.

It may be necessary to transmit a partially filled ATM cell to limit packet emission delay. In such a case, the remainder of the cell is padded with all zero bytes. A cell whose payload contains only the STF and 47 padding bytes can also be transmitted to meet other needs, such as serving a keep-alive function and satisfying a traffic contract.

A.5.3   AAL Types 3 and 4 (AAL3/4)

AAL3 was designed for class C services, and AAL4 for class D services. During the standardization process, the two AALs were merged and are now the same. The CPCS of AAL type 3/4 plays the role of transporting, variable-length information units through the SAR sublayer. Figure A.27 shows the structure of AAL3/4 CPCS–PDU. The CPCS–PDU header includes the fields *common part identifier* (CPI), *beginning tag* (BTA), and

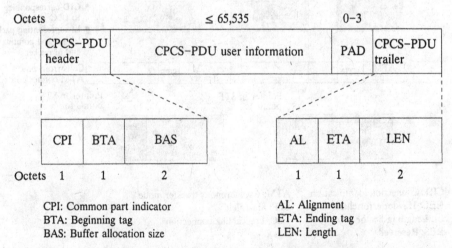

Fig. A.27   Structure of the AAL3/4 CPCS–PDU.

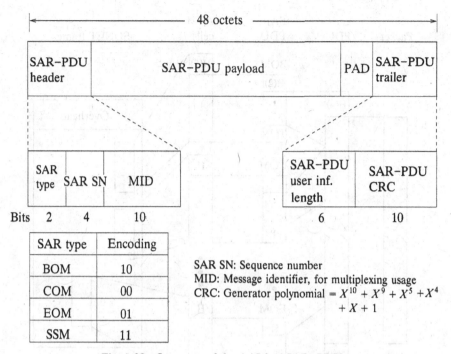

Fig. A.28 Structure of the AAL3/4 SAR–PDU.

buffer allocation size (BAS), whereas the trailer includes the fields *alignment* (AL), *ending tag* (ETA) and length (LEN). CPI is used to interpret the subsequent fields in the CPCS–PDU header and trailer, for example the counting units of the subsequent fields BAS and LEN. BTA and ETA are equal and consistent within the same CPCS–PDU. Different bytes are used in general for different CS–PDUs, and the receiver checks the equality of BTA and ETA. BAS indicates to the receiver the number of bytes required to store the whole CPCS–PDU. AL is used to make the trailer a 4-byte field, and LEN indicates the actual content of the PDU's payload, whose length is up to 65,535 bytes. A padding field (PAD) is also used to make the payload an integer multiple of 4 bytes, which could simplify the receiver design. The current specification of CPI is limited to the interpretation just described for the BAS and LEN fields.

Figure A.28 shows the structure of the AAL3/4 PDU. The segment type (ST) is a 2-bit field that indicates a SAR–PDU as containing the beginning of message (BOM), continuation of a message (COM), or end of a message (EOM), or a single-segment message (SSM). The 4-bit sequence number (SN) field is used to number the SAR–PDUs modulo 16 (i.e., 0 to 15). The multiplexing identification (MID) field is 10 bits long and is used to multiplex different connections into a single ATM connection. All SAR–PDUs of a particular connection are assigned the same MID value. Using this field, it is

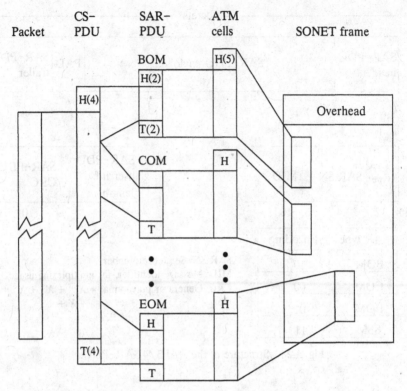

Fig. A.29   Packet-cell conversion (AAL3/4).

possible to interleave and reassemble SAR–PDUs of different sources that are transported using the same VPI or VCI. Cells of different sources are distinguished by carrying different MID values. The 6-bit user information length indicates the number of bytes in the SAR–PDU, and its value ranges from 4 to 44. A 10-bit CRC is used over the entire SAR–PDU to detect bit errors. Figure A.29 shows how a data packet is converted to ATM cells (using AAL3/4), which are then embedded in SDH frames.

A.5.4   AAL Type 5 (AAL5)

AAL5 is the most widely used AAL type to date. For instance, AAL5 is the adaptation layer for ILMI, UNI signaling, PNNI signaling, and IP packets.

Figure A.30 shows the structure of AAL5 CPCS–PDU. Its payload length can be from 1 to 65,535 bytes. A PAD field, ranging 0 to 47 bytes, is added to make the CPCS–PDU length an integer multiple of 48 bytes. Thus, it can completely fill in the SAR–PDU payloads. The CPCS user-to-user indication (UU) is one byte long and may be used to communicate between two AAL5

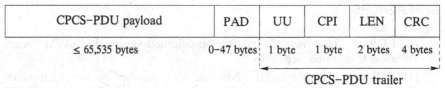

UU: User-to-user indication
CPI: Common part indicator
LEN: Length
CRC: Cyclic redundancy check

Fig. A.30 AAL5 CPCS–PDU format.

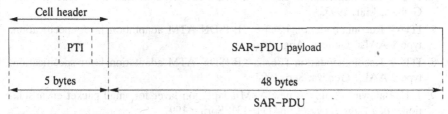

PTI: Payload type identifier

Fig. A.31 AAL5 SAR–PDU format.

entities. The common part indicator (CPI), one byte long, is not defined yet and must be set to 0. LEN indicates the actual length of the CPCS–PDU payload, so as to identify the PAD size. A 32-bit CRC (cyclic redundancy check), following the FDDI standards, is used to detect bit errors in the CPCS–PDU. The CPCS–PDU is segmented into 48-byte SAR–PDUs and passed to the ATM layer, where 5-byte cell headers are inserted to the front of the data units, as shown in Figure A.31.

The main reasons that AAL5 is the most widely used are its high utilization (the entire 48-byte payload is used to carry user's information) and less processing overhead (one CRC calculation for the entire CPCS–PDU). To have all 48 bytes carry user's information, the ATM layer is required to indicate the last cell of a data packet to facilitate packet reassembly, which in turn violates the protocol stack. When a cell's PTI is 000, it can be either a beginning of a message (BOM) or a continuation of a message (COM). When it is 001, the cell can be either an end of a message (EOM) or a single-segment message (SSM). A receiver will start to reassemble arriving cells when it receives the very first cell until it receives a cell with the PTI set to 001. Once an EOM or SSM is received, the receiver will start the reassembly process for the following packet.

REFERENCES

1. ATM Forum, "User–network interface specification version 3.1," ATM Forum Technical Committee, Sep. 1996.

2. ITU-T Recommendation G.707, "Network node interface for the synchronous digital hierarchy (SDH)," Geneva, Nov. 1995.

3. ITU-T Recommendation I.432, "B-ISDN user network interface specification," Geneva, Aug. 1996.

4. R. Onvural, *Asynchronous Transfer Mode Networks: Performance Issues*, Artech House, Norwood, MA, 1995.

5. ITU-T Recommendation I.361, "B-ISDN ATM layer specification," Geneva, Nov. 1995.

6. ITU-T Recommendation I.363, "B-ISDN ATM adaptation layer specification," Geneva, Mar. 1993.

7. ITU-T Recommendation I.363.1, "B-ISDN ATM adaptation layer specification: type 1 AAL," Geneva, Aug. 1996.

8. ITU-T Recommendation I.363.2, "B-ISDN ATM adaptation layer specification: type 2 AAL," Geneva, Sep. 1997.

9. J. H. Baldwin, et al., "A new ATM adaptation layer for small packet encapsulation," *Bell Labs Tech. J.*, pp. 111–131, Spring 1997.

10. ITU-T Recommendation I.363.3, "B-ISDN ATM adaptation layer specification: type 3/4 AAL," Geneva, Aug. 1996.

11. ITU-T Recommendation I.363.5, "B-ISDN ATM adaptation layer specification: type 5 AAL," Geneva, Aug 1996.

INDEX

Absolute dropper, differentiated services (DS/Diffserv) architecture, 343

Acceptance region, integrated services internet (ISI), call admission control (CAC), controlled-load service (CLS), 51

Access flow control, backlog balancing flow control, 253

Acknowledgment mechanism:
new RENO TCP, flow control, congestion control, 286
TCP flow control, congestion control, 278–280, 282–284
fast retransmit and recovery, 284–286

Action elements, differentiated services (DS/Diffserv) architecture, 342–343

ActiveList, packet scheduling, round-robin scheduling, 112–113

Active queue management, drop on full disciplines, 215

Actual total buffer usage, backlog balancing flow control, simulation results, 258–263

Adaptability, multiprotocol label switching (MPLS), traffic engineering, 367

Adaptive clock, SONET protocol, ATM adaptation layer, 402

Add-drop mux (ADM), SONET protocol, sublayer structure, 380–382

Address summarization and reachability, PNNI routing protocols, 314–315

Admission control:

ATM VBR services, call admission control (CAC):
effective bandwidth, 24–25
Lucent's CAC, 25–27
NEC CAC, 27–30
lossless multiplexing, 28
statistical multiplexing, 28–30
tagged-probability-based CAC, 30–43
cell loss rate (CLR), 30–32
modified effective bandwidth, 34
modified statistical multiplexing algorithm, 34–35
performance evaluation, 35–43
TAP algorithm, 32–34
worst-case traffic model, 23–24
characteristics and techniques, 17–18
deterministic bound, 18–19
integrated services internet (ISI):
application, 4
admission control (CAC), 43–54
admission control algorithms, 50–52
controlled-load service (CLS), 49–54
measurement mechanisms, 52–54
quality of service (QoS) guarantee:
delay/buffer bounds, 45–47
guaranteed service (GS) application, 48–49
probabilistic bound, equivalent bandwidth, 19–23
Bernoulli trials and binomial distribution, 20